S0-DFG-515

Advances in Physical Organic Chemistry

Advances in Physical Organic Chemistry

Volume 35

Edited by

T. T. TIDWELL

Department of Chemistry
University of Toronto
Toronto
Ontario M5S 3H6, Canada

San Diego San Francisco New York
Boston London Sydney Tokyo

This book is printed on acid-free paper.

Academic Press
A Harcourt Science and Technology Company
Harcourt Place, 32 Jamestown Road, London NW1 7BY, UK
http://www.academicpress.com

Academic Press
A Harcourt Science and Technology Company
525 B Street, Suite 1900, San Diego, California 92101-4495, USA
http://www.academicpress.com

ISBN 0-12-033535-2

A catalogue record for this book is available from the British Library

This serial is covered by The *Science Citation Index*.

Typeset by Paston Prepress Ltd, Beccles, Suffolk

Printed and bound in Great Britain by
MPG Books Ltd, Cornwall, UK

00 01 02 03 04 05 MP 9 8 7 6 5 4 3 2 1

Contents

Editor's preface

This volume is the first since Donald Bethell served as Editor of the series, and he set a very high standard both for the quality of the contributions presented and for the excellence of the editorial work. It is my hope as the new Editor to maintain this high level, and continue to provide the chemical community with authoritative and critical assessments of different aspects of the field of physical organic chemistry. The chapters in the previous volumes provide a lasting record that is widely cited and used, and will continue to serve for decades to come. Because this series has maintained such a high level of quality and utility there is little need for change, and one of the few innovations is the adoption of the numerical system of reference citation now used by almost all chemical journals.

The four chapters in this volume are intimately related to the study of carbocations and of free radicals, which are two classes of intermediates that were both recognized as discrete reactive intermediates just at the beginning of the twentieth century. The first chapter, on excess acidities, is a lucid exposition of the current understanding of a field that has been relevant to many of the great triumphs of physical organic chemistry throughout the century. The second chapter, on the behavior of carbocations in solution, demonstrates the exquisite detail with which these processes may now be understood. Two chapters concern electron transfer, and thus involve not only free radicals but charged species as well.

I wish to extend my thanks to the authors of the chapters in this volume for the uniformly high quality and timeliness of their contributions. The Advisory Board has been generous with their suggestions, and the success of this series is due in no small part to their efforts. Regretably the Board has suffered the loss of the services of Lennart Eberson, who died in February, 2000, and will be remembered as a distinguished chemist, a longtime contributor to this series, and a valuable member of the Board.

The new century is a time of great opportunity for physical organic chemistry, which in recent decades has expanded far beyond its traditional boundaries. This now encompasses fields ranging from the purely theoretical to the largely applied, and includes chemistry in the gas, liquid, and solid phases, and many aspects of biological, medicinal, and environmental chemistry. It is our intention to cover as many of these areas as possible.

It is also a time for reflection, for as I have discussed elsewhere (*Pure and Applied Chemistry* (1997), **69**: 211–213), the history of the field of physical organic chemistry belongs almost completely in the twentieth century. Thus the seminal recognition of reactive intermediates including carbocations, free radicals, and carbenes came very early in the century, along with the mechanistic and theoretical tools needed for understanding and interpreting the behavior of these species. Throughout the century the achievements of physical organic chemistry have been widely recognized, not least by the award of the Nobel Prize. In the past decade these prizes have honoured the theory of electron transfer reactions (R. Marcus, 1992), the direct observation of carbocations (G. A. Olah, 1994), molecular orbital calculations (J. Pople and W. Kohn, 1998), and the study of transition states (A. Zewail, 1999). All of these areas are central to the modern practice of physical organic chemistry, and the other prizes in this same period all show the influence of physical organic thinking. The twenty-first century offers even more opportunities, and *Advances in Physical Organic Chemistry* will aspire to bring the best of these to the chemical public.

As Editor I feel an obligation to continue the enviable record of this series, and to provide at reasonable cost a service to the users. Suggestions for prospective fields or authors are welcome.

T. T. Tidwell

Contributors to Volume 35

Tina L. Amyes Department of Chemistry, University at Buffalo, SUNY, Buffalo, NY 14260, USA

Robin A. Cox Department of Chemistry, University of Toronto, 80 St George Street, Toronto, Ontario M5S 3H6, Canada

Jay K. Kochi Department of Chemistry, University of Houston, Houston, Texas 77204-5641, USA

Shrong-Shi Lin Department of Chemistry, University at Buffalo, SUNY, Buffalo, NY 14260, USA

AnnMarie C. O'Donoghue Department of Chemistry, University at Buffalo, SUNY, Buffalo, NY 14260, USA

Rajendra Rathore Department of Chemistry, University of Houston, Houston, Texas 77204-5641, USA

John P. Richard Department of Chemistry, University at Buffalo, SUNY, Buffalo, NY 14260, USA

Jean-Michel Savéant Laboratoire d'Electrochimie Moléculaire, Unité Mixte de Recherche Université – CNRS No 7591, Université de Paris 7 – Denis Diderot, 2 place Jussieu, 75251 Paris, Cedex 05, France

Maria M. Toteva Department of Chemistry, University at Buffalo, SUNY, Buffalo, NY 14260, USA

Yutaka Tsuji Department of Chemistry, University at Buffalo, SUNY, Buffalo, NY 14260, USA

Kathleen B. Williams Department of Chemistry, University at Buffalo, SUNY, Buffalo, NY 14260, USA

Excess Acidities

Robin A. Cox

Department of Chemistry, University of Toronto, Toronto, Ontario, Canada

1 Introduction

Large numbers of reactions of interest to chemists only take place in strongly acidic or strongly basic media. Many, if not most, of these reactions involve proton transfer processes, and for a complete description of the reaction the acidities or basicities of the proton transfer sites have to be determined or estimated. These quantities are also of interest in their own right, for the information available from the numbers via linear free energy relationships (LFERs), and for other reasons.

It is therefore necessary to have methods of dealing with kinetic and equilibrium data obtained in these media. Many chemists are convinced that

ADVANCES IN PHYSICAL ORGANIC CHEMISTRY
VOLUME 35 ISBN 0-12-033535-2

trying to use information obtained in strong acid or strong base media for the determination of the mechanisms of reactions is fraught with difficulty and complication. The most notorious example of this is R. P. Bell leaving out of the second edition of his book *The Proton in Chemistry*[1] the chapter on concentrated solutions of acids and bases that had been included in the first edition,[2] "partly because the interpretation of reaction velocities in these concentrated solutions has become more rather than less confused with the passage of time". It is my intention to present a simple, unified method of dealing with these systems, called the "excess acidity" method; to show that reasonably reliable thermodynamic acidity constants can be obtained by using it; and to show that the method leads to mechanistic information that is difficult if not impossible to obtain in any other way when used with kinetic data.

Space considerations permit only the consideration of strong aqueous acid media in this review, primarily H_2SO_4, $HClO_4$ and HCl. An equivalent technique (the "excess basicity" method) can be applied to strongly basic media; for instance, it has been applied to weak acidity determinations in aqueous dimethyl sulfoxide mixtures,[3–6] and used for kinetic studies in this system.[7] Sulfamide ionizations have been studied,[8] and ground and excited state acidities in other aqueous media have been determined.[9] The excess basicities of methanolic methoxide solutions have been examined,[10] and kinetics in these solutions have been looked at.[11,12] However, by and large strong bases have not been studied to nearly the same extent that strong acids have.

2 Determination of weak basicities

AQUEOUS SOLUTION

Since the days of Brønsted[13] the strengths of acids in aqueous solution have been defined in terms of the equilibrium constant K_a for the ionization of HA, equation (1):

$$\mathrm{HA + H_2O \overset{K_a}{\rightleftharpoons} H_3O^+ + A^-} \tag{1}$$

$$K_a = \frac{C_{H_3O^+}C_{A^-}}{C_{HA}} \tag{2}$$

with the mathematical definition of K_a being that of equation (2) and C being molar concentration, which can be used in dilute solution. (According to Bell,[1] Lowry[14] and Lewis[15] were proposing somewhat similar ideas at the same time but did not give the actual definition according to equation (2).) Other states of protonation are possible in equation (1), for instance $H_2A^+ \rightarrow HA + H^+$, or $HA^- \rightarrow A^{2-} + H^+$, but for simplicity equation (2) will be used exclusively for $HA \rightarrow A^- + H^+$.

In equation (2), by convention (since it is in large excess when HA is in dilute solution) the concentration of water, C_{H_2O}, is *left out* of the definition. This leads to problems. For instance, when hydronium ion itself is the acid, for consistency equation (3) must be written:

$$H_3O^+ + H_2O \overset{K_a}{\rightleftharpoons} H_2O + H_3O^+ \tag{3}$$

$$K_a = \frac{C_{H_2O}C_{H_3O^+}}{C_{H_3O^+}} = C_{H_2O} = 55.34\,\text{M at } 25^\circ\text{C} \tag{4}$$

and the K_a of water is defined according to equation (4), which leads to the pK_a of water being −1.743 rather than 0.000. (Similarly its pK_b is 15.743, not 14.000.)

In strong acids the convention is to write the protonation equilibrium of a weak base B as equation (5); the species H_3O^+ in equation (1) (or such higher proton solvates as may be present) is just written as "H^+" for simplicity, without indicating its structural environment:

$$B + H^+ \overset{K_{BH^+}}{\rightleftharpoons} BH^+ \tag{5}$$

The mathematical definition of K_{BH^+} is like that of K_a (now right-to left, see equation (5)); writing a for activities and f for molar activity coefficients, as is commonly done in strong acid work, equation (6) is obtained:

$$K_{BH^+} = \frac{a_B a_{H^+}}{a_{BH^+}} = \frac{C_B C_{H^+}}{C_{BH^+}} \cdot \frac{f_B f_{H^+}}{f_{BH^+}} \tag{6}$$

Note that the water activity is left out of the definition of K_{BH^+} in just the same way as it is for the definition of K_a in dilute aqueous solution. This is not necessarily a good idea in these non-ideal strong acid media, since the water activity can vary drastically as the medium changes from dilute to concentrated acid;[16–19] by no means does it remain constant, as the dilute solution definition implies, and so from this point of view it would be a good idea to include it. However, there are several practical problems involved if this is done. First, it would be necessary to alter either all listed pK_a values or all listed pK_{BH^+} values by 1.743 as given above, to allow comparison between values determined in aqueous buffers and those determined in strong acid. This seems an unnecessary complication. Secondly, the activities of water (and of the acid) have all been measured using the mole fraction activity scale,[16–19] which has the standard state (where $f = 1$) defined as being the pure solvent. Now this is different from the standard state used in equation (6), where $f = 1$ is defined as being infinite dilution in the reference medium, which is a hypothetical ideal solution 1 M in the acid being used, the same reference medium as that used for pH measurements.[20] Thus, before water activities can be included in the definition it is necessary to convert the listed mole fraction-based water activities to concentration-based ones, which is not a trivial operation. (The standard state for water now looks a bit strange, being an infinitely dilute solution of water in itself!) By and large it seems a lot easier to stay with the definition of equation (6). Nevertheless, there is one case where it is necessary to have these molarity-based water activities for equilibrium measurements, for the determination of pK_{R^+} values, and they are also needed in kinetic studies, see below.

STRONG ACID MEDIA

$$pK_{BH^+} = \log I - \log C_{H^+} - \log \frac{f_B f_{H^+}}{f_{BH^+}} \tag{7}$$

Writing equation (6) in logarithmic form results in equation (7). Again by convention, the log ionization ratio, $\log I = \log(C_{BH^+}/C_B)$, is defined, with the ionized form on top. Equations (6) and (7) are thermodynamically exact; the problem with them has always been what to do about the unknown activity coefficient ratio term. The first person to tackle this problem was Hammett,[21,22] who defined an acidity function, H_0, as in equation (8):

$$H_0 = pK_{BH^+} - \log I = -\log C_{H^+} - \log \frac{f_B f_{H^+}}{f_{BH^+}} \tag{8}$$

H_0 is defined so as to be similar to pH, and to reduce to it in dilute solution, i.e. to $\mathrm{pH} = pK_a - \log I$. The idea is that versions of equation (8) can be written for weak base indicators that protonate to different extents in the same acid solutions (overlapping indicators; indicators because they indicate the solution acidity); subtracting two of these (say for indicators A and B) leads to equation (9), and if the activity coefficients for A and B, and for AH^+ and BH^+, approximately cancel, the value of pK_{BH^+} can be calculated from the measured ionization ratios for A and B if pK_{AH^+} is known:

$$pK_{BH^+} - \log \frac{C_{BH^+}}{C_B} - pK_{AH^+} + \log \frac{C_{AH^+}}{C_A} = \log \frac{f_A f_{BH^+}}{f_{AH^+} f_B} = 0 \tag{9}$$

A is an *anchor compound*, one whose ionization ratios are measurable in dilute aqueous acid; $pK_{BH^+} = 1.00$ for *p*-nitroaniline is used for H_0.[22,23] Equation (9) is known as the *cancellation assumption*; using it on a series of overlapping weak base indicators of similar type (primary aromatic amines in the case of H_0) leads to H_0 values covering a wide acidity range according to equation (8), once all the pK_{BH^+} values are known.

It was soon realized that there are problems with this approach.[24,25] Log ionization ratios for weak bases that are not primary aromatic amines, while linear in H_0, do not give the unit slope required by equation (8). This soon led to many other acidity functions, defined for other types of weak base, H_A for amides,[24] H_0''' for tertiary aromatic amines,[25] C_0 or H_R for carbocations,[26,27] and so on. In a recent review of acidity functions,[28] 28 different ones were listed for aqueous sulfuric acid mixtures alone! So H_0 is not a universal function, although it can still be used to obtain values for the pK_{BH^+} of other types of compound. There are two ways of doing this in common use; the first, sometimes referred to as the Yates–McClelland method,[29] is simply to accept a slope m other than unity and use equation (10); the pK_{BH^+} is then (mH_0) when $\log I$ is zero (half-protonation):

$$\log I = m(-H_0) + pK_{BH^+} \quad (10)$$

The second way, called the Bunnett–Olsen method,[30] makes the less drastic assumption that log activity coefficient ratios such as those in equation (7) are linear functions of one another, rather than cancelling out. From the definition of H_0 in equation (8) we can write equation (11), where Am refers to the primary aromatic amines used in the determination of H_0, and then any specific activity coefficient ratio, say for the weak base B, is assumed to be linear in this according to equation (12):

$$H_0 + \log C_{H^+} = -\log \frac{f_{Am} f_{H^+}}{f_{AmH^+}} \quad (11)$$

$$\log \frac{f_B f_{H^+}}{f_{BH^+}} = (1 - \phi_e) \log \frac{f_{Am} f_{H^+}}{f_{AmH^+}} \quad (12)$$

Equation (12) is a linear free-energy relationship, since activity coefficients f can be represented as ΔG° values. The reason for defining the slope parameter as in equation (12) (subscript e for equilibrium) is that a little rearranging of equations (11) and (12) leads to the easy-to-use Bunnett–Olsen equation for equilibria, equation (13):[30]

$$\log I + H_0 = pK_{BH^+} + \phi_e(H_0 + \log C_{H^+}) \quad (13)$$

This linear plot works very well, giving pK_{BH^+} values as intercepts (and slopes ϕ_e); thus only one acidity function (H_0) is needed for the purpose of estimating weak basicities. In the Bunnett–Olsen method C_{H^+} is simply the acid molarity. The terms m from the Yates–McClelland method and $(1 - \phi_e)$ from the Bunnett–Olsen method are, for all practical purposes, equivalent: $m = \sim 1$, $\phi_e = \sim 0$ for primary nitroanilines; $m = \sim 0.6$, $\phi_e = \sim 0.4$ for amides; and so on.

3 The excess acidity method

A philosophical problem remains, however. The Bunnett–Olsen method, which assumes the linearity of activity coefficient ratios in one another, still uses H_0, and H_0 values are derived using the cancellation assumption! The cancellation assumption is eliminated altogether in the *excess acidity method* (also called the Marziano–Cimino–Passerini and Cox–Yates methods, which is unfortunate since both are the same – the term "excess acidity method" is preferred).

For an anchor compound B* (say *p*-nitroaniline), whose ionization ratios are measurable in dilute acid, we can write the thermodynamically exact equation (14):

$$\log I_{B^*} - \log C_{H^+} = \log \frac{f_{B^*} f_{H^+}}{f_{B^*H^+}} + pK_{B^*H^+} \tag{14}$$

For an overlapping indicator B′ we can write equation (15):

$$\begin{aligned} \log I_{B'} - \log C_{H^+} &= \log \frac{f_{B'} f_{H^+}}{f_{B'H^+}} + pK_{B'H^+} \\ &= n_1 \log \frac{f_{B^*} f_{H^+}}{f_{B^*H^+}} + pK_{B'H^+} \\ &= n_1(\log I_{B^*} - \log C_{H^+} - pK_{B^*H^+}) + pK_{B'H^+} \end{aligned} \tag{15}$$

giving n_1 and $pK_{B'H^+}$ from the resulting linear plot, and enabling further values of the activity coefficient ratio for B* to be calculated. This process can be continued into stronger and stronger acid media. This technique was originally formulated by Marziano, Cimino and Passerini,[31] who abbreviated the activity coefficient ratio term for B* as M_C (equation 16), and the slopes as n, later $n_{i,j}$.[32] These authors provided several scales for aqueous H_2SO_4 and $HClO_4$ media.[31]

$$\log \frac{f_{B^*} f_{H^+}}{f_{B^*H^+}} = M_C \quad \text{or} \quad X \tag{16}$$

Subsequently the calculation of M_C (now called $M_C f(x)$) was improved by mathematical treatment.[33] The assumption upon which the method is based, linearity according to equation (15), has been thoroughly tested for aqueous $HClO_4$[32] and H_2SO_4[34] media. Cox and Yates[20] computerized the calculation of these functions, preferring the simpler terminology X for "excess acidity", since the activity coefficient ratio represents the difference between the actual solution acidity and the stoichiometric acid concentration, and m^* for the slopes, as in equation (17). The term excess acidity was first used by Perrin,[35] although he defined it in Bunnett–Olsen terms as being $(-H_0 - \log[H^+])$, see equation (11), which is somewhat different from the current equation (16) definition.

EXCESS ACIDITY SCALES

$$\log I - \log C_{H^+} = m^* X + pK_{BH^+} \tag{17}$$

Equation (17) is the heart of the excess acidity method for the determination of unknown pK_{BH^+} values in strongly acidic media. Without going into detail (which is tedious) polynomial coefficients have been calculated that enable the calculation of X for 0–99.5 wt% H_2SO_4 and 0–80 wt% $HClO_4$. These are used with equation (18) and are given in Table 1. The form of equation (18) was

Table 1 Polynomial coefficients giving X as a function of wt% acid at 25°C for aqueous sulfuric and perchloric acid mixtures.[a]

Polynomial coefficient	Gives X for aq. H_2SO_4	Gives X for aq. $HClO_4$
a_1	−1.2192412	−0.74507718
a_2	1.7421259	1.0091461
a_3	−0.62972385	−0.30591601
a_4	0.11637637	0.049738522
a_5	−0.010456662	−0.0040517065
a_6	0.00036118026	0.00012855227

[a] From ref. 20. Use with equation (18).

chosen in order to have $X = 0$ in pure water, and to provide polynomial coefficients near to unity:

$$X = a_1(z-1) + a_2(z^2-1) + a_3(z^3-1) + \cdots$$
$$z = \text{antilog}(\text{wt\%}/100) \text{ for } H_2SO_4; \quad z = \text{antilog}(\text{wt\%}/80) \text{ for } HClO_4 \tag{18}$$

Subsequently this computer method was investigated in more detail,[36] and it was found that it was not necessary to be as elaborate as equation (18). The polynomial coefficients given in Table 2 for HCl and $HClO_4$ are used with the much simpler equation (19):

$$X = a_1(\text{wt\%}) + a_2(\text{wt\%})^2 + a_3(\text{wt\%})^3 \tag{19}$$

The X_0 scale for $HClO_4$ that can be obtained from Table 2 is derived using H_0 indicators only (primary aromatic amines), rather than the broad mix of indicators of different type used in deriving X. Values of X calculated from these polynomial coefficients are given for H_2SO_4 in Tables 3 and 4, for $HClO_4$ in Tables 5 and 6 (with X_0), and for HCl in Tables 7 and 8, as a function of wt% acid (odd-numbered tables) and of the acid molarity (even-numbered tables).

Other information is provided in Tables 3–8. This includes values of log C_{H^+} for use with equation 17; for $HClO_4$ and HCl these are simply the log acid molarity, assuming the acid to be fully dissociated. The maximum acid strength is 80 wt% for $HClO_4$, at which point the acid mixtures become solid at 25°C, and 40% for HCl, at which point the aqueous solution is saturated with the

Table 2 Polynomial coefficients giving X as a function of wt% acid at 25°C for aqueous hydrochloric acid mixtures, and X_0 at 25°C for aqueous perchloric acid mixtures.[a]

Polynomial coefficient	Gives X for aq. HCl	Gives X_0 for aq. $HClO_4$
a_1	0.0527767	0.0335096
a_2	0.00190497	−0.000745044
a_3	−0.0000197423	0.0000222391

[a] From ref. 36. Use with equation (19).

Table 3 Values of X, and of $\log C_{H^+}$, $\log a_{H_2O}$, $\log C_{HSO_4^-}$ and $\log a_{H_2SO_4}$ in molarity units, for aqueous sulfuric acid at 25°C, at different values of wt% acid.

Wt% H_2SO_4	X	$\log C_{H^+}$	$\log a_{H_2O}$	$\log C_{HSO_4^-}$	$\log a_{H_2SO_4}$
0.2	0.004	−1.534	1.741	−1.940	–
0.5	0.009	−1.178	1.739	−1.449	–
1.0	0.019	−0.897	1.734	−1.108	–
2.0	0.038	−0.609	1.726	−0.780	–
3.0	0.059	−0.433	1.717	−0.596	–
4.0	0.080	−0.307	1.708	−0.467	–
5.0	0.103	−0.206	1.699	−0.368	–
7.5	0.164	−0.019	1.677	−0.192	–
10.0	0.231	0.118	1.654	−0.066	–
15.0	0.387	0.315	1.603	0.113	–
20.0	0.573	0.461	1.546	0.242	–
25.0	0.790	0.577	1.481	0.346	–
30.0	1.038	0.673	1.406	0.434	−7.623
35.0	1.317	0.756	1.317	0.514	−7.062
40.0	1.628	0.828	1.213	0.588	−6.479
45.0	1.969	0.891	1.088	0.658	−5.871
50.0	2.345	0.947	0.939	0.728	−5.234
53.0	2.590	0.977	0.834	0.769	−4.837
56.0	2.853	1.005	0.716	0.810	−4.427
59.0	3.138	1.031	0.581	0.851	−4.003
62.0	3.450	1.056	0.427	0.892	−3.565
65.0	3.795	1.078	0.250	0.933	−3.111
68.0	4.178	1.099	0.045	0.973	−2.641
71.0	4.606	1.118	−0.193	1.012	−2.156
74.0	5.080	1.136	−0.472	1.050	−1.657
77.0	5.598	1.151	−0.800	1.086	−1.149
80.0	6.150	1.163	−1.184	1.117	−0.641
82.0	6.528	1.165	−1.475	1.132	−0.310
84.0	6.906	1.160	−1.792	1.137	0.008
86.0	7.277	1.140	−2.134	1.127	0.303
88.0	7.637	1.098	−2.492	1.093	0.564
90.0	7.985	1.024	−2.854	1.024	0.781
92.0	8.340	0.919	−3.197	0.919	0.942
94.0	8.743	0.800	−3.504	0.800	1.049
95.0	8.989	0.743	−3.646	0.743	1.085
96.0	9.285	0.684	−3.796	0.684	1.117
97.0	9.656	0.604	−3.990	0.604	1.149
98.0	10.132	0.462	−4.303	0.462	1.187
99.0	10.754	0.150	−4.894	0.150	1.232
99.5	11.136	−0.171	−5.421	−0.171	1.252

gaseous HCl. H_2SO_4 can used up to 100 wt%, and past this point by going into oleum mixtures. The estimation of $\log C_{H^+}$ is more complicated for H_2SO_4, however. Operationally defined values of $\log C_{H^+}$ are provided in Tables 3 and 4; these are updates of the ones originally given.[20] These take into account the

Table 4 Values of X, and of $\log C_{H^+}$, $\log a_{H_2O}$, $\log C_{HSO_4^-}$ and $\log a_{H_2SO_4}$ in molarity units, for aqueous sulfuric acid at 25°C, at different values of the acid molarity.

$[H_2SO_4]$ (M)	X	$\log C_{H^+}$	$\log a_{H_2O}$	$\log C_{HSO_4^-}$	$\log a_{H_2SO_4}$
0.01	0.002	−1.803	1.742	−2.372	—
0.10	0.018	−0.907	1.734	−1.119	—
0.25	0.047	−0.527	1.722	−0.693	—
0.50	0.097	−0.228	1.701	−0.389	—
0.75	0.152	−0.048	1.681	−0.219	—
1.00	0.210	0.080	1.661	−0.100	—
1.50	0.337	0.263	1.619	0.066	—
2.00	0.477	0.393	1.575	0.182	—
2.50	0.628	0.494	1.530	0.272	—
3.00	0.789	0.577	1.481	0.346	—
3.50	0.960	0.646	1.430	0.409	−7.788
4.00	1.137	0.705	1.375	0.464	−7.419
4.50	1.321	0.757	1.316	0.515	−7.056
5.00	1.509	0.802	1.253	0.561	−6.697
5.50	1.702	0.843	1.186	0.604	−6.344
6.00	1.899	0.879	1.115	0.645	−5.993
6.50	2.099	0.911	1.038	0.683	−5.648
7.00	2.302	0.941	0.956	0.720	−5.305
7.50	2.509	0.967	0.869	0.756	−4.966
8.00	2.721	0.992	0.775	0.790	−4.629
8.50	2.940	1.014	0.675	0.823	−4.295
9.00	3.166	1.034	0.567	0.855	−3.962
9.50	3.402	1.052	0.451	0.886	−3.630
10.00	3.649	1.069	0.326	0.916	−3.298
10.50	3.909	1.085	0.190	0.945	−2.967
11.00	4.182	1.099	0.043	0.973	−2.637
11.50	4.470	1.112	−0.116	1.000	−2.307
12.00	4.771	1.125	−0.288	1.026	−1.978
12.50	5.086	1.136	−0.476	1.051	−1.651
13.00	5.414	1.146	−0.680	1.074	−1.326
13.50	5.753	1.155	−0.904	1.096	−1.004
14.00	6.100	1.162	−1.148	1.115	−0.127
14.50	6.453	1.165	−1.415	1.129	−0.375
15.00	6.810	1.162	−1.708	1.137	−0.072
15.50	7.168	1.148	−2.030	1.133	0.218
16.00	7.528	1.114	−2.381	1.070	0.488
16.50	7.889	1.047	−2.755	1.047	0.726
17.00	8.267	0.942	−3.131	0.942	0.915
17.50	8.711	0.809	−3.483	0.809	1.043
18.00	9.376	0.666	−3.842	0.666	1.125
18.50	10.780	0.133	−4.925	0.133	1.233

partial dissociation of bisulfate ion at concentrations less than mole fraction 0.5 (84.46 wt%), and the fact that at acidities higher than this H_2SO_4 is only partially dissociated, since there is insufficient water to accommodate the protons.[20] In other words, these values are for protons solvated by water only,

Table 5 Values of X and X_0, and of $\log C_{H^+}$ and $\log a_{H_2O}$ in molarity units, for aqueous perchloric acid at 25°C at different values of wt% acid.

Wt% $HClO_4$	X	X_0	$\log C_{H^+}$	$\log a_{H_2O}$
0.20	0.003	0.007	−1.702	1.741
0.50	0.008	0.017	−1.303	1.739
1.00	0.016	0.033	−1.001	1.734
2.00	0.033	0.064	−0.697	1.726
3.00	0.051	0.094	−0.519	1.717
4.00	0.070	0.124	−0.391	1.708
5.00	0.091	0.152	−0.292	1.700
7.50	0.147	0.219	−0.110	1.677
10.00	0.212	0.283	0.022	1.654
15.00	0.371	0.410	0.211	1.605
20.00	0.571	0.550	0.349	1.551
25.00	0.819	0.720	0.460	1.490
30.00	1.116	0.935	0.554	1.419
35.00	1.468	1.214	0.637	1.334
40.00	1.879	1.572	0.711	1.226
43.00	2.155	1.831	0.752	1.145
46.00	2.458	2.130	0.792	1.050
49.00	2.790	2.470	0.830	0.936
52.00	3.156	2.855	0.867	0.799
55.00	3.561	3.289	0.903	0.635
58.00	4.011	3.776	0.938	0.438
61.00	4.508	4.320	0.971	0.204
64.00	5.051	4.923	1.004	−0.075
67.00	5.629	5.589	1.035	−0.405
70.00	6.220	6.323	1.064	−0.794
72.00	6.607	6.851	1.084	−1.089
74.00	6.983	7.412	1.102	−1.417
76.00	7.356	8.006	1.120	−1.779
78.00	7.766	8.635	1.138	−2.179
80.00	8.311	9.299	1.155	−2.619

with no account being taken of free H_2SO_4 molecules; this was done to maintain communication with the aqueous standard state, to make an attempt, at least, to have equation (17) provide thermodynamic pK_{BH^+} values. The values in Tables 3 and 4 stop at 99.5 wt%, because autoprotolysis of H_2SO_4 becomes important very near 100 wt% acid,[37] giving the very strong acid $H_3SO_4^+$.

The X-functions are illustrated in various ways in Figs 1–3. It can be seen that the relative acid strengths of the three acids depend on the concentration scale used. $HClO_4$ is always stronger than H_2SO_4, their molecular weights being very similar. HCl, however, can be the strongest acid of the three (wt% acid scale, Fig. 1), intermediate (mole fraction acid scale, Fig. 3), or the weakest (acid molarity scale, Fig. 2). In Fig. 1 the $\log C_{H^+}$ values for H_2SO_4 are illustrated; according to what was said above the values drop off above about 85 wt% H_2SO_4, since they represent protons solvated by water. For the determination of

Table 6 Values of X and X_0, and of $\log C_{H^+}$ and $\log a_{H_2O}$ in molarity units, for aqueous perchloric acid at 25°C at different values of the acid molarity.

$[HClO_4]$ (M)	X	X_0	$\log C_{H^+}$	$\log a_{H_2O}$
0.10	0.016	0.033	−1.000	1.734
0.20	0.033	0.064	−0.699	1.726
0.30	0.050	0.094	−0.523	1.717
0.40	0.069	0.122	−0.398	1.709
0.50	0.089	0.149	−0.301	1.701
0.75	0.141	0.212	−0.125	1.680
1.00	0.200	0.271	−0.001	1.659
1.50	0.333	0.382	0.175	1.616
2.00	0.489	0.494	0.300	1.572
2.50	0.667	0.615	0.397	1.527
3.00	0.863	0.751	0.476	1.479
3.50	1.076	0.905	0.543	1.429
4.00	1.305	1.081	0.601	1.374
4.50	1.547	1.280	0.652	1.314
5.00	1.801	1.501	0.698	1.247
5.50	2.065	1.746	0.740	1.172
6.00	2.340	2.012	0.777	1.088
6.50	2.624	2.298	0.812	0.994
7.00	2.919	2.605	0.844	0.889
7.50	3.227	2.931	0.874	0.771
8.00	3.548	3.276	0.902	0.640
8.50	3.884	3.639	0.929	0.495
9.00	4.234	4.019	0.953	0.335
9.50	4.597	4.417	0.977	0.160
10.00	4.971	4.833	0.999	−0.032
10.50	5.353	5.266	1.020	−0.243
11.00	5.739	5.720	1.040	−0.473
11.50	6.121	6.196	1.060	−0.725
12.00	6.496	6.695	1.078	−1.001
12.50	6.859	7.222	1.096	−1.305
13.00	7.216	7.780	1.114	−1.639
13.50	7.589	8.372	1.131	−2.010
14.00	8.047	9.005	1.148	−2.422
14.50	8.742	9.683	1.164	−2.881

pK_{BH^+} values via equation (17) in sulfuric acid, it is important to use $\log C_{H^+}$ values that are compatible with the X scales used. The scales provided here are compatible. With the M_C scales $\log C_{H^+}$ values given by Robertson and Dunford[38] are used. With the X scale given by Bagno, Scorrano and More O'Ferrall[39] the $\log C_{H^+}$ scale given there must be used, since it combines both H_3O^+ and H_2SO_4 concentrations (and thus their values depart from the strictly aqueous standard state).

The X-functions in Tables 3–8 are quite general; they were derived using data for 165 weak bases in H_2SO_4, 76 in $HClO_4$, and 29 in HCl, and include

Table 7 Values of X and $-H_0$, and of $\log C_{H^+}$ and $\log a_{H_2O}$ in molarity units, for aqueous hydrochloric acid at 25°C at different values of wt% acid.

Wt% HCl	X	$\log C_{H^+}$	$\log a_{H_2O}$	$-H_0$
0.20	0.011	−1.262	1.741	−1.251
0.50	0.027	−0.863	1.739	−0.836
1.00	0.055	−0.561	1.735	−0.506
2.00	0.113	−0.258	1.726	−0.145
3.00	0.175	−0.080	1.716	0.095
4.00	0.240	0.047	1.707	0.287
6.00	0.381	0.228	1.686	0.609
8.00	0.534	0.357	1.662	0.891
10.00	0.699	0.458	1.636	1.157
12.00	0.874	0.541	1.606	1.415
14.00	1.058	0.612	1.572	1.670
16.00	1.251	0.674	1.534	1.925
18.00	1.452	0.729	1.492	2.181
20.00	1.660	0.779	1.446	2.439
22.00	1.873	0.824	1.394	2.697
24.00	2.091	0.866	1.338	2.957
26.00	2.313	0.905	1.276	3.218
28.00	2.538	0.941	1.208	3.479
30.00	2.765	0.975	1.134	3.740
32.00	2.993	1.006	1.055	3.999
34.00	3.221	1.036	0.969	4.257
36.00	3.448	1.065	0.878	4.513
38.00	3.673	1.092	0.782	4.765
40.00	3.896	1.117	0.685	5.013

protonation at N, C, O and S atoms, and negatively, positively and dipositively charged species as well as neutral ones.[20,36] The computer derivation technique used has several advantages. Overlap between adjacent indicators is unnecessary when using it (although the scales given here involve a great deal of overlap), and anchoring with a known compound is not necessary either, since X, being an activity coefficient ratio, automatically becomes zero in the standard state. Acidity functions as such become unnecessary when using X, but one specifically applicable to the substrate (S) being studied, H_S, can easily be obtained if required by using equation (20). H_S scales of this type have already seen some use, for instance in studies of the kinetics of the aquation of some inorganic cobaloxime complexes in aqueous H_2SO_4.[40] An H_0 scale is given in Tables 7 and 8 for HCl, obtained by simply adding X and $\log C_{H^+}$.[36] H_0 scales for the other acids can be obtained in the same way.

$$-H_S = \log C_{H^+} + m^* X \tag{20}$$

Another advantage to the excess acidity approach comes from the realization

Table 8 Values of X and $-H_0$, and of $\log C_{H^+}$ and $\log a_{H_2O}$ in molarity units, for aqueous hydrochloric acid at 25°C at different values of the acid molarity.

[HCl] (M)	X	$\log C_{H^+}$	$\log a_{H_2O}$	$-H_0$
0.10	0.020	−1.000	1.740	−0.980
0.20	0.039	−0.699	1.737	−0.660
0.30	0.060	−0.523	1.734	−0.463
0.40	0.081	−0.398	1.731	−0.317
0.50	0.102	−0.301	1.727	−0.199
0.75	0.156	−0.125	1.719	0.031
1.00	0.213	0.000	1.711	0.213
1.50	0.333	0.176	1.693	0.509
2.00	0.461	0.301	1.674	0.762
2.50	0.595	0.398	1.653	0.993
3.00	0.735	0.477	1.630	1.212
3.50	0.880	0.544	1.605	1.424
4.00	1.029	0.602	1.577	1.631
4.50	1.182	0.653	1.548	1.835
5.00	1.337	0.698	1.517	2.035
5.50	1.495	0.740	1.483	2.235
6.00	1.654	0.778	1.447	2.432
6.50	1.814	0.812	1.409	2.626
7.00	1.976	0.845	1.368	2.821
7.50	2.137	0.874	1.325	2.199
8.00	2.299	0.902	1.280	3.201
8.50	2.460	0.929	1.232	3.389
9.00	2.621	0.954	1.182	3.575
9.50	2.782	0.977	1.129	3.759
10.00	2.941	0.999	1.073	3.940
10.50	3.098	1.021	1.016	4.119
11.00	3.255	1.041	0.956	4.296
11.50	3.409	1.060	0.894	4.469
12.00	3.561	1.078	0.830	4.639
12.50	3.711	1.096	0.766	4.807
13.00	3.858	1.113	0.701	4.971

that it is based on a linear free energy relationship. If this is so, then equation (21) is true, and values of X at any absolute temperature T can be derived by simply multiplying the values of X measured at 25°C by the absolute temperature ratio, as in equation (22):[41,42]

$$X = \frac{-\Delta G^*}{RT \ln 10} \tag{21}$$

$$X_T = X_{25^\circ}\left(\frac{298.15}{T}\right) \tag{22}$$

Thus it is not necessary to derive X scales at different temperatures. This has, nevertheless, been done,[43] although the scales (now called Mc) or polynomial coefficients enabling their calculation are not given.[43] Values of $\log C_{H^+}$ for HCl

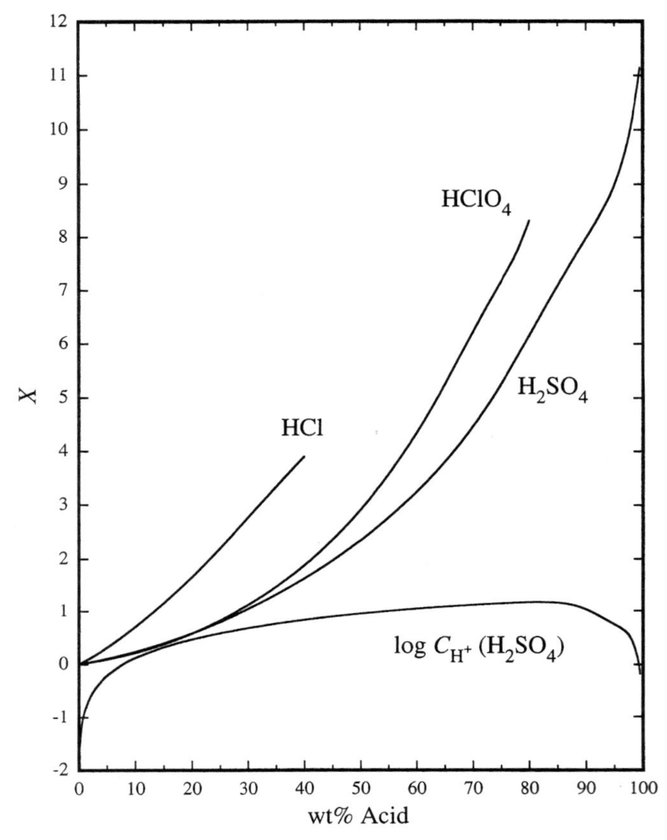

Fig. 1 Plots of X as a function of wt% acid for aqueous H_2SO_4, $HClO_4$ and HCl media at 25°C, with $\log C_{H^+}$ values for H_2SO_4.

and $HClO_4$ can be adjusted to temperature via the available solution densities, but this is more difficult for H_2SO_4. The densities can be used, but the degree of dissociation of bisulfate ion also changes with temperature,[44] and, frankly, good data on bisulfate and sulfate concentrations in aqueous sulfuric acid mixtures as a function of temperature are lacking. To a first approximation, the 25°C values, adjusted via the solution densities, work reasonably well. Values of $\log C_{HSO_4^-}$ at 25°C, needed for kinetic studies, are listed in Tables 3 and 4.

Tables 3 and 4 contain values of the log water activity and log sulfuric acid activity in molarity units. These can be obtained at any temperature by using the polynomial coefficients supplied by Zeleznik,[45] which are based on all of the pre-existing thermodynamic data obtained for this medium. The numbers were converted to the molarity scale using the conversion formula given in Robinson and Stokes.[46] Molarity-based water activities are given for $HClO_4$ in Tables 5 and 6. These are calculated from data obtained at 25°C by Pearce and Nelson,[17]

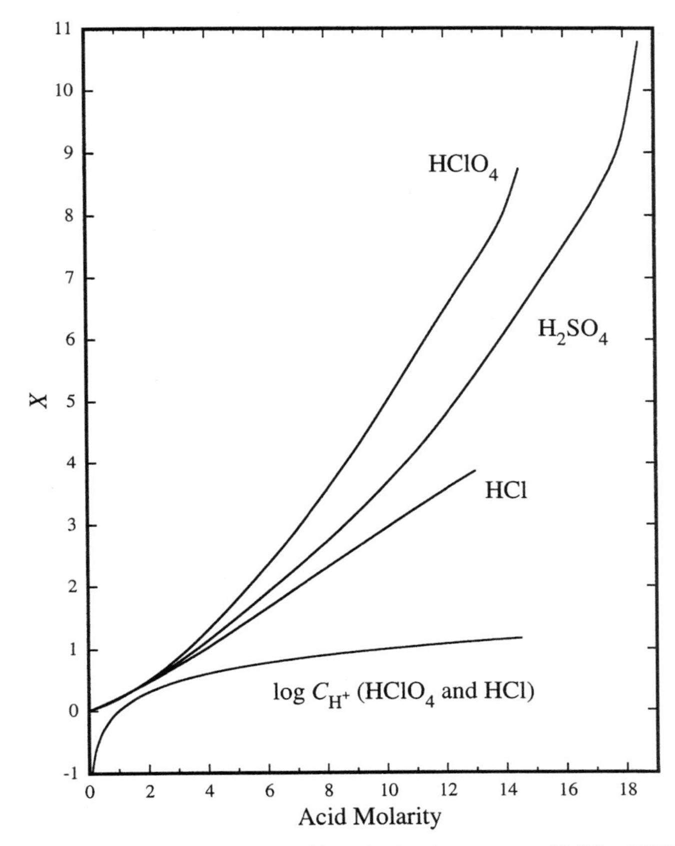

Fig. 2 Plots of X as a function of the acid molarity for aqueous H_2SO_4, $HClO_4$ and HCl media at 25°C, with $\log C_{H^+}$ values for HCl and $HClO_4$.

Robinson and Baker,[18] and Wai and Yates;[19] below 71 wt% they can be converted to other temperatures using partial molal enthalpies calculated from the relative enthalpies given by Bidinosti and Biermann.[47] Similar values for HCl in Tables 7 and 8 are obtained from Randall and Young[48] and Åkerlöf and Teare,[49] quoted by Bunnett,[50] and can be converted to other temperatures using coefficients supplied by Liu and Grén.[51] The author uses computer programs, written in Microsoft Basic for the Macintosh, that compute these quantities and others.

Plenty of other X and M_C scales are available, for these acid media and others. We have published an X scale for aqueous HBr and an X_0 scale for HCl,[36] and an X scale for $DClO_4$,[52] although different scales for deuterated acids are unnecessary; the scales for the protonated acids can be used, along with the assumption of equal acidity at equal molarity or mole fraction for protonated and deuterated acids,[53] which works very well. Marziano, Tomasin

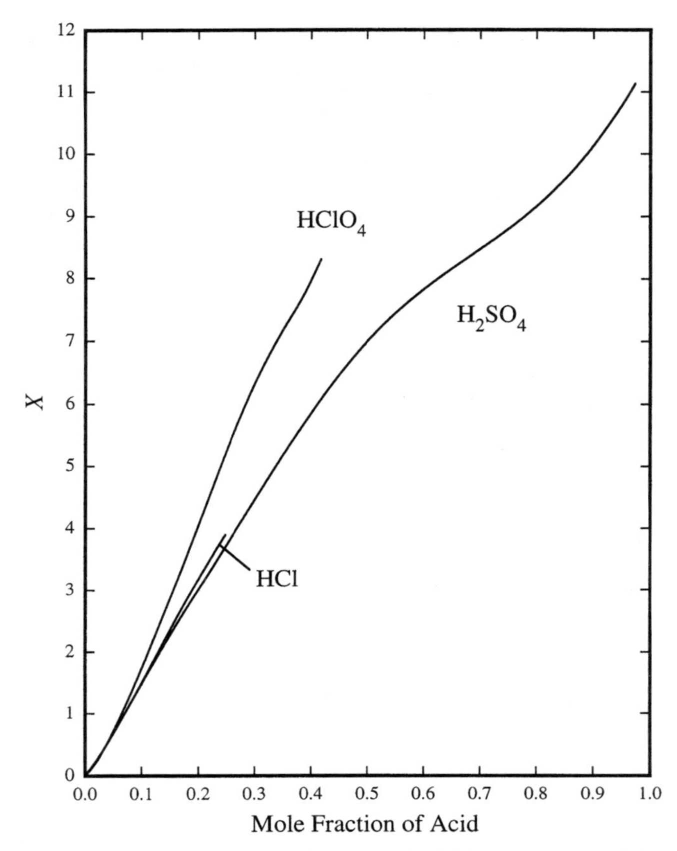

Fig. 3 Plots of X as a function of mole fraction of acid for aqueous H_2SO_4, $HClO_4$ and HCl media at 25°C.

and Traverso[54] recomputed M_C scales in aqueous H_2SO_4 and $HClO_4$ assuming that the same indicator has the same pK_{BH^+} value in both media. Analysis shows little if any difference between the various computed M_C or X scales in H_2SO_4 and $HClO_4$.[36,55]

$$Mc(\mathrm{s}) = \log \frac{f_{\mathrm{CH_3SO_3^-}} f_{\mathrm{H^+}}}{f_{\mathrm{CH_3SO_3H}}} \tag{23}$$

Degrees of dissociation of aqueous methanesulfonic acid mixtures and an M_C scale in these media are reported.[56] Mc(s) and Ac(s) scales in aqueous CH_3SO_3H, HNO_3, CF_3SO_3H, $HClO_4$ and HBr have been developed, Mc(s) referring to an activity coefficient ratio for the acid solvent itself rather than for organic indicators, e.g. equation (23) for CH_3SO_3H, and Ac(s) being acidity functions for them based on equation (20).[57] These have been used to obtain pK_{BH^+} values and n_{ij} slopes in these media, and the results have been analyzed.[58]

Most recently, aqueous H_2SO_4 and CF_3SO_3H mixtures have been studied.[59] The latest terminology[59] seems to be: *Mc*(s), referring to an activity coefficient ratio for the solvent itself, equation (23); *Mc*(i), referring to an activity coefficient ratio based on various indicators (i.e. an *X* scale), equation (16); and *Ac*(s) and *Ac*(i) for equation (20)-type acidity functions. This author finds all this terminology, with its various changes, quite confusing, and will stick to *X* and m^*. Any of the *X* or M_C scales, if used with the correct $\log C_{H^+}$ values for the medium and scale in question, will give reliable pK_{BH^+} and m^* values.

DETERMINATION OF BASICITIES

The actual using of the excess acidity method to obtain weak basicities and slope parameters is quite easy, and is now the method of choice. The following examples are chosen at random on the basis of not having been published before elsewhere, or of having been recalculated, and to illustrate the different possibilities and different acid systems.

Carpentier and Lemetais[60] have published ionization ratios for the protonation of some secondary aromatic amines in aq. H_2SO_4, and these are illustrated as an excess acidity plot ($\log I - \log C_{H^+}$ vs. *X*, using the values in Table 3) in Fig. 4. In all the calculations in this section the data have been weighted using an

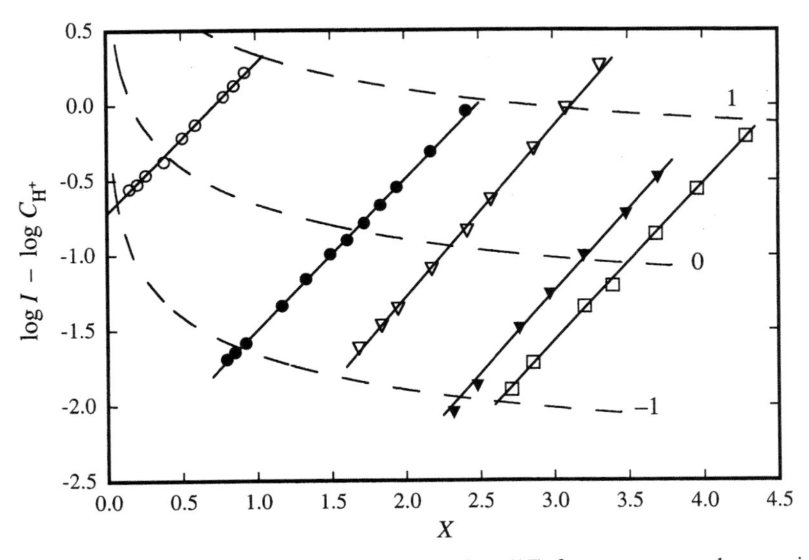

Fig. 4 Excess acidity plots according to equation (17) for some secondary amines in aqueous sulfuric acid at 25°C. Open circles, *N*-ethyl-4-chloro-2-nitroaniline; closed circles, 4-nitrodiphenylamine; open triangles, 2-nitrodiphenylamine; closed triangles, *N*-ethyl-2,4-dinitroaniline; open squares, 5-chloro-2-nitrodiphenylamine. Data from ref. 60; numbers to the right of the dashed lines are $\log I$ values.

Table 9 Intercept pK_{BH^+} values and m^* slopes for some secondary aromatic amines in aqueous sulfuric acid at 25°C.[a]

Substrate	AF value[a]	pK_{BH^+} [b]	m^* [b]	N [c]	r [d]
N-Ethyl-4-chloro-2-nitroaniline	−0.69	−0.7218 ± 0.0052	1.003 ± 0.010	9	0.9997
4-Nitrodiphenylamine	−2.48	−2.513 ± 0.019	1.009 ± 0.011	12	0.9994
2-Nitrodiphenylamine	−3.28	−3.569 ± 0.038	1.141 ± 0.016	9	0.9993
N-Ethyl-2,4-dinitroaniline	−4.40	−4.537 ± 0.082	1.097 ± 0.026	7	0.999
5-Chloro-2-nitrodiphenylamine	−4.69	−4.748 ± 0.096	1.057 ± 0.028	7	0.998

[a] Data from ref. 60.
[b] Intercepts and slopes of Fig. 4. The errors are standard deviations.
[c] Number of points.
[d] Linear correlation coefficient.

error function provided by Kresge and Chen,[61] which allows data between the $\log I$ values of +1 and −1 to contribute more to the slope and intercept determination; the $\log I$ values are indicated on the figures. The results obtained are given in Table 9, and they illustrate very well the accuracy and limitations of the method. As can be seen, the farther away from the standard state, the higher the errors and the poorer the correlation coefficients (although the values in Table 9 would all be regarded as very good). The more the slopes differ from one another, the greater the difference between the excess acidity values and those obtained by deriving an acidity function from the $\log I$ values.

Ionization ratios obtained as a function of temperature can be analyzed to obtain enthalpies and entropies of ionization, by adjusting X to the appropriate temperature using equation (22); if this is done, the resulting m^* values are temperature independent. Many values obtained in aqueous H_2SO_4 in this way for a considerable variety of compounds are already available, e.g. for nitroanilines, pyridines, pyridine oxides, thioamides and amides.[42,62,63] In Fig. 5 are presented excess acidity plots for three amides studied by Attiga and Rochester[64] in aqueous $HClO_4$ at 15, 25, 35 and 45°C, of which the 15 and 45°C data are shown. Amides have much smaller m^* values than do aromatic amines,[20] so the lines in Fig. 5 are less steep than the ones in Fig. 4. The pK_{BH^+} values at 25°C, m^* and the enthalpies and entropies of ionization are given in Table 10. The latter bear little relation to the numbers given by Attiga and Rochester,[64] however, probably owing to the fact that these authors used an H_A acidity function for $HClO_4$,[65] now known to have been mis-anchored.[66]

$$\mathrm{ROH} + \mathrm{H}^+ \underset{}{\overset{K_{\mathrm{R}^+}}{\rightleftharpoons}} \mathrm{R}^+ + \mathrm{H_2O} \qquad (24)$$

$$K_{\mathrm{R}^+} = \frac{a_{\mathrm{ROH}}a_{\mathrm{H}^+}}{a_{\mathrm{R}^+}a_{\mathrm{H_2O}}} = \frac{C_{\mathrm{ROH}}}{C_{\mathrm{R}^+}} \cdot C_{\mathrm{H}^+} \cdot \frac{f_{\mathrm{ROH}}f_{\mathrm{H}^+}}{f_{\mathrm{R}^+}} \frac{1}{a_{\mathrm{H_2O}}} \qquad (25)$$

$$\log I - \log C_{\mathrm{H}^+} + \log a_{\mathrm{H_2O}} = \mathrm{p}K_{\mathrm{R}^+} + m^* X \qquad (26)$$

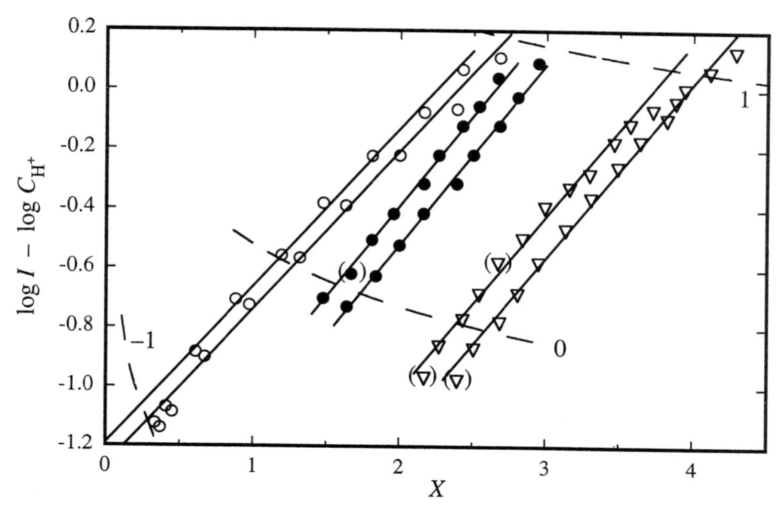

Fig. 5 Excess acidity plots according to equation (17) for three amides in aqueous perchloric acid at 45°C (upper lines) and 15°C (lower lines). Open circles, 4-methoxybenzamide; closed circles, 3,4,5-trimethoxybenzamide; open triangles, 3-nitrobenzamide. Data from ref. 64; numbers to the right of the dashed lines are $\log I$ values.

For protonation–dehydration processes, such as trityl cation formation from triphenylcarbinols, equation (24), the water activity has to be included if the formulation of the activity coefficient ratio term is to be the same as that in equation (7), which it should be if linearity in X is to be expected; see equation (25). The excess acidity expression in this case becomes equation (26); this is a slightly different formulation from that used previously for these processes,[36] the one given here being more rigorous. Molarity-based water activities must be used, or else the standard states for all the species in equation (26) will not be the same, see above. For consistency this means that all values of pK_{R^+} listed in the literature will have to have 1.743 added to them, since at present the custom

Table 10 pK_{BH^+} values, m^* slopes, enthalpies and entropies of ionization for some amides in aqueous perchloric acid.[a]

Substrate	pK_{BH^+} at 25°C	m^*	$\Delta H^{\circ\, b}$	$\Delta S^{\circ\, c}$
4-Methoxybenzamide	-1.241 ± 0.014	0.5298 ± 0.0095	1.06 ± 0.19	9.23 ± 0.65
3,4,5-Trimethoxy-benzamide	-1.688 ± 0.014	0.6065 ± 0.0068	1.808 ± 0.092	13.79 ± 0.31
3-Nitrobenzamide	-2.276 ± 0.026	0.5874 ± 0.0087	1.92 ± 0.14	16.87 ± 0.50

[a] Data from ref. 64.
[b] In kcal mol^{-1}.
[c] In cal mol^{-1} K^{-1}.

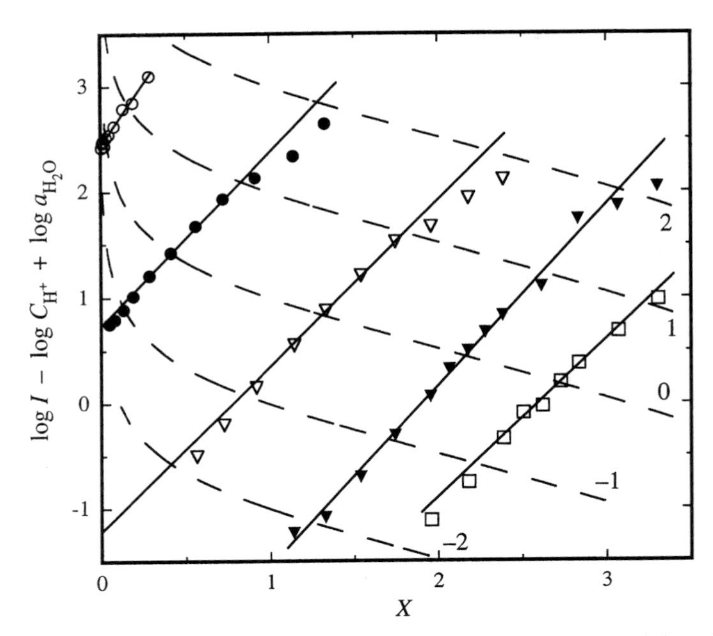

Fig. 6 Excess acidity plots according to equation (26) for some triphenylcarbinols (TPCs) in aqueous HCl at 26°C. Open circles, 4,4′,4″-trimethoxy-TPC; closed circles, 4,4′-dimethoxy-TPC; open triangles, 4-methoxy-TPC; closed triangles, 4-methyl-TPC; open squares, TPC. Data from ref. 67; numbers to the right of the dashed lines are $\log I$ values.

is to leave out the molecule of water formed in equation (24) *as well as* the one solvating the H^+, discussed above.

Plots according to equation (26) for some triphenylcarbinols at 26°C in aqueous HCl[67,68] are shown in Fig. 6; the slopes and pK_{R^+} values (on the basis discussed above) are listed in Table 11. The values of X, $\log C_{H^+}$ and $\log a_{H_2O}$

Table 11 Intercept pK_{R^+} values and m^* slopes for some triphenylcarbinols in aqueous hydrochloric acid at 26°C.[a]

Substrate	AF value[a]	pK_{R^+}[b]	m^{*b}	N^c	r^d
4,4′,4″-tri-OMe-TPC	2.56	2.4348 ± 0.0072	2.34 ± 0.13	10	0.990
4,4′-di-OMe-TPC	0.68	0.733 ± 0.028	1.641 ± 0.062	11	0.994
4-OMe-TPC	−1.34	−1.218 ± 0.085	1.566 ± 0.063	10	0.994
4-Me-TPC	−3.19	−3.26 ± 0.13	1.713 ± 0.058	13	0.994
Triphenylcarbinol (TPC)	−4.38	−3.90 ± 0.21	1.502 ± 0.078	9	0.991

[a] Data from refs 67 and 68.
[b] Intercepts and slopes of Fig. 6. The errors are standard deviations.
[c] Number of points.
[d] Linear correlation coefficient.

given in Table 7 were used; as can be seen, the linearity of the resulting plots and the errors on the results in Table 11 are quite acceptable. Carbocations have much larger m^* values than aromatic amines,[36] so the lines in Fig. 6 are steeper than the ones in Fig. 4. Recently the pK_{R^+} of triphenylcarbinol has been measured by an entirely different method than the one under discussion here, from the ratio of the directly measured rates of the forward and reverse reactions in equation (24) ($R = Ph_3C$).[69] The value thus determined (-4.72 ± 0.03) is in good agreement with most of the values found by using the excess acidity and acidity function methods (it is poorest for the ones measured in aqueous HCl, see Table 11). This increases considerably the confidence one can place on basicity values obtained in strong acids.

The excess acidity method is now the method of choice for the determination of pK_{BH^+} and pK_{R^+} values, being used by 29 different research groups, at last count (end of 1999). Space considerations preclude citing every reference, so only the most recent work of those using it will be mentioned. Chandler and Lee[70] have studied the protonation of alcohols using ^{13}C NMR spectroscopy. Freiberg and his group[71] have determined the basicities of many ketones and other carbonyl compounds, mainly using ^1H NMR spectroscopy; some oximes have also been studied.[72] Géribaldi *et al.* have looked at many α,β-unsaturated ketones.[73] Haldna and his group have studied ketones[74] and hydroxybenzoic acids,[75] and Lee and Sadar have looked at aliphatic carboxylic acids.[76] Some thiophene-2-carboxamides are discussed by Alberghina *et al.*;[77] thiophenes[78] and many other compounds have been studied by the Spinelli group, among them benzamides[79] and benzoic acids.[80] García *et al.* have examined many equilibria in strong acids, also including benzamides.[81] Perisic-Janjic *et al.* have examined the protonation of some phthalimides and other nitrogen-containing compounds using UV spectroscopy,[82] and hydroxamic acids have been looked at by Ghosh and Ghosh[83] and by Sharma *et al.*[84] Many sulfur-containing compounds have been studied by the Edward group.[85] Moodie's group have observed a switch from *N*-protonation at low acidities to *O*-protonation at high acidities for 1-methoxycarbonylpiperidine.[86]

Marziano's group have determined the basicity of mesitylene in aqueous CF_3SO_3H media, and estimated the basicities of some other aromatics.[87] Bagno and Scorrano have looked at many equilibria by various techniques, among them the protonation–dehydration equilibria involved in the formation of carbocations.[88] The latter process has also been studied by Dorion and Gaboriaud.[89] Andonovski *et al.* have studied indoles,[90] as have Bálon's group[91] and Gut and Wirz.[92] DNA bases and related molecules have received much attention.[93,94] Pincock and Redden have used X values for some excited-state protonation studies.[95] Even inorganic molecules in strong acids have been examined by the excess acidity technique.[96–98] Many authors have also determined pK_{BH^+} values as part of a kinetic study, and some of these will be discussed below.

MEDIUM EFFECTS

The measurement of pK_{BH^+} and m^* values is sometimes complicated by the fact that the spectroscopic methods used (generally ^{1}H and ^{13}C NMR spectroscopy, and UV-vis) are subject to medium effects. Water and (say) 80 wt% H_2SO_4 are very different media, and it is not very surprising that spectral peaks for the same species in the two media can occur at different wavelengths or different chemical shifts. Several methods have been devised for handling this problem;[24,99,100] the excess acidity method lends itself to dealing with medium effects quite well.[101]

It is well known[23] that ionization ratios I can be determined from spectra using the expression $I = (D - D_B)/(D_A - D)$, where D is the measured UV-vis optical density or NMR chemical shift and D_A and D_B refer to the values for the pure acid form (i.e. BH^+) and pure base form (i.e. B), respectively. Recasting this in terms of D gives equation (27):

$$D = \frac{D_B + ID_A}{I + 1} \tag{27}$$

Using modern computer curve-fitting techniques,[102] equations (17) and (27) can be fitted simultaneously with variables X, $\log C_{H^+}$ and D, obtaining pK_{BH^+}, m^*, D_B and D_A as coefficients, which is a more precise technique than calculating $\log I$ values first and using equation (17) and the error function[61] discussed above.

$$D_A = D_{Aaq} + \delta D_A I'_A \quad \text{or} \quad D_B = D_{Baq} + \delta D_B I'_B \tag{28}$$

$$I'_A = \frac{I_A - 1}{1 + I_A}, \quad \text{where} \quad \log I_A = \Delta m^*_A X \tag{29}$$

With medium effects present, D_B or D_A (or both) also become functions of acidity, as shown in equation (28). I' is a function of X according to equation (29) (written in terms of A); the additional coefficient Δm^*_A often takes values near 0.1.[101,103] In fact, with values near 0.1, I' proves to be linear in the substrate acidity function H_S,[101] in agreement with the empirical observations of Katritzky, Waring and Yates,[24] lending considerable support to this analysis. Suitable substitutions of equations (27)–(29) into equation (17) will give all the coefficients with considerable accuracy, provided that sufficient data points are available (values of D as a function of acidity, in our hands at least 20 values are needed). This technique has proved to be of great value in determining the pK_{BH^+} and m^* values for benzamides, acetamides and lactams, measured using UV-vis, ^{1}H NMR and ^{13}C NMR spectroscopy.[103]

Another technique that has been used in recent years to deal with medium effects on basicity determinations is factor analysis,[104] also known as characteristic vector analysis.[105] This technique, first developed by Reeves,[100] can be used for correcting for medium effects, but only if used with considerable care. It has been shown that the basic technique developed by Edward and Wong[105] does not work as such,[101] but that the coefficients of the first vector can be used with the excess acidity technique described above to obtain accurate acidity constants.[101] Alternatively, the target testing approach described by Malinowski and Howery[104] can be used to provide reliable pK_{BH^+} and m^* values, as has been shown in some detail by Haldna and his group.[75,106–112] The latter technique has been shown to give essentially the same pK_{BH^+} and m^* values for amides as does the excess acidity method.[113]

SLOPE VALUES; OTHER MEDIA

The excess acidity slope values, m^*, have been associated with solvation differences between the unprotonated B form and the protonated BH^+ form in acid media. This aspect has been thoroughly reviewed[39] and will not be considered further here. However, it is interesting to speculate as to what the maximum possible value of m^* might be. To this end a plot against X for H_2SO_4, and X and X_0 for $HClO_4$, of values of $\log f^*_{H^+}$, obtained using an electrochemical technique and the TEA^+PCP^- assumption,[114] is given in Fig. 7.

This highly interesting plot shows good linearity for the X scales in both media, with slopes that are, within experimental error, the same: 2.105 ± 0.012. The intercepts are small, although not quite zero (0.069 ± 0.021 for H_2SO_4 and 0.260 ± 0.044 for $HClO_4$) and the correlation coefficients are very high at 0.9998 and 0.9991. Thus one could guess that pK_{BH^+} values obtained in $HClO_4$ media might be more error-prone than those obtained in H_2SO_4. The close similarity of the two X scales is good evidence that they are, in fact, measuring the same thing in the different acid media. However, the X_0 scale in $HClO_4$ does not correlate nearly so well; the plot looks curved and the correlation coefficient is only 0.996.

One could thus speculate that the slope value for $\log f_{H^+}$, 2.1, is the largest value possible for m^*; no larger values are reported (except for the error-prone value of 2.34 for 4,4′,4″-trimethoxytriphenylcarbinol reported here in Table 11). On the other hand, Marziano's group[115] have found that the $M_C(s)$ function, which describes the dissociation of the acid itself, is linear in the normal indicator-based $M_C(i)$ function with slopes of 3.30 for CH_3SO_3H, 2.80 for HBr, 2.35 for $HClO_4$, 2.10 for CF_3SO_3H and 1.55 for HNO_3.[57] One could guess that these were the largest possible m^* values in these solvent systems. At present there is no way to tell which, if either, speculation is correct.

The $M_C(s)$ approach leads to the conclusion that the m^* values for a series of indicators increase as the indicator pK_{BH^+} values become less basic.[43,58,59] This is interesting, insofar as we have observed a similar phenomenon for rate constants. Slope values give linear plots against Hammett σ^+ values for some acylal and thioacylal hydrolyses and also for the hydrolyses of some other sulfur-containing molecules,[116] and for some aromatic hydrogen exchange processes,[117] slopes mostly increasing as reactivity decreases. However, for

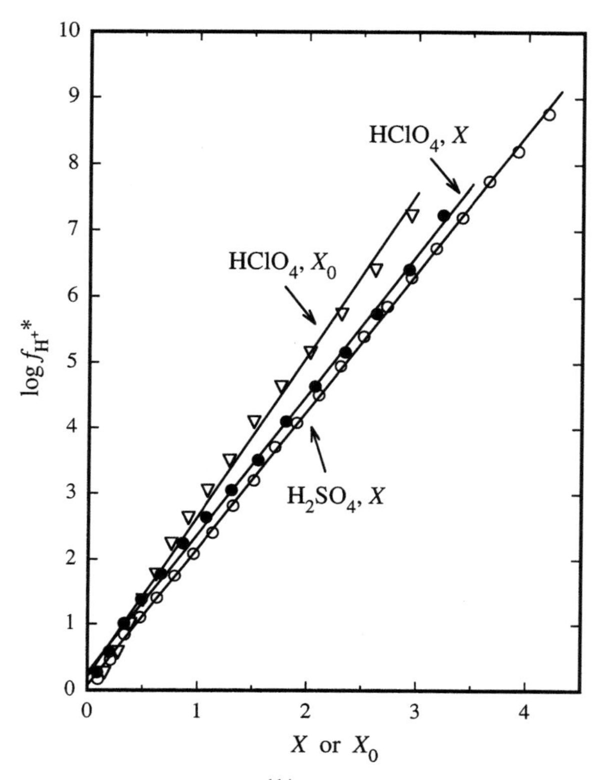

Fig. 7 Plots of $\log f^*_{H^+}$ values at 25°C,[114] defined relative to the standard ion TEA^+, against X for aqueous H_2SO_4 and $HClO_4$, and against X_0 for aqueous $HClO_4$.

some phenylacetylene hydrations the slopes mostly stay the same or slowly *de*crease with decreasing reactivity,[118] and this is also true for some alkyl-nitramine decompositions (plots against σ^*).[119] For styrene hydrations and related reactions the slopes stay the same with changing substrate reactivity.[120] What, if anything, this means, or what the slopes of the linear plots mean is not known at present.

It is an observational fact that m^*, m from the Yates–McClelland method and $(1 - \phi_e)$ from the Bunnett–Olsen method are all roughly equivalent, although they are defined in quite different ways. We made a comparison and pointed this out;[121] for sulfuric acid a plot of $-(H_0 + \log C_{H^+})$ against X is accurately linear (correlation coefficient 0.9996) over the 0–99.5 wt% concentration range, slope 0.983 (although this is not the case for perchloric acid). Thus for acid media without measured X or M_C scales (which are becoming fewer all the time) the Bunnett–Olsen method can be used as a good second-best, since almost every acid system has a measured H_0 scale. However, the philosophical objection to using it, mentioned above, still holds.

Many workers in the field like to use the X_0 scale for aqueous perchloric acid media, since Kresge's group showed that it offered the best extrapolation from concentrated to dilute aqueous acids.[122] This may be a problem in dilute acid (see the small intercepts in Fig. 7); for instance, García and Leal have pointed out that the pH scale and the excess acidity scale do not correspond well in dilute solution.[123] Examination of Fig. 7 shows that it might not be such a good idea to use X_0, for instance if numbers obtained in H_2SO_4 and $HClO_4$ are to be compared. However, as was pointed out above, it does not matter which scale is used, provided it is used consistently, if comparisons of results obtained in different experiments are to be made; this seems to be particularly true for kinetic measurements. It is used for this purpose by the Kresge and Lajunen groups, among others; for instance, Ortiz *et al.* use it to model behavior in complex media.[124,125]

Many reactions of interest involve substrates that are not very soluble in water, and for this reason organic solvents are often added to the aqueous acid system used. These can be treated simply as inert diluents as long as they do not undergo reaction with the acid medium (e.g. not alcohols, which esterify). Dioxane is commonly used for this purpose, at concentrations up to 60 vol% (which may be a little high to regard it as a simple diluent), by the groups of Ghosh,[126,127] Satchell,[128,129] and others. Satchell's group measured H_0 and X functions in the 10 vol% dioxane–aqueous perchloric acid solvent system and found them to be very similar to those found in the purely aqueous medium,[130] so it seems that regarding non-reactive ethers as simple diluents is reasonable.

OTHER MATTERS

The excess acidity method was originally developed as a general tool for the study of the mechanisms of reactions occurring in strong aqueous acid media, which is the author's main interest, since the H_0 and other acidity function methods had obvious deficiencies (see below). It has also proved to be of considerable utility in the determination of pK_{BH^+} values, dealing with medium effects, and so on, but this was not really intended to be its prime function. Like all general methods, it sometimes gives results that are not as good as those obtained by using more specific, purpose-designed techniques. For instance, the pK_{BH^+} values for some simple aliphatic amides reported by McTigue's group,[131] using a conductimetric technique, are undoubtedly highly reliable, as a later comparison of values obtained using conductimetry and ones obtained using UV-vis spectroscopy and the excess acidity method would appear to indicate.[132]

Johnson and Stratton[133] have compared pK_{BH^+} values obtained using the Hammett acidity function method and the excess acidity method, and they consider that results obtained using the former may correspond more nearly to the true thermodynamic quantities. This is only going to be true if the Hammett cancellation approximation, equation (9), applies to the acidity function used with some exactitude. Criteria for deciding this have been given;[3,20] in general it is remarkable how poorly this approximation works for most published acidity functions. "Good" acidity functions are the H_0 (primary aromatic amines) and H_I (indoles) scales in H_2SO_4, and the H_A (amides) scale in $HClO_4$. That is all; in $HClO_4$ even the H_0 scale is not very good.[20]

Values of pK_{BH^+} obtained using the Hammett acidity function, Bunnett–Olsen and excess acidity methods have been compared by Johnson and Stratton[134] and by Kresge's[122] and Spinelli's groups.[135] Values obtained using the M_C and X functions have been compared with those resulting from the application of the Bunnett–Olsen method (using different acidity functions) in a series of papers by Haldna.[136–139] In H_2SO_4 the results given by the different techniques are in "rather good" agreement,[137] but for very weak bases (half-protonated in > 70 wt% H_2SO_4 or > 50 wt% $HClO_4$) there can be considerable disagreement between results obtained using the three methods. This is not particularly surprising, as the required extrapolations to the aqueous standard state are very long for these compounds. On the other hand, Kresge *et al.*[122] suggest that extrapolations using the X_0 function are the most reliable for the stronger bases. The differences in pK_{BH^+} values for very weak bases obtained in aqueous H_2SO_4 and $HClO_4$ media are also apparent from the comparison in Cox and Yates;[20] the reason for it is not clear at this time. Traverso suggests that the confidence limits on the pK_{BH^+} and m^* values obtained for very weak bases are quite large.[140] Forcing the same indicator base to have the same pK_{BH^+} values, or the same pK_{BH^+} and m^* values, in aqueous H_2SO_4, $HClO_4$ and HCl media does not change the calculated M_C or X functions much, if at all.[36,54,55] For this review I contemplated deriving new, up-to-date excess acidity scales incorporating the latest data, but I decided against it, since the resulting X-functions were, within experimental error, the same as the ones quoted in Tables 3–8, dating from 1978 and 1981! In practical terms, as mentioned, the Bunnett–Olsen method is quite useful in other acids.

Wojcik has offered the criticism that basicity values determined by the acidity function or excess acidity methods are contaminated by an "indeterminate constant", and may be inaccurate because of this, particularly for very weak bases.[141,142] It is difficult to know what to do about this (Wojcik does not offer any means of overcoming the problem), but it is likely that the constant is zero, or at any rate small, since comparison between values obtained in these ways and those obtained by totally different methods, where possible,[69,131] shows that the excess acidity-derived basicities compare very well. If the indeterminate constant were large, noticeable effects would surely be apparent somewhere, considering the vast amount of work that has gone into basicity determinations in strong acid media over the years, but there is no evidence of it. The suggestion has been made[28] that excess acidity basicity determinations could be improved by adding a quadratic term to equation (17), $m^{**}X^2$, based on an observation by

Edward *et al.*,[143] but in practice this has not proved to be necessary, m^{**} values being indistinguishable from zero.[55]

4 Reactions in strong acids

A1 MECHANISM

$$\mathrm{S} + \mathrm{H}^+ \underset{\text{fast}}{\overset{K_{\mathrm{SH}^+}}{\rightleftharpoons}} \mathrm{SH}^+ \xrightarrow[\text{slow}]{k_0} \text{Products} \tag{30}$$

$$\text{trioxane} + \mathrm{H}^+ \rightleftharpoons \text{O-protonated trioxane} \xrightarrow[\text{slow}]{k_0} \mathrm{HO{-}CH_2{-}O{-}CH_2{-}O^+{=}CH_2} \xrightarrow{\text{fast}} 3\,\mathrm{H_2C{=}O} \tag{31}$$

This consists of protonation in a fast pre-equilibrium process followed by rate-determining reaction of the protonated substrate SH^+, equation (30); a typical example is the much-studied acid-catalyzed depolymerization of trioxane, shown in equation (31).[144]

$$\frac{\mathrm{d}C_\mathrm{P}}{\mathrm{d}t} = -\frac{\mathrm{d}(C_\mathrm{S} + C_{\mathrm{SH}^+})}{\mathrm{d}t} = k_\psi(C_\mathrm{S} + C_{\mathrm{SH}^+}) = k_0 a_{\mathrm{SH}^+}\frac{1}{f_\ddagger} = k_0 C_{\mathrm{SH}^+}\frac{f_{\mathrm{SH}^+}}{f_\ddagger} \tag{32}$$

Kinetic theory indicates that equation (32) should apply to this mechanism. Since the extent of protonation as well as the rate constant will vary with the acidity, the sum of protonated and unprotonated substrate concentrations, $(C_\mathrm{S} + C_{\mathrm{SH}^+})$, must be used. The observed reaction rate will be pseudo-first-order, rate constant k_ψ, since the acid medium is in vast excess compared to the substrate. The medium-independent rate constant is k_0, and the activity coefficient of the transition state, $f_\ddagger$, has to be included to allow equation of concentrations and activities.[145] We can use the antilogarithmic definition of h_0 in equation (33) and the definition of K_{SH^+} in equation (34):

$$h_0 = \frac{a_{\mathrm{H}^+} f_\mathrm{Am}}{f_{\mathrm{AmH}^+}} \tag{33}$$

$$\begin{aligned} K_{\mathrm{SH}^+} &= \frac{a_\mathrm{S} a_{\mathrm{H}^+}}{a_{\mathrm{SH}^+}} = \frac{C_\mathrm{S}}{C_{\mathrm{SH}^+}} \cdot a_{\mathrm{H}^+} \cdot \frac{f_\mathrm{S}}{f_{\mathrm{SH}^+}} \\ &= \frac{C_\mathrm{S}}{C_{\mathrm{SH}^+}} \cdot h_0 \cdot \frac{f_{\mathrm{AmH}^+} f_\mathrm{S}}{f_{\mathrm{SH}^+} f_\mathrm{Am}} \end{aligned} \tag{34}$$

Substituting for C_{SH^+} in equation (32), rearranging, cancelling and taking logs gives a usable rate equation, equation (35):[23]

$$\log k_\psi - \log\frac{C_S}{C_S + C_{SH^+}} = \log\frac{k_0}{K_{SH^+}} - H_0 + \log\frac{f_{AmH^+}f_S}{f_{Am}f_\ddagger} \quad (35)$$

If the substrate is predominantly unprotonated in the acidity range covered, the protonation correction term (second on the left in equation (35)) will be zero, and if the activity coefficient term on the right cancels to zero, $\log k_\psi$ values will be linear in $-H_0$ with unit slope (the Zucker–Hammett hypothesis).[146] In practice, linearity is usually observed,[23] e.g. for the trioxane depolymerization in equation (31),[144] although the slopes are seldom exactly unity. For more strongly basic substrates that are predominantly protonated in the acidity range covered, equation (36) is easily derived; this is acidity independent and should have zero slope against $-H_0$ if the substrate is fully protonated and the last term cancels.[145]

$$\log k_\psi - \log\frac{C_{SH^+}}{C_S + C_{SH^+}} = \log k_0 + \log\frac{f_{SH^+}}{f_\ddagger} \quad (36)$$

A-S_E2 MECHANISM

This involves rate-determining proton transfer, equation (37); in principle it should show general acid catalysis, but in practice this usually cannot be seen as the catalyzing acid is simply "H_3O^+". A typical example would be aromatic hydrogen exchange, such as the detritiation of tritiated benzene shown in equation (38):[147]

$$S + H^+ \xrightarrow[\text{slow}]{k_0} SH^+ \xrightarrow[\text{fast}]{} \text{Products} \quad (37)$$

$$C_6H_5\text{–}T + H^+ \xrightarrow[\text{slow}]{k_0} [C_6H_5(H)(T)]^+ \xrightarrow[\text{fast}]{-T^+} \text{Exchanged product} \quad (38)$$

Here the relevant rate equation is equation (39), which gives equation (40) if h_0 is used to substitute for a_{HA}:

$$k_\psi C_S = k_0 C_S a_{HA}\frac{f_S}{f_\ddagger} \quad (39)$$

$$\log k_\psi = \log k_0 - H_0 + \log\frac{f_{AmH^+}f_S}{f_{Am}f_\ddagger} \quad (40)$$

Equation (40) also predicts linearity in $-H_0$ with unit slope if the activity coefficient ratio term cancels, and thus is indistinguishable from equation (35) for A1 reactions.[145] Again, in practice plots according to equation (40) are often linear, but the slope is not unity.[23]

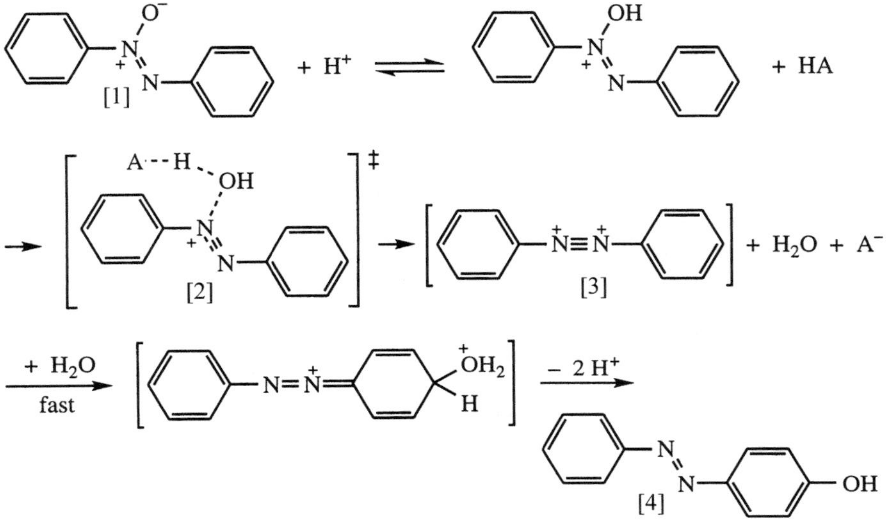

Scheme 1

One way around this difficulty is to exploit different acid media, e.g. to compare rate constants measured in sulfuric, perchloric, nitric and phosphoric acids.[148,149] Another is to work in sulfuric acid in the acidity region above 80 wt% H_2SO_4, where undissociated H_2SO_4 molecules (and $H_3SO_4^+$ in the strongest acids) are present in addition to H_3O^+, and catalysis by these different acid species can be seen. For instance, the Wallach rearrangement has been found to have the mechanism shown in Scheme 1;[150–154] the log reaction rate constants give linear correlations with $\log a_{H_2SO_4}$, and with $\log C_{H_3SO_4^+}$ in the strongest acids, but not with H_0. For azoxybenzene itself, [1], the first proton transfer is a fast pre-equilibrium, but the second is part of the rate-determining step, with HA in the transition state [2] being an undissociated H_2SO_4 molecule, or an $H_3SO_4^+$ ion in the strongest acids, but not H_3O^+. The symmetrical dicationic intermediate, which can be represented as [3], undergoes fast nucleophilic attack. Water is shown in Scheme 1, but the nucleophile is equally likely to be HSO_4^- or even H_2SO_4 in the strong acid media used; the 4-hydroxyazobenzene product [4] is obtained on workup.[150]

A2 MECHANISM

$$\mathrm{S} + \mathrm{H}^+ \underset{\text{fast}}{\overset{K_{\mathrm{SH}^+}}{\rightleftharpoons}} \mathrm{SH}^+ + \mathrm{Nu} \xrightarrow[\text{slow}]{k_0} \text{Products} \qquad (41)$$

$$\mathrm{R{-}C({=}O)OMe} + \mathrm{H^+} \underset{\text{fast}}{\overset{K_{SH^+}}{\rightleftharpoons}} \mathrm{R{-}C^{+}(OH)OMe} + 2\,\mathrm{H_2O} \xrightarrow[\text{slow}]{k_0} \mathrm{R{-}C(OH)_2{-}OMe}\ [5] + \mathrm{H_3O^+} \xrightarrow{\text{fast}} \text{Carboxylic acid and methanol} \qquad (42)$$

This is like the A1 mechanism except that the protonated substrate SH^+, instead of reacting alone, reacts with something else in the rate-determining step, a nucleophile or a base, written as Nu in equation (41). A typical example of this very common mechanism is ester hydrolysis, say of the methyl esters in equation (42), where SH^+ reacts with two water molecules in the slow step,[29] giving the neutral tetrahedral intermediate [5] directly. Reasonably enough, the rate equations obtainable are very similar to those for the A1 mechanism, with the addition of an extra a_{Nu} term, see equation (43):

$$\log k_\psi - \log \frac{C_S}{C_S + C_{SH^+}} = \log \frac{k_0}{K_{SH^+}} - H_0 + \log a_{Nu}$$

and (43)

$$\log k_\psi - \log \frac{C_{SH+}}{C_S + C_{SH^+}} = \log k_0 + \log a_{Nu}$$

Plots against H_0 according to equation (43) will be *curved*, so the A2 mechanism can be distinguished from the A1 and A-S_E2 mechanisms this way. (The original Zucker–Hammett hypothesis[146] had plots against $\log C_{H^+}$ being linear for an A2 process, but this has since been shown to be wrong,[145] and will not be covered further here.)

The predominant nucleophile/base in acid systems is *water*. Thus we can write equation (44):

$$\log k_\psi - \log \frac{C_S}{C_S + C_{SH^+}} + H_0 = \log \frac{k_0}{K_{SH^+}} + \log a_{H_2O}$$

and (44)

$$\log k_\psi - \log \frac{C_{SH+}}{C_S + C_{SH^+}} = \log k_0 + \log a_{H_2O}$$

and plots against the log water activity according to equation (44) should be linear for an A2 process.[23,50] They often are, up to a point anyway (slope = w); but the suggestion made at the time that these plots were first proposed was that plots for *all* types of reaction should be linear in $\log a_{H_2O}$, not just those for A2 reactions: reactions not involving water, $w \leq 0$; water acting as a nucleophile, $w = 1.2$–3.3; water acting as a proton-transfer agent, $w > 3.3$.[50] However, this is simplistic; for instance, it makes little sense to plot log rate constants against the log water activity if water is not involved in the reaction.

THE BUNNETT–OLSEN METHOD

This method can also be applied to reactions in strong acid media:[155]

$$\log \frac{f_S f_{H^+}}{f_\ddagger} = (1 - \phi_\ddagger) \log \frac{f_{Am} f_{H^+}}{f_{AmH^+}} \tag{45}$$

$$\log k_\psi + H_0 = \log \frac{k_0}{K_{SH^+}} + \phi_\ddagger (H_0 + \log C_{H^+}) \tag{46}$$

$$\log k_\psi = \log k_0 + (\phi_\ddagger - \phi_e)(H_0 + \log C_{H^+}) \tag{47}$$

By analogy with equation (12), the assumption made regarding the linearity of activity coefficient ratios is equation (45) (slope parameter $\phi_\ddagger$), and the resulting Bunnett–Olsen equations that apply to kinetic measurements are equations (46) and (47) for unprotonated and protonated substrates, respectively.[156] These apply to the A1 and A-S_E2 mechanisms; for the A1 and A2 mechanisms they may require correction for partial substrate protonation as in equations (25) and (26) above. For A2 reactions an additional term such as the log water activity has to be added as in equation (33). These equations have been widely tested and work quite well.[155–160] The difference between the Bunnett–Olsen and the excess acidity kinetic methods (discussed below) is that the Bunnett–Olsen method features an *additive* combination of the slope parameters ϕ_e and $\phi_\ddagger$, whereas the excess acidity method features a *multiplicative* one. There seems to be no theoretical justification for the former. Also the Bunnett–Olsen method still uses H_0, whereas acidity functions are not needed for the excess acidity approach; see above.

5 Application of excess acidities to kinetics

This is now the most widely used method for analyzing reaction kinetics and determining reaction mechanisms in aqueous strong acid media, being used routinely not only by ourselves but by at least twenty other research groups (as of the end of 1999). The three main mechanisms occurring in strong acid media will be covered first and others subsequently – substitutions, ring closures, condensations, enolizations, cyclopropane reactions, the benzidine rearrangement, and others.

A1 MECHANISM

This is given in equation (30), and the relevant rate equation is equation (32) above. Substituting the equation (48) definition of K_{SH^+} into equation (32), rearranging and taking logs gives equation (49):[145,161]

$$a_{SH^+} = \frac{a_S a_{H^+}}{K_{SH^+}} = \frac{C_S C_{H^+} f_S f_{H^+}}{K_{SH^+}} \tag{48}$$

$$\log k_\psi - \log \frac{C_S}{C_S + C_{SH^+}} - \log C_{H^+} = \log \frac{k_0}{K_{SH^+}} + \log \frac{f_S f_{H^+}}{f_\ddagger} \tag{49}$$

Now we make the excess acidity assumption that activity coefficient ratios such as that in equation (49) are linear in one another. The best assumption to make is that the term with the activity coefficient of the transition state is linear in the activity coefficient ratio for the *same* substrate S,[162,163] since this is as closely similar as possible to the one in equation (49), see equation (50). We already know that the latter terms are linear in X. Thus for unprotonated substrates the relevant rate equation becomes equation (51).[145,161]

$$\log \frac{f_S f_{H^+}}{f_\ddagger} = m^\ddagger \log \frac{f_S f_{H^+}}{f_{SH^+}} = m^\ddagger m^* X \tag{50}$$

$$\log k_\psi - \log \frac{C_S}{C_S + C_{SH^+}} - \log C_{H^+} = \log \frac{k_0}{K_{SH^+}} + m^\ddagger m^* X \tag{51}$$

For substrates that are predominantly protonated under the reaction conditions, the relevant rate equation is equation (52). A little algebra[162] shows that equation (53) is true, leading to equation (54):[145,161]

$$\log k_\psi - \log \frac{C_{SH^+}}{C_S + C_{SH^+}} = \log k_0 + \log \frac{f_{SH^+}}{f_\ddagger} \tag{52}$$

$$\log \frac{f_{SH^+}}{f_\ddagger} = (m^\ddagger - 1) \log \frac{f_S f_{H^+}}{f_{SH^+}} = (m^\ddagger - 1) m^* X \tag{53}$$

$$\log k_\psi - \log \frac{C_{SH^+}}{C_S + C_{SH^+}} = \log k_0 + (m^\ddagger - 1) m^* X \tag{54}$$

The excess acidity method can also be used to obtain the protonation correction terms, if needed, in equations (51) and (54); dividing by C_S shows them to be $\log(1/(I+1))$ and $\log(I/(I+1))$, for the unprotonated and protonated cases, respectively, and $\log I$ values are available from equation (17).

These equations have been tested extensively and work well; the results show that for A1 reactions $m^\ddagger > 1$ is always true.[145,161] For example, the results of Bell's group for the hydrolysis of trioxane in aqueous H_2SO_4, $HClO_4$ and HCl media, equation (31),[144] are shown as an excess acidity plot according to equation (51) in Fig. 8. As can be seen, the results for all three acid media fall on the same line, increasing the likelihood that the derived excess acidities are in fact measuring the same thing in the different acid media. The line in Fig. 8 has a slope of 1.091 ± 0.010, an intercept of -7.4597 ± 0.0083, and a correlation coefficient of 0.998 over 54 points, three of which were rejected as being off the line by the program used.[117] An equivalent plot for the hydrolysis of par-

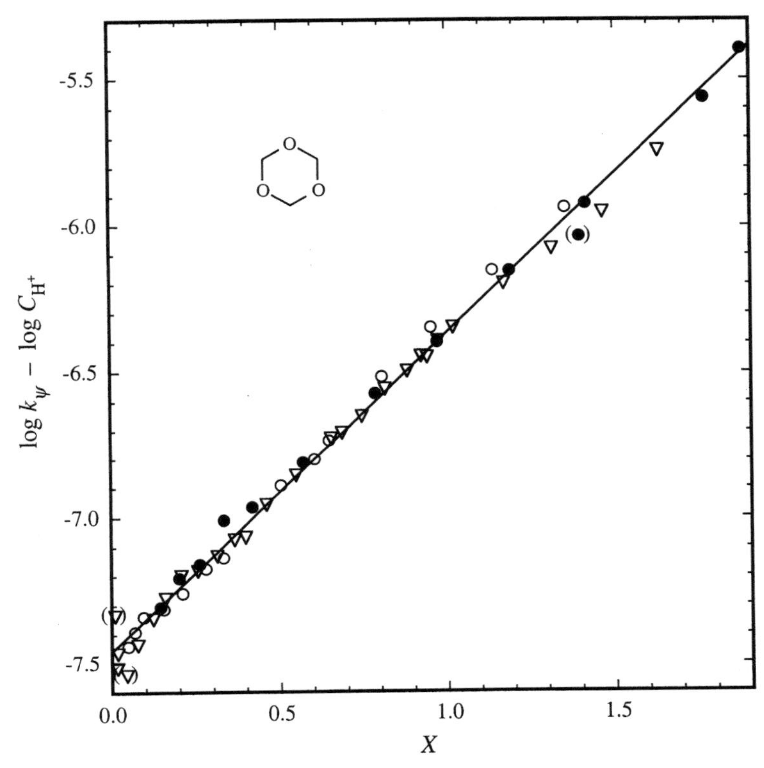

Fig. 8 Excess acidity plot against X of A1 hydrolysis rate constants for trioxane in aqueous H_2SO_4 (open circles), aqueous $HClO_4$ (closed circles) and aqueous HCl (open triangles) at 25°C. Data from ref. 144.

aldehyde (acetaldehyde trimer) using the data of Bell and Brown,[164] covering the same three acid media, has a slope of 1.271 ± 0.054, an intercept of -3.892 ± 0.018 (a much faster reaction, by a factor of 3700), and a correlation coefficient of 0.97 over 35 points, one of which was rejected. The slopes are $m^{\ddagger}m^{*}$; m^{*} values for these acetals are not known but are certainly less than 0.5,[39] giving $m^{\ddagger} > 2$.

Several of the reactions studied by the Satchell group are A1 processes. For instance, in aqueous $HClO_4$ and aqueous $HClO_4$/dioxane mixtures the hydrolyses of benzaldehyde diethyl thioacetals,[165] of benzaldehyde diaryl thioacetals and *S*-aryl, *S*-ethyl acetals,[166] and of benzophenone diethyl and diphenyl thioacetals,[167] give linear A1 excess acidity plots. Benzoic acid anhydrides generally exhibit an A1 hydrolysis in water, as determined from rate constant data obtained in aqueous $HClO_4$ and in aqueous $HClO_4$/dioxane mixtures,[128,168] although an A2 hydrolysis is possible in the dioxane mixtures.

Acyl fluorides also hydrolyze by an A1 mechanism as long as the water content of the medium is high.[129,169]

The excess acidity method has been used to show that the hydrolysis of several aromatic sulfonyl chlorides in aqueous H_2SO_4 media follow an A1 pathway.[170–172] Many of the ether hydrolysis reactions studied by the Lajunen group in aqueous $HClO_4$ are A1 reactions,[173] as are the alcohol dehydrations they looked at.[174] Nitronic acid formation from a nitroalkane precursor has been shown to be an A1 process.[175] Some of the oxidations by chromium(VI) in aqueous $HClO_4$ studied by Alvarez Macho and Montequi Martin, e.g. of dimethyl sulfoxide[176] and propanal,[177] have been shown to have an A1 mechanism. A plot of $\log k_\psi - \log C_{H^+}$ against X is linear for the reaction between HOCl and HCl in aqueous H_2SO_4, and it is suggested that this reaction involves initial protonation of HOCl, followed by an attack on H_2OCl^+ by Cl^- to give the products chlorine and water.[178]

Many hydrolysis reactions, e.g. of esters and some amides, switch from an A2 mechanism to an A1 mechanism as the acidity is increased, and we will see several of these under the A2 reaction and amide hydrolysis headings.

A-S_E2 MECHANISM

This mechanism is given in equation (37). Absolute rate theory leads to equation (55), and making the same assumption as for the A1 case, equation (50), leads to the relevant rate equation, equation (56).[145,161] This equation is derived on the assumption that all the acidity of the medium comes from "solvated protons", H_3O^+; in sulfuric acid it will require modification above 80 wt% acid as the medium acidity begins to be due to the presence of undissociated H_2SO_4 molecules as well, see above.[179]

$$k_\psi C_S = k_0 a_S a_{H^+} \frac{1}{f_\ddagger} = k_0 C_S C_{H^+} \frac{f_S f_{H^+}}{f_\ddagger} \tag{55}$$

$$\log k_\psi - \log C_{H^+} = \log k_0 + m^\ddagger m^* X \tag{56}$$

Equation (56) predicts linearity against X, as for the A1 case. However, the $m^\ddagger$ value now has physical meaning, being identical with the term α_A previously defined for A-S_E2 reactions by the Kresge group,[162,163] which was one reason for adopting the definition of $m^\ddagger$ given in equation (50) above in the first place. This α_A is equivalent to a Brønsted α, giving the extent of proton transfer at the transition state. An examination of equation (57) makes this clear, the transition state having to be somewhere between the initial and product states.

$$f_\ddagger = (f_S f_{H^+})^{1-\alpha} (f_{SH^+})^\alpha \tag{57}$$

$$\underset{H_3C}{\overset{Ph}{\diagdown}}C{=}CH_2 + H^+ \xrightarrow[\text{slow}]{k_0} \underset{H_3C}{\overset{Ph}{\diagdown}}\overset{+}{C}{-}CH_3 + H_2O \xrightarrow[\text{fast}]{-\,H^+} Ph(HO)C(CH_3)CH_3 \qquad (58)$$

Thus for these reactions $m^‡$ is necessarily less than unity, a result that has now been widely observed in practice,[117,118,120,161,180,181] and thus the $m^‡$ value offers a clear distinction between the A1 and A-S_E2 mechanisms, which is not the case with the H_0 correlations discussed above. A number of different excess acidity plots according to equation (56), covering a wide reactivity range, are shown in Fig. 9. These are for the hydration of α-methylstyrene,[120] equation (58), and the mechanistically similar hydration of phenylacetylene;[118] for the isomerization of *cis*-stilbene;[120] and for the detritiation of tritiated benzene, equation (28) above.[117] As can be seen, all four plots are good straight lines; the references cited may be consulted for the details. The slopes look steep, but m^* values for carbon protonation approximate 1.8,[36] and the $m^‡$ values are all calculated to be

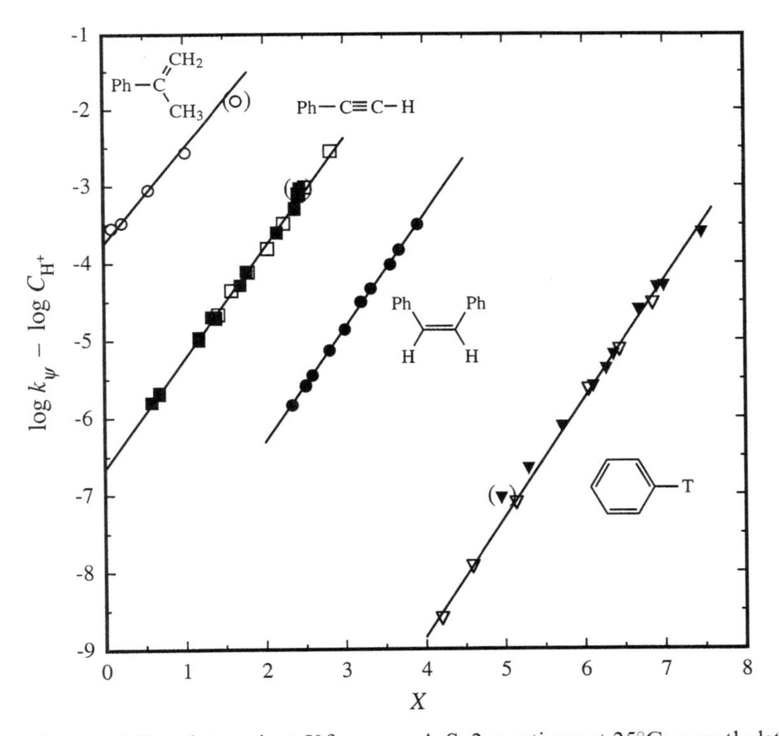

Fig. 9 Excess acidity plot against X for some A-S_E2 reactions at 25°C: α-methylstyrene hydration (open circles);[120,182] phenylacetylene hydration (open and closed squares);[118,160,183,184] *cis*-stilbene isomerization (closed circles);[120,185] and tritiated benzene T-exchange (open and closed triangles).[117,147,186,187]

< 1: 0.77 for α-methylstyrene, 0.76 for phenylacetylene, 0.74 for *cis*-stilbene and 0.86 for benzene.

Excess acidity correlations have been used to show that some aromatic sulfonic acid desulfonations have an A-S_E2 mechanism.[188,189] This mechanism (alternative terminologies are Ad-E2 and A(E) + A(N))[190] has also been found to apply to the hydration of acetylene itself,[191] to ynamines[192] and to many other alkynes,[193–195] as well as to many different alkenes[196–199] and vinyl ethers.[200–203] The excess acidity method has been used to evaluate α_A values for several alkene hydrations.[204,205]

Some cyclic thioacetals have an A-S_E2 hydrolysis mechanism,[206] as do some 2-aryl-2-methyl-1,3-dithianes, except for the 4-NO_2 derivative, which looks more A2-like.[207] In 10 vol% dioxane/aqueous $HClO_4$ mixtures, reactive 2-aryl-2-phenyl-1,3-dithianes are believed to have an A-S_E2 hydrolysis mechanism, whereas the least reactive ones have an A2 mechanism.[130] Isothiocyanates are believed to hydrolyze by a mechanism that involves simultaneous proton transfer to nitrogen and attack of water at carbon in a cyclic transition state.[208,209]

A2 MECHANISM

$$\log k_\psi - \log \frac{C_S}{C_S + C_{SH^+}} - \log C_{H^+} = \log \frac{k_0}{K_{SH^+}} + m^\ddagger m^* X + \log a_{Nu} \qquad (59)$$

$$\log k_\psi - \log \frac{C_{SH^+}}{C_S + C_{SH^+}} = \log k_0 + (m^\ddagger - 1) m^* X + \log a_{Nu} \qquad (60)$$

The mechanism is given in equation (41) above, and the corresponding derived excess acidity rate equations are equations (59) and (60) for substrates that are predominantly unprotonated and protonated, respectively, under the reaction conditions.[145,161]

Plots of the left-hand side of these equations against X are curved, allowing easy distinction of an A2 mechanism. Excess acidity plots using equation (59) are shown for some ester hydrolyses in sulfuric acid at 25°C in Fig. 10, for *t*-butyl acetate,[29] and for methyl 2,6-dimethylbenzoate and methyl benzoate.[41] The first ester (leftmost line in Fig. 10) clearly undergoes an A1 hydrolysis, specifically $A_{Al}1$;[29] the parameters of the line are slope 1.552 ± 0.027; intercept −3.983 ± 0.015, correlation coefficient 0.999. The second ester (middle line) does too, but now the mechanism is $A_{Ac}1$:[41] slope 1.253 ± 0.033; intercept −10.78 ± 0.17 (a much slower reaction, by nearly seven powers of ten); correlation coefficient 0.997. Methyl benzoate, however, undergoes an A2 reaction at the lower acidities, since the plot is curved, only switching over to an $A_{Ac}1$ reaction at ≈80 wt% H_2SO_4.[41] The parameters of the A1 process were obtained by curve-fitting: slope 1.245 ± 0.038; intercept −14.48 ± 0.28, nearly

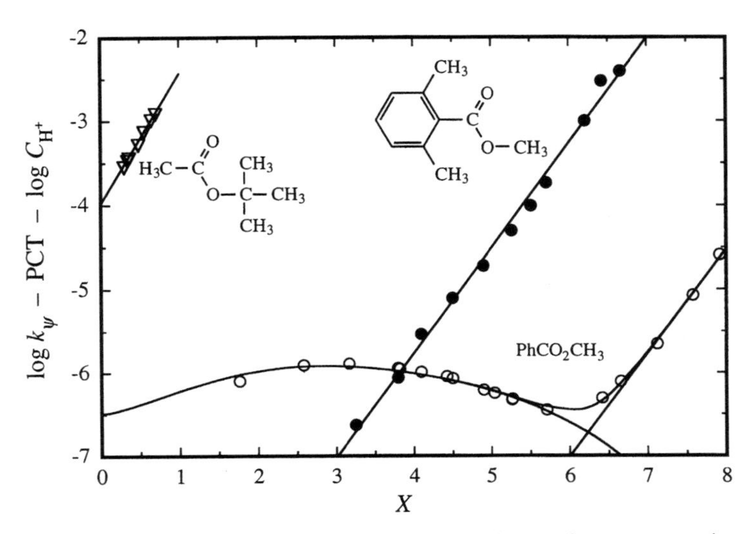

Fig. 10 Excess acidity plots against X for the hydrolyses of some esters in aqueous H_2SO_4 at 25°C; for methyl benzoate (open circles) and methyl 2,6-dimethylbenzoate (closed circles), data from ref. 41, and for *t*-butyl acetate (open triangles), data from ref. 29. PCT, protonation correction term.

four powers of ten slower than methyl 2,6-dimethylbenzoate, explaining the more favorable A2 process at the lower acidities for methyl benzoate.

Now a very useful feature of the excess acidity method comes into play; likely nucleophiles or bases can be tested by subtracting their log activities or concentrations from the left-hand side of equations (59) and (60), and the species reacting with SH^+ is uniquely identified when linearity of the result against X is achieved.[145,161] For instance, subtraction of twice the water activity is required to attain linearity in ester hydrolysis processes such as equation (42), as shown in Fig. 11 for methyl benzoate[41] and ethyl benzoate.[210] The water activities given in Table 3 were used. The parameters of the lines in Fig. 11, obtained by curve-fitting, are: methyl benzoate, slope 0.921 ± 0.010, intercept -9.974 ± 0.044; and ethyl benzoate, slope 0.9098 ± 0.0089, intercept -9.802 ± 0.032. As would be expected, the slopes are almost identical; the intercept difference shows that methyl benzoate reacts about 1.5 times as fast as does ethyl benzoate in the standard state, a result easily attributable to the slight increase in steric crowding to the equation (42) hydrolysis in the latter case. The order of the A2 ester hydrolysis reaction in water is thus two, a result quite difficult to obtain in other ways, even in dilute solution, perhaps requiring a proton inventory study of a reaction that is very slow in water.

The excess acidity method has been used to show that some reactions do have pure A2 mechanisms, for instance the hydrolysis of some benzohydroxamic acids,[126,211] the hydrolysis of sultams,[212] and the oxidation of lactic acid in

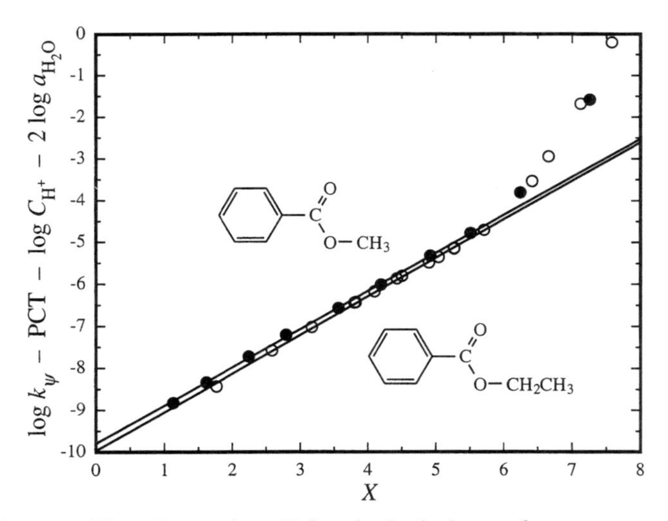

Fig. 11 Excess acidity plot against X for the hydrolyses of some esters in aqueous H_2SO_4 at 25°C; for methyl benzoate (closed circles), data from ref. 41, and for ethyl benzoate (open circles), data from ref. 210, showing the reaction with two water molecules. PCT, protonation correction term.

strong acid media.[213] The hydrolysis of the relatively unreactive 2-(4-methoxyphenyl)-1,3-dithiane in aqueous perchloric acid is suggested to have an A2 mechanism.[214] The decomposition of coenzyme B_{12} and related compounds in aqueous acids may have an A1 or an A2 mechanism depending on the substrate.[215]

However, the switchover from an A2 to an A1 hydrolysis is a very common mechanistic pathway in strong acid media, probably more common than the pure A2 mechanism. Excess acidity analyses have shown that thioacetic acid, several thiobenzoic acids, and many thiolbenzoate and thionbenzoate esters show this sort of mechanism switch.[179] Acylals and thioacylals also show this behavior,[116] with thioacylals using two water molecules and acylals one. Many hydroxamic acids react this way,[127,216] as do esters of various types,[41,217,218] episulfoxides[219] and aryloxatriazoles.[220] Acylhydrazines can also show a mechanism switch of this sort, although with these substrates the situation is somewhat more complex.[221]

OTHER MECHANISMS

Many of the reactions that take place in strong acid media cannot really be defined as being simple A1, A-S_E2 or A2 processes, and some of these will be described here. Space considerations preclude detailed discussion of all of the many different reactions that have been studied mechanistically using the excess acidity technique, and so some will only be mentioned in passing.

Substitutions; N-*nitro amines*

N-nitro amines, $RNHNO_2$, decompose to alcohols and nitrous oxide in strong acid media. Rate constants obtained for R = methyl in sulfuric acid[222–224] are illustrated as excess acidity plots in Fig. 12.[119] This shows multiple curvature, but analysis according to equation (59) shows that one water molecule is involved in the reaction up to about 80 wt% H_2SO_4, and one bisulfate ion above this point, see Fig. 13. The proposed mechanism is shown in Scheme 2.[119]

$$RHN{-}\overset{+}{N}(=O)O^- \underset{}{\overset{K_T}{\rightleftharpoons}} R{-}N{=}\overset{+}{N}(OH)O^- + H^+ \underset{K_{SH^+}}{\rightleftharpoons} R{-}N{=}\overset{+}{N}(\overset{+}{O}H_2)O^-$$

$$Nu^- + R{-}N{=}\overset{+}{N}(\overset{+}{O}H_2)O^- \xrightarrow[\text{slow}]{k_0} Nu{-}R + N_2O + H_2O$$

Scheme 2

This is an acid-catalyzed S_N2 displacement from the protonated *aci*-nitro tautomer. The substrate is not protonated in even the strongest acids used; the protonation site shown in Scheme 2 is not the most basic, but the only one that results in decomposition. Bisulfate is a poorer nucleophile than water by a factor of about 1000,[119] but above 80 wt% in acidity its concentration is at least 1000 times that of water.[150] Twelve compounds were studied; the intercept

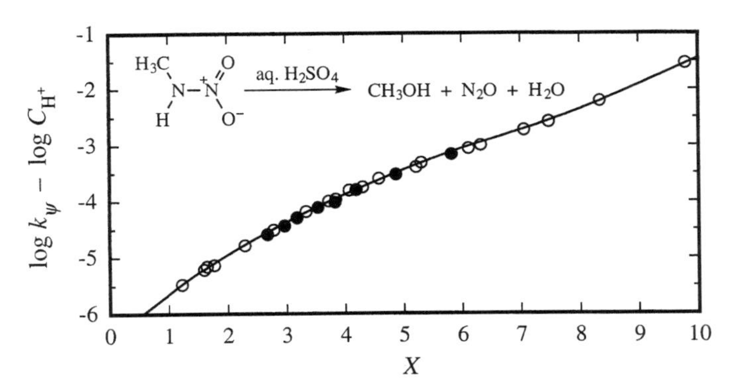

Fig. 12 Excess acidity plot against *X* for the decomposition of *N*-nitromethylamine in aqueous H_2SO_4 at 25°C, assuming no nucleophile involvement. Data from refs 222 and 223, open circles, and data obtained from rate constants for the decomposition of methyldinitroamine given in ref. 224, closed circles.

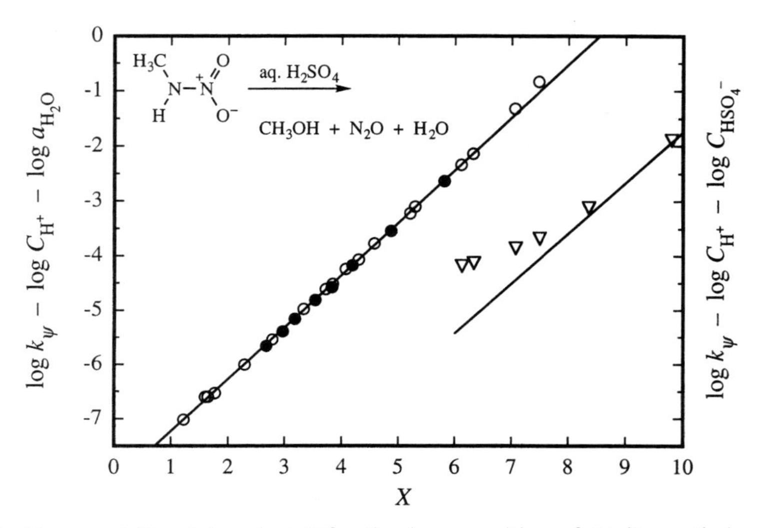

Fig. 13 Excess acidity plot against X for the decomposition of *N*-nitromethylamine in aqueous H_2SO_4 at 25°C. Data of Fig. 12, assuming the involvement of one water molecule (left-hand line) at low acidities, and of one bisulfate ion (right-hand line, triangles) at high acidities.

$\log k_0$ values were well correlated by the σ^* values of the alkyl groups.[119] The parent compound NH_2NO_2, nitramide, was studied in aqueous $HClO_4$;[225] an excess acidity analysis shows that it reacts by a similar mechanism to the alkyl compounds in the more concentrated acids, but has a non-acid-catalyzed route available to it in the more dilute acids and water, perhaps involving two water molecules as shown in Scheme 3.[226]

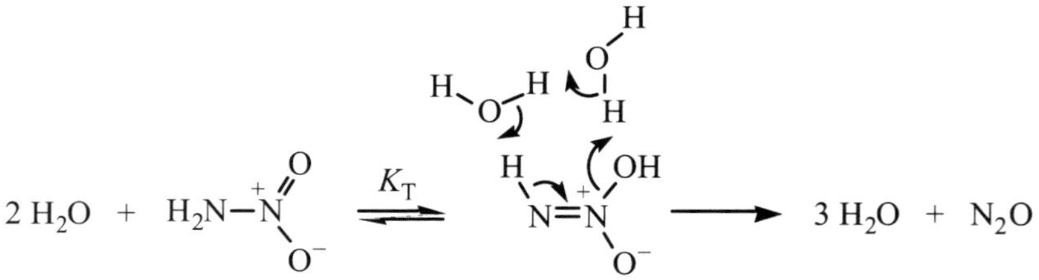

$$2\,H_2O + H_2N{-}\overset{+}{N}(=O)O^- \overset{K_T}{\rightleftharpoons} (H_2O)_2 \cdots HN{=}\overset{+}{N}(OH)O^- \longrightarrow 3\,H_2O + N_2O$$

Scheme 3

Aromatic nitration

Studies of the reaction between various substituted benzenes and nitric acid, which can take place in a number of different acid systems, have been reported in an extensive series of papers by the Marziano group, using correlations with the M_C activity coefficient function. This reaction is a two-stage process, equations (61) and (62), either of which may be rate-determining under the reaction conditions.

$$HO-\overset{+}{N}(=O)-O^- + H^+ \rightleftharpoons H_2\overset{+}{O}-\overset{+}{N}(=O)-O^- \rightleftharpoons {}^+NO_2 + H_2O \qquad (61)$$

$$Y-C_6H_5 + {}^+NO_2 \longrightarrow Y-[C_6H_5(H)(NO_2)]^+ \xrightarrow[\text{fast}]{-H^+} Y-C_6H_4-NO_2 \qquad (62)$$

Thus knowledge of $^+NO_2$ concentrations as a function of acidity is very useful. To this end, the M_C scale has been used to determine the basicity of nitric acid in concentrated H_2SO_4, using Raman and UV spectroscopy.[227] These techniques have also been used to study all of the nitric acid equilibria over the entire available concentration ranges in H_2SO_4 and $HClO_4$,[228] and in concentrated CF_3SO_3H media.[229] The thermodynamics of the $HNO_3 \rightarrow {}^+NO_2$ equilibrium have also been examined in H_2SO_4.[43]

The treatment of reactions according to the M_C scale was formulated for aromatic nitrations by Traverso, Marziano and Passerini,[230] and the resulting linear description of nitration rate profiles in the 80–98 wt% H_2SO_4 range was given by Marziano and Sampoli[231] and Marziano *et al.*[232] Correction for the reactant concentration results in rate constants that are independent of the medium acidity,[233] and the reasonable conclusion is drawn that, for substrates capable of being protonated by the medium, only the unprotonated species undergoes nitration.[233] Among the nitration reactions that have been examined using the M_C technique are those of benzene itself with some other compounds in H_2SO_4 and $HClO_4$,[234] of some cinnamic acids and other styrene derivatives in aqueous H_2SO_4,[235] of various aromatic compounds in aqueous CF_3SO_3H,[236] and of some sulfonic acids in H_2SO_4, with the equilibria involved in the reaction also being investigated.[237] Some aromatics have been studied at several temperatures.[43]

The hydrolysis of azophenyl ethers

Several azophenyl ether hydrolyses have been analyzed using the excess acidity method.[238–241] The mechanisms of these reactions are interesting and quite different from the usual A1-type acid-catalyzed ether hydrolysis processes.[173] As an example, the two methoxy groups in [6] (Scheme 4) hydrolyze by quite different mechanisms. The *p*-methoxy group reacts in an A2 type process with more than one water molecule, probably three, since the excess acidity plot of $\log k_\psi - \log C_{H^+} - 3\log a_{H_2O}$ is linear, with a break-point where the slope changes. This is interpreted in terms of a change of rate-determining step with acidity, from the k_1 step to the k_2 step in Scheme 4 as the acidity increases.[240]

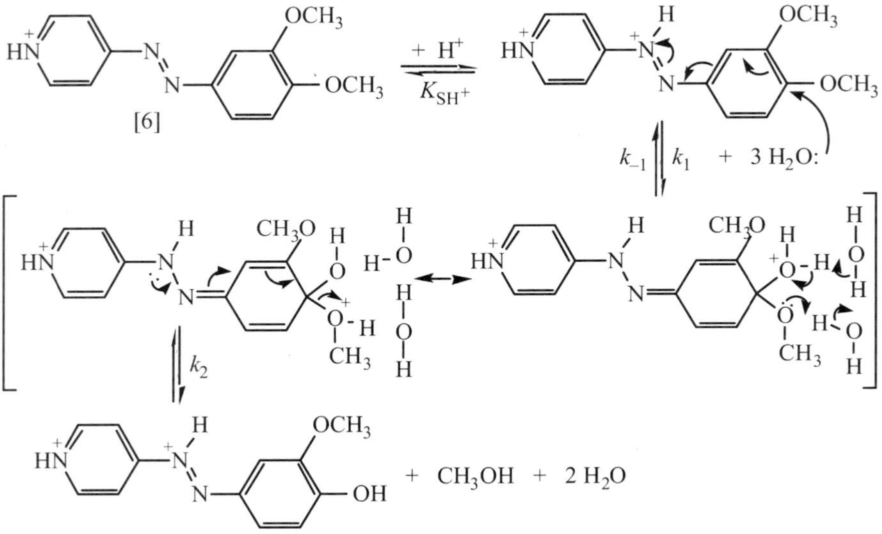

Scheme 4

However, the *m*-methoxy group in [6] is deactivated, reacting some 7000 times more slowly than the *p*-methoxy group and by a quite different A-S_E2-type mechanism; the plot of $\log k_\psi - \log C_{H^+}$ (corrected for the unproductive protonation shown in Scheme 5) is linear in X with a slope of only 0.25. Scheme 5 shows the more likely of two possible mechanisms.[240]

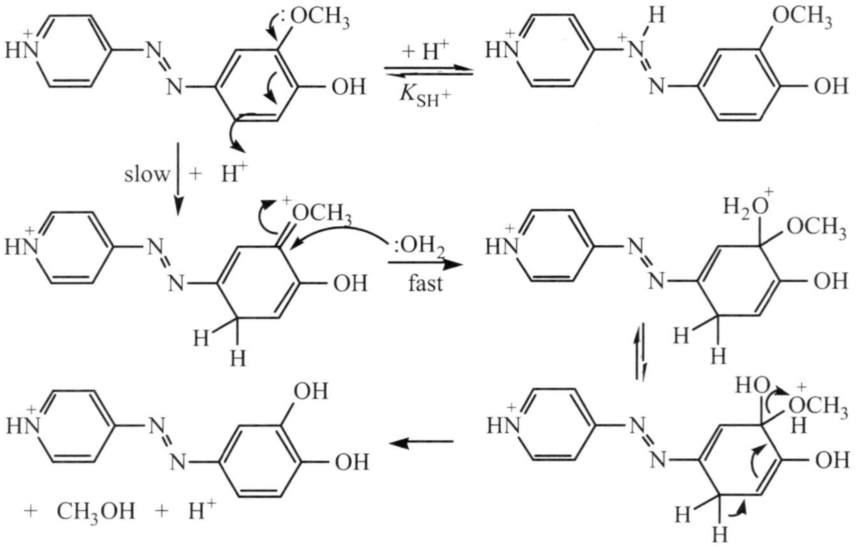

Scheme 5

Enolization

Not every excess acidity mechanistic analysis has been an outstanding success. For instance, several enolization studies have used this technique. The enolization of acetophenone was one of the reactions originally studied by Zucker and Hammett;[146] their sulfuric acid rate constant data, obtained by iodine scavenging (the reaction is zero-order in halogen), was used in an excess acidity analysis,[242] together with additional results obtained for some substituted acetophenones (using bromine scavenging).[243]

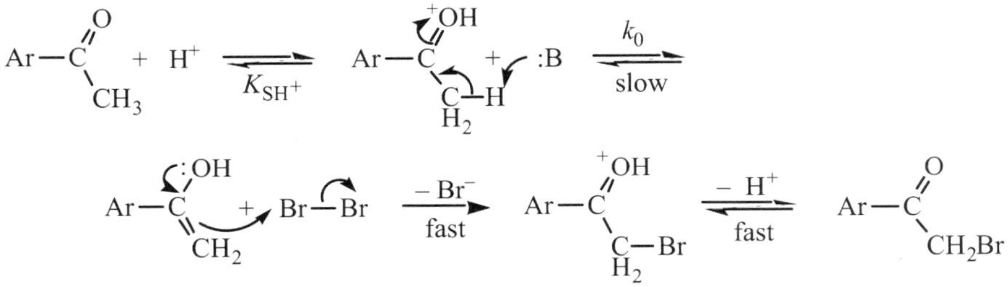

$$\text{ArC(=O)CH}_3 + \text{H}^+ \underset{K_{\text{SH}^+}}{\rightleftharpoons} \text{ArC(=}\overset{+}{\text{O}}\text{H)CH}_2\text{–H} + \text{:B} \xrightarrow[\text{slow}]{k_0} \text{ArC(OH)=CH}_2 + \text{Br–Br} \xrightarrow[\text{fast}]{-\text{Br}^-} \text{ArC(=}\overset{+}{\text{O}}\text{H)CH}_2\text{Br} \underset{\text{fast}}{\overset{-\text{H}^+}{\rightleftharpoons}} \text{ArC(=O)CH}_2\text{Br}$$

Scheme 6

The resulting kinetic plots were curved, resembling the low acidity region of Fig. 10, consistent with the base B removing the proton in the slow step being two water molecules, see Scheme 6. The behavior of 4-nitroacetophenone was anomalous, however, giving a linear A1-type plot.[242] Later it was found that for the enolization of acetophenone the standard-state rate constant extrapolated using rate constant data obtained in sulfuric acid media was 35% greater than that found using perchloric acid. This result was attributed to catalysis by the bisulfate ion in H_2SO_4, difficult to see in other ways.[244] Analysis of rate constant data for acetone[245] also indicated attack by two water molecules at acidities below 80% H_2SO_4, and by one bisulfate ion at acidities higher than this.[242] For the enolization of acetaldehyde in H_2SO_4 (see Scheme 7) subtracting twice the water activity from the A2 excess acidity plot for the k_1 step also gave a straight line.[246]

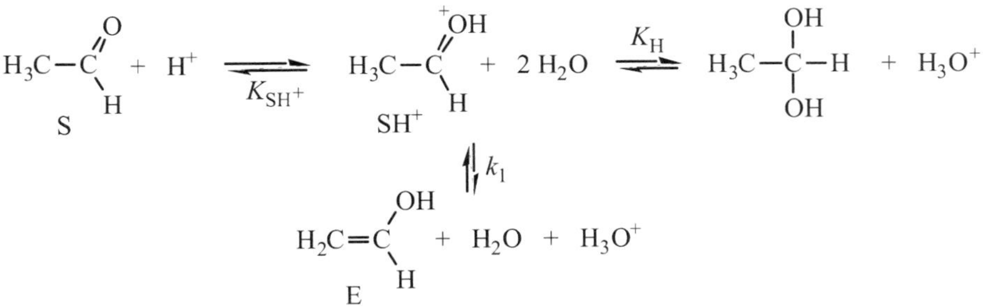

$$\underset{\text{S}}{\text{H}_3\text{C–C(=O)H}} + \text{H}^+ \underset{K_{\text{SH}^+}}{\rightleftharpoons} \underset{\text{SH}^+}{\text{H}_3\text{C–C(=}\overset{+}{\text{O}}\text{H)H}} + 2\,\text{H}_2\text{O} \overset{K_\text{H}}{\rightleftharpoons} \text{H}_3\text{C–C(OH)}_2\text{–H} + \text{H}_3\text{O}^+$$

$$\text{SH}^+ \overset{k_1}{\rightleftharpoons} \underset{\text{E}}{\text{H}_2\text{C=C(OH)H}} + \text{H}_2\text{O} + \text{H}_3\text{O}^+$$

Scheme 7

However, it has since been found that the curvature observed in these excess acidity plots is an artifact, caused by insufficiently effective halogen scavenging in the original experiments,[247] and that when corrected for this the excess acidity plots of $\log k_\psi - \log C_{H^+}$ against X for these reactions are linear, not curved. This is a very interesting result, as it means that the base, **B** in Scheme 6, does not appear in the rate law, unlike the nucleophile in equation (42), see Figs 10 and 11. Why this should be is, at present, not clear; further work seems necessary.

Condensation; ring closure

In contrast to the above, other reactions have been found to require base assistance by water in the rate-determining step, i.e. the water activity does appear in the rate law. The mechanism formulated for the condensation of acetaldehyde in sulfuric acid is given in equation (63), following on from the enolization of Scheme 7, subsequent dehydration to crotonaldehyde occurring as shown in Scheme 8. The k_1, k_2, k_3 and k_{-3} steps shown were all studied.[246]

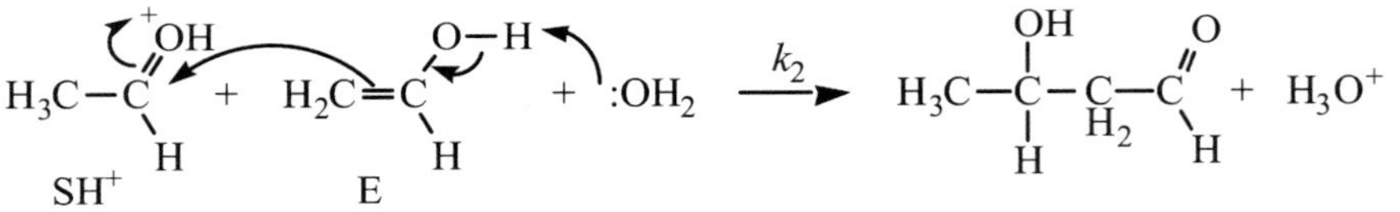

$$\underset{SH^+}{H_3C-CH=\overset{+}{O}H} + \underset{E}{H_2C=CH-O-H} + :OH_2 \xrightarrow{k_2} H_3C-CH(OH)-CH_2-CH=O + H_3O^+ \tag{63}$$

$$H_3C-CH(OH)-CH_2-CH=O \rightleftharpoons H_3C-CH(OH)-CH=CH-OH + H^+ \rightleftharpoons$$

$$H_3C-CH(\overset{+}{O}H_2)-CH=CH-\ddot{O}H \underset{k_{-3}}{\overset{k_3}{\rightleftharpoons}} H_2O + H_3C-CH=CH-CH=\overset{+}{O}H \rightleftharpoons H_3C-CH=CH-CH=O + H^+$$

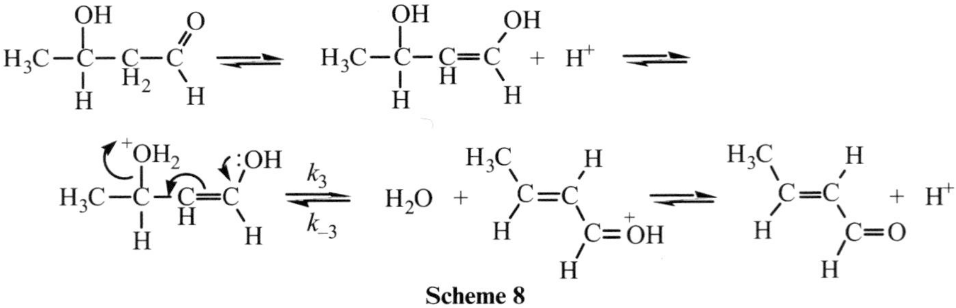

Scheme 8

The k_2 step in equation (63) is interesting because it is bimolecular. The resulting rate expression is equation (64). Substituting equation (65) into this and taking logs gives equation (66):

$$k_\psi C_S^2 = k_2 a_{SH^+} a_E a_{H_2O} \frac{1}{f_\ddagger} \tag{64}$$

$$K_E = \frac{a_S}{a_E} \quad \text{and} \quad K_{SH^+} = \frac{a_S a_{H^+}}{a_{SH^+}} \tag{65}$$

$$\log k_\psi - \log C_{H^+} - \log a_{H_2O} = \log \frac{k_2 K_E}{K_{SH^+}} + \log \frac{f_S^2 f_{H^+}}{f_\ddagger} \tag{66}$$

The activity coefficient term in equation (66) contains an extra f_S term, so exact linearity of the left-hand side in X would not be expected. Nevertheless, the resulting plot was almost linear, with a correlation coefficient of 0.999,[246] meaning that the excess acidity treatment does in fact apply. Note that the water molecule in equation (63) is acting as a base. For the k_3 step in Scheme 8, $\log k_\psi - \log C_{H^+}$ against X was found to be linear, and for the k_{-3} step $\log k_\psi - \log C_{H^+} - \log a_{H_2O}$ was linear; here the water molecule is acting as a nucleophile.[246]

In aqueous H_2SO_4 mixtures 2-substituted imidazolines [7] cyclize, giving [8] or [9] as products, depending on whether the ring nitrogen is methylated or not;[248] see equation (67).

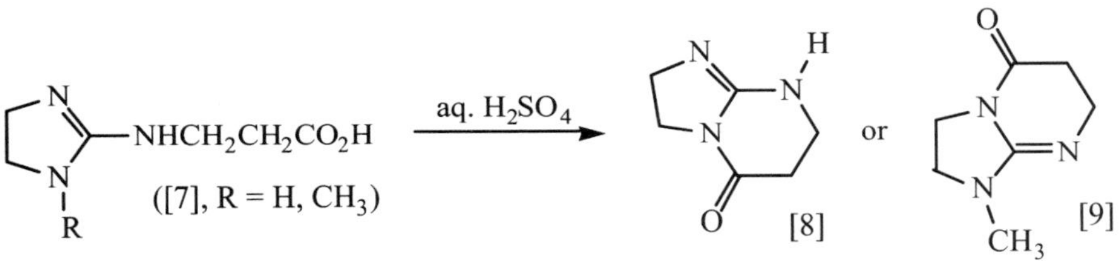

$$\text{[7]-NHCH}_2\text{CH}_2\text{CO}_2\text{H} \;([7],\ R = H,\ CH_3) \xrightarrow{\text{aq. } H_2SO_4} [8] \text{ or } [9] \qquad (67)$$

When the log rate constants were subjected to an excess acidity analysis, $\log k_\psi - \log C_{H^+} - \log a_{H_2O}$ was found to be linear in X in both cases, whereas $\log k_\psi - \log C_{H^+}$ was not. This again suggests the involvement of a water molecule in the rate-determining step, probably acting as a base, reacting with protonated [7] as shown in Scheme 9.[248]

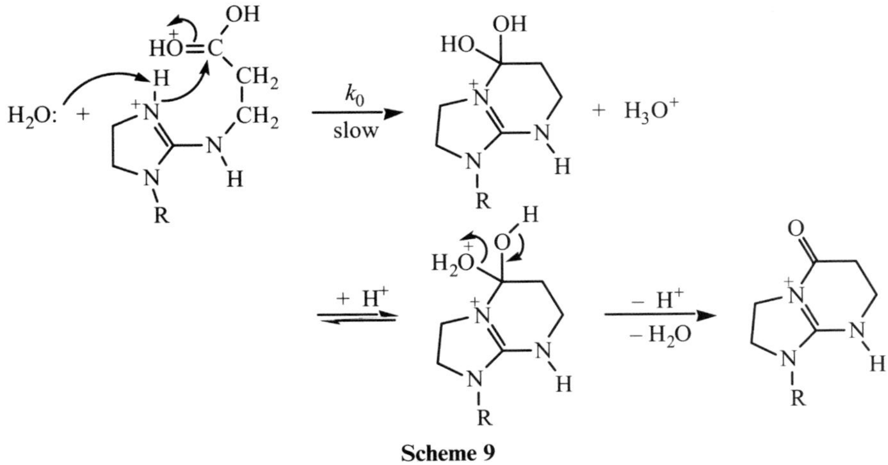

$$H_2O: + \text{[protonated 7]} \xrightarrow[\text{slow}]{k_0} \text{[intermediate]} + H_3O^+ \underset{}{\overset{+\,H^+}{\rightleftharpoons}} \text{[protonated intermediate]} \xrightarrow[-\,H_2O]{-\,H^+} \text{[product]}$$

Scheme 9

Other ring closure reactions that have been studied include those of 5-phenylhydantoic acids to 3-phenylhydantoins in aq. H_2SO_4 solutions at 70°C.[249] The low $m^\ddagger$ values observed suggest a mechanism that involves C–N bond formation being concerted with proton transfer from nitrogen.[249] The ring closures of 3-(3′-methylureido)propanoic acid and 3-(3′-phenyluriedo)-2-methylpropanoic acid to hexahydropyrimidine-2,4-diones, which also take place in aq. H_2SO_4 solutions, have a similar mechanism.[250] The pK_{BH^+} values for protonation of the ureido group were determined in both cases, and also for carbonyl group protonation of the products in the latter case. The excess acidity plots for the reverse ring-opening process, observable in the latter case, are consistent with slow C–N bond scission concerted with proton transfer,[250] which seems reasonable.

Acylimidazoles – non-acid-catalyzed processes

Having seen that the excess acidity method works for second-order as well as for first-order acid-catalyzed processes, it is of interest to see whether it extends to reactions that are not acid catalyzed. The hydrolysis of acylimidazoles, equation (68), takes place in aqueous acids; the substrate is protonated on the ring nitrogen in the pH range, and in acid media the reaction rate constants decrease steadily with increasing acidity.[251,253]

$$\mathrm{R{-}C({=}O){-}N\langle imidazolium\rangle NH^+} \xrightarrow[\text{or } H_2SO_4]{\text{aq. HCl, HClO}_4} \mathrm{RCO_2H} + \mathrm{HN\langle imidazolium^+\rangle NH} \qquad (68)$$

$R = CH_3$, Ph

Several different aqueous acid media can be used for the reaction. Figure 14 refers to acetylimidazole in sulfuric acid; $\log a_{H_2O}$ is subtracted from the

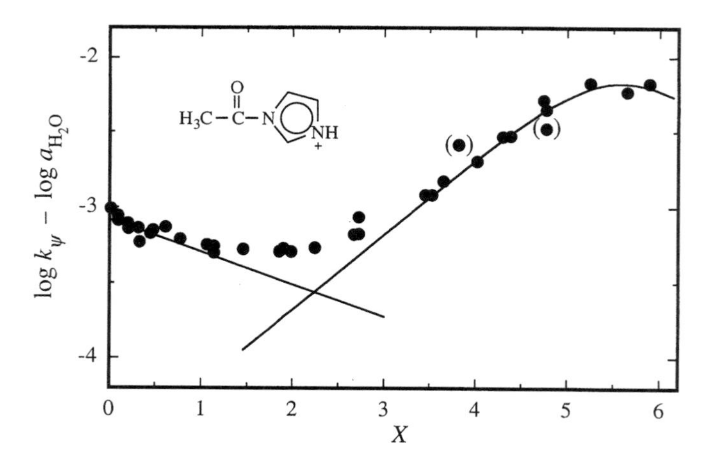

Fig. 14 Plot of $\log k_\psi - \log a_{H_2O}$ against X for the hydrolysis of acetylimidazole in aqueous sulfuric acid at 25°C. Data from refs 251 and 252. Reprinted with permission from ref. 254, © 1997 NRC Canada.

observed log reaction rate constants and the result is plotted against X.[254] As can be seen, there are two regions, an initial linear decrease followed by an upward region that starts linear but curves downward in strong acid. The proposed reaction mechanism is shown in Scheme 10; hydration is reversible, and either the k_1 or the k_2 step may be rate-determining, depending on the acidity.[254]

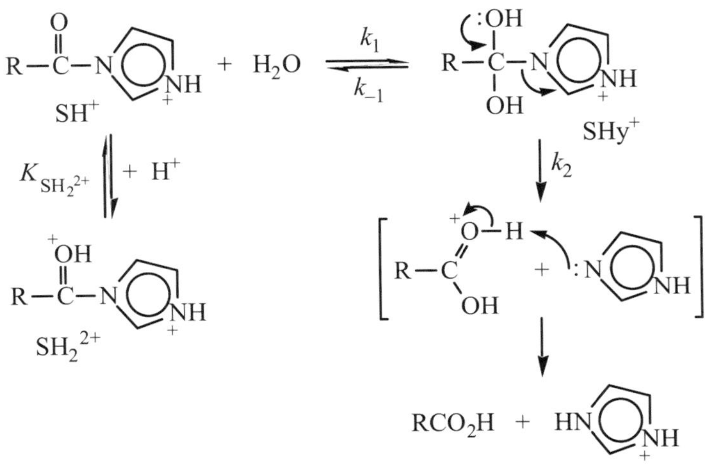

Scheme 10

At low acidity, where the k_2 step is rate-determining (see Scheme 10), we can write equation (69); substituting the definitions of equation (70), rearranging and taking logs gives equation (71):[254]

$$k_\psi C_{SH^+} = \frac{k_2 a_{SHy^+}}{f_\ddagger} \tag{69}$$

$$K_{Hy} = \frac{k_1}{k_{-1}} = \frac{a_{SHy^+}}{a_{SH^+} a_{H_2O}} \quad \text{and} \quad a_{SH^+} = C_{SH^+} f_{SH^+} \tag{70}$$

$$\log k_\psi - \log a_{H_2O} = \log \frac{k_1 k_2}{k_{-1}} + \log \frac{f_{SH^+}}{f_\ddagger} \tag{71}$$

It was shown in Fig. 7 that $\log f^*_{H^+}$ is linear in X, and if this is so then activity coefficient ratio terms like that in equation (71) will also be linear in X, see equation (50); since the substrate here is SH^+, f_{SH^+} in equation (71) corresponds to f_S in equation (50). Hence the first linear region in Fig. 14.

At high acidity (i.e. low water activity), where the k_1 step is rate limiting, we have to take the substrate concentration as the sum $(C_{SH^+} + C_{SH_2^{2+}})$, to take account of the unproductive second protonation shown in Scheme 10. Equation (72) is easily derived:[254]

$$\log k_\psi - \log \frac{C_{SH^+}}{C_{SH^+} + C_{SH_2^{2+}}} - \log a_{H_2O} = \log k_1 + \log \frac{f_{SH^+}}{f_\ddagger} \tag{72}$$

which predicts the second upward curving region of Fig. 14. Enough data points were present to enable the calculation of the $pK_{SH_2^{2+}}$ value by curve-fitting; a value of -6.77 ± 0.12 was obtained in sulfuric acid. Values of k_2/k_{-1} are easily derived; in water, hydrolysis is more than ten times as likely as reversion to starting materials.[254] Thus, the excess acidity method does indeed also apply to reactions that are not acid catalyzed in these aqueous acid media.

Cyclopropane reactions

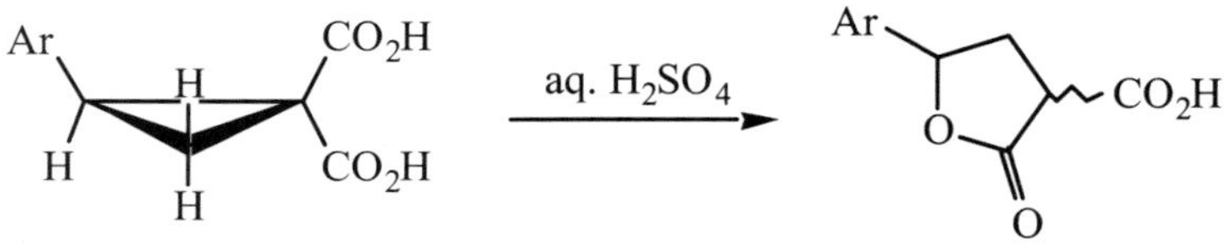

$$\text{Ar(H)C-CH}_2\text{-C(CO}_2\text{H)}_2 \xrightarrow{\text{aq. H}_2\text{SO}_4} \gamma\text{-lactone (Ar, CO}_2\text{H)} \quad (73)$$

Cyclopropane 1,1-dicarboxylic acids rearrange to γ-lactones in sulfuric acid, see equation (73). The excess acidity plots of $\log k_\psi - \log C_{H^+}$ and $\log k_\psi - \log C_{H^+} - \log a_{H_2O}$ against X are both linear, with the latter giving slightly better correlation coefficients.[255] The proposed mechanism is shown in Scheme 11, with either the k_1 or the k_2 step being rate-determining, the edge going to k_2 on the basis of activation parameters and $m^\ddagger$ values. The observed ρ value was -3.49 ± 0.13. The $m^\ddagger$ values found are too large for the last step, involving carbon protonation, to be rate-determining, and they are also incompatible with an alternative mechanism in which ring carbon protonation is the rate-determining first step.[255]

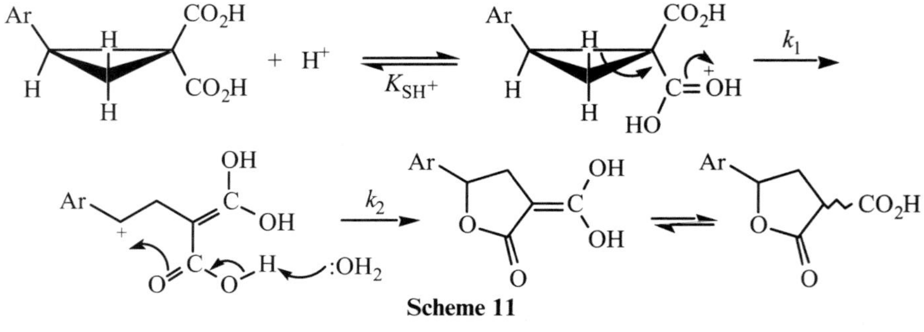

Ar, H, H, H, CO₂H, CO₂H + H⁺ ⇌ (K_{SH^+}) Ar, H, H, H, CO₂H, C=OH⁺, HO → (k_1) Ar, OH, C, OH, O=C–O–H, :OH₂ → (k_2) Ar, O, O, =C(OH)OH ⇌ Ar, O, O, CO₂H

Scheme 11

The cyclopropane acetals [10] and [11] also hydrolyze in aqueous sulfuric acid, but the reaction mechanisms for the two are not the same. Reaction of [10] involves pre-equilibrium oxygen protonation followed by rate-determining A1 ring opening, whereas that of [11] involves carbon protonation concurrent with

Ph, OCH₃, OCH₃ [10] Ph, O, O [11]

cyclopropane ring cleavage in an A-S_E2-type process.[255] This is reminiscent of the different hydrolysis mechanisms found for different thioacetal structural types by the Satchell group, discussed above.

Phenylazopyridines and phenylhydrazopyridines

These compounds undergo quite complex reactions in aqueous sulfuric acid media, which have recently been studied in some detail and are now well understood.[256–258] Actually it is somewhat surprising that these compounds react at all in acid, the parent azobenzenes being so stable that they can be used as H_0 indicators,[259] but the protonated pyridinium ring activates them toward nucleophilic attack. Typically, 2 moles of 4-(phenylazo)pyridine [12] give the hydroxylated (or oxidized) product 4-(4-hydroxyphenylazo)pyridine [13] and the reduction products 4-aminophenol [14] and 4-aminopyridine [15]. The corresponding [*N′*-(4-hydroxyphenylhydrazino)]pyridine [16] was surmised to be a reaction intermediate, and this was confirmed by synthesizing it separately and subjecting it to the reaction conditions; it gave the same products.[257]

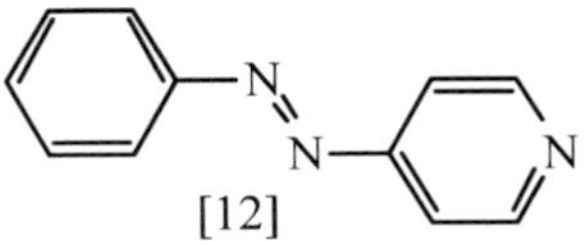

[12] 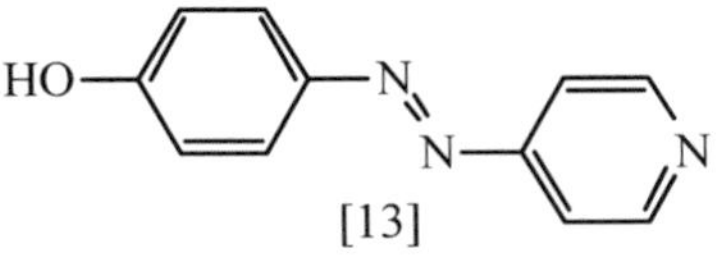 [13]

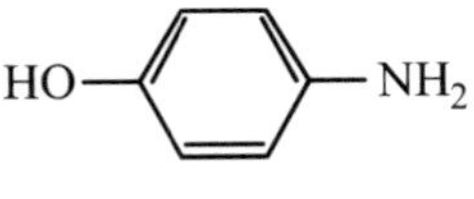

HO— —NH$_2$ H$_2$N—

[14] 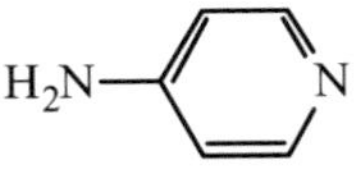 [15] 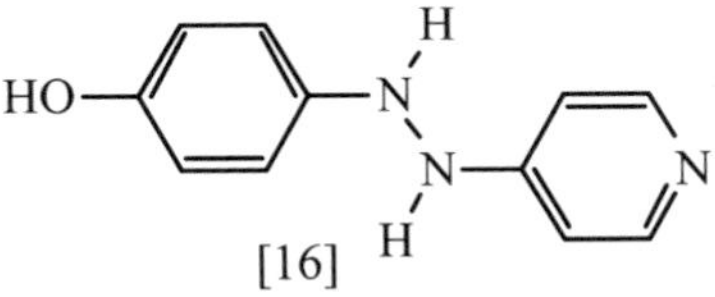 [16]

For the excess acidity analysis, $pK_{SH_2^{2+}}$ values for protonation of the substrates and products ([12], [13] and some others) on the azo-group were determined; the first protonation, on the pyridine ring nitrogen, occurs in the pH region.[257] Analysis of the kinetics indicated an A2 process, with the diprotonated substrate and either one water molecule or one bisulfate ion in the activated complex, depending on the acidity.[257] For the 4-chloro derivative, which is less reactive, only the reaction involving bisulfate is visible.[258] The first reaction, formation of intermediate [16] from protonated [12], is formulated as being a nucleophilic attack according to Scheme 12.[257]

Rate constants were obtained for the subsequent reaction of [16] and for two other derivatives, those with H and Cl replacing the OH group on [16]. These can be quite fast reactions, and for all substrates are faster than the initial hydroxylation shown in Scheme 12. Excess acidity plots indicated that the subsequent reaction involved a second proton transfer, followed by an A1-type process. The substrates were too reactive for the basicities to be determined directly, but there were enough kinetic points in most cases to enable accurate

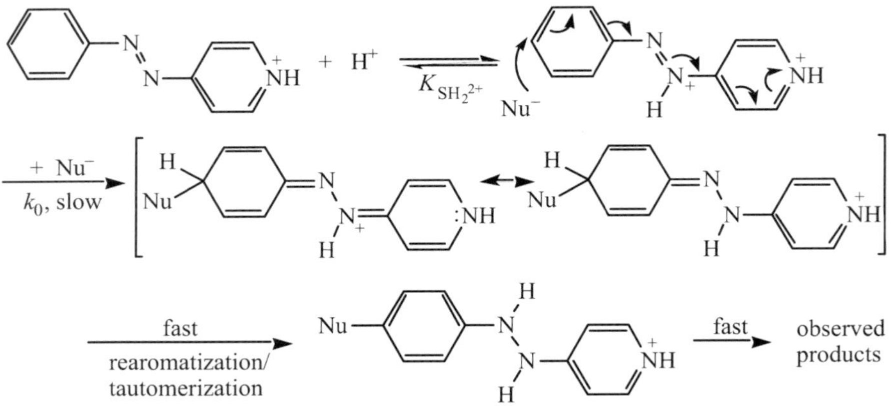

Scheme 12

$pK_{SH_2^{2+}}$ values to be calculated from the kinetic plots by curve-fitting;[258] for instance, the quite reasonable value of -1.038 ± 0.090 was obtained for [16]. The resulting linear kinetic plots according to equation (54), i.e. assuming full substrate protonation, are shown in Fig. 15.

Other evidence, for instance the observation of a semidine as one of the reaction products,[256] led to the realization that the reaction is a benzidine disproportionation, such as those observed when benzidines with two *p*-substituents are subjected to benzidine rearrangement conditions;[260,261] this also very conveniently explains the products formed. The mechanism is given in Scheme 13.[258]

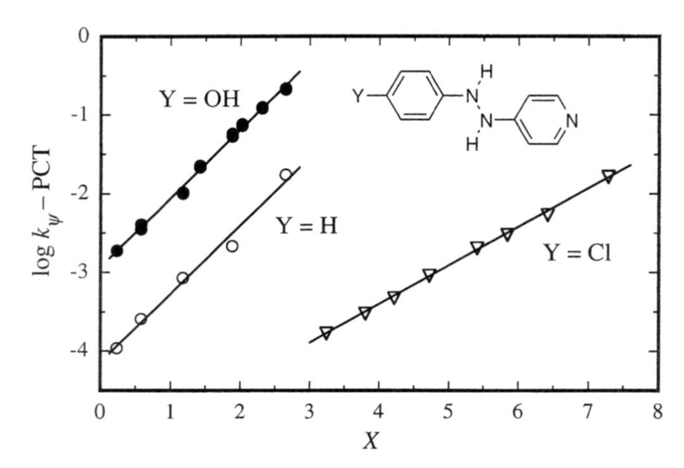

Fig. 15 Excess acidity plot of $\log k_\psi - \log(C_{SH_2^{2+}}/(C_{SH^+} + C_{SH_2^{2+}}))$, i.e. the protonation correction term PCT, against X for the reactions of some [N'-(arylhydrazino)]pyridines in aqueous H_2SO_4 at 25°C. Reprinted with permission from ref. 258, © 1998 NRC Canada.

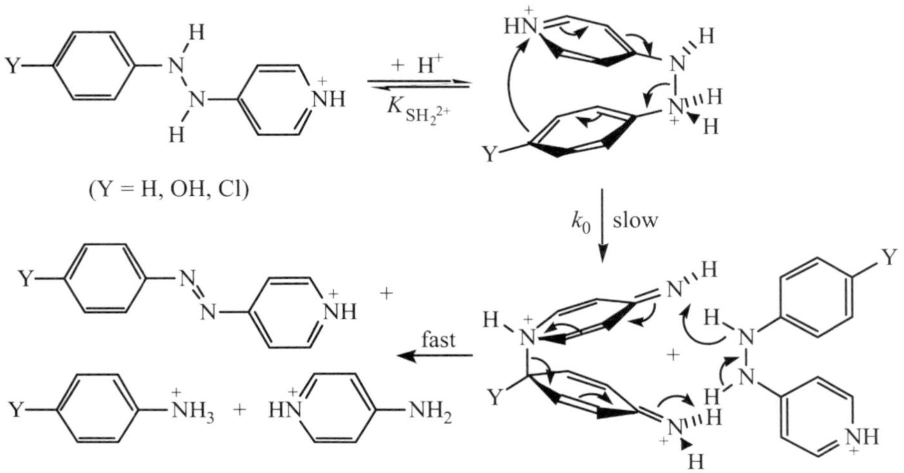

(Y = H, OH, Cl)

Scheme 13

Scheme 13 may look unfavorable on the face of it, but in fact the second two reactions are thermally allowed 10- and 14-electron electrocyclic reactions, respectively. The aromatic character of the transition states for these reactions is the major reason why the benzidine rearrangement is so fast in the first place.[261] The second bimolecular reaction is faster than the first rearrangement (bimolecular kinetics were not observed); it is downhill energetically because the reaction products are all aromatic, and formation of three molecules from two overcomes the entropy factor involved in orienting the two species for reaction.[258]

AMIDE HYDROLYSIS

This reaction has its own section for a number of reasons. It is a very important reaction in the sense that many physiologically active and other biologically interesting molecules, and all proteins, are amides. Consequently, many groups have studied this reaction over many years, so there is a great deal of rate constant and other data. Despite this, the mechanism remains poorly understood in detail, and different mechanistic proposals have been made by different groups (and sometimes by the same groups at different times); amides of different structural types have different hydrolysis mechanisms. There is agreement that the reaction involves protonation as the first step, but even the position of protonation was controversial for many years;[262] only fairly recently has the bulk of accumulated evidence led to oxygen protonation being preferred.[101,103] Of course this does not mean that reaction has to occur from this form, although it does make it more likely; it could occur from the nitrogen-protonated form if this is very reactive, even if it is present in much smaller amounts, and this does happen for some amides. The consensus of today's textbooks would favor the mechanism shown in Scheme 14.

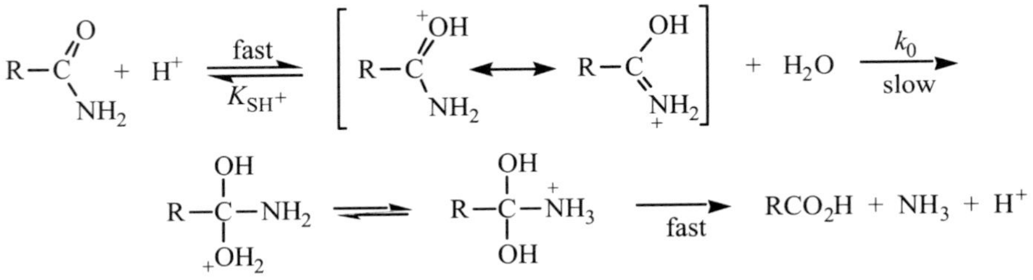

$$\mathrm{R{-}C({=}O)NH_2} + \mathrm{H^+} \underset{K_{\mathrm{SH^+}}}{\overset{\text{fast}}{\rightleftharpoons}} \left[\mathrm{R{-}C({=}\overset{+}{O}H)NH_2} \longleftrightarrow \mathrm{R{-}C(OH){=}\overset{+}{N}H_2}\right] + \mathrm{H_2O} \xrightarrow[\text{slow}]{k_0}$$

$$\mathrm{R{-}C(OH)(\overset{+}{O}H_2){-}NH_2} \rightleftharpoons \mathrm{R{-}C(OH)_2{-}\overset{+}{N}H_3} \xrightarrow[\text{fast}]{} \mathrm{RCO_2H + NH_3 + H^+}$$

Scheme 14

Benzamides, benzimidates and lactams

This is not the forum in which to review all of the work that has been done on acid-catalyzed amide hydrolysis[263,264] (this would take an equally long chapter in its own right), so I will concentrate on the studies that have used the excess acidity method. This has mainly involved benzamides, the presence of the aryl group making it possible to follow the reaction using UV spectroscopy. In many cases the number of water molecules reacting with the protonated amide had previously been found to be three according to the r-parameter treatment,[265–267] which is an update of the Bunnett w-parameter treatment discussed above,[50] using the amide acidity function H_A instead of H_0.[265]

The excess acidity method has been applied to a considerable amount of data obtained in aqueous sulfuric acid at several different temperatures:[268] for the hydrolysis of benzamide, *N*-methylbenzamide and *N,N*-dimethylbenzamide,[269–271] for the corresponding methylbenzimidates and methylbenzimidatium ions, used as model compounds for the *O*-protonated benzamides,[253] for methyl and ethyl 2,6-dimethylbenzimidate, as a check for the S_N2 displacement at the *O*-methyl group which can occur in methylbenzimidates,[272] and for several lactams of different ring sizes, using data obtained by ^{1}H NMR spectroscopy.[273] Reliable values for the pK_{BH^+} values for these compounds were available.[101,103] To summarize, the results were found to be consistent with a three-water-molecule process in relatively dilute acid media, with a one-water-molecule mechanism taking over in stronger acid; at the very highest acidities, a one-bisulfate-ion mechanism was apparent.[268] The three-water-molecule process was attractive because in this way the *N*-protonated tetrahedral intermediate can be formed in one step with little or no charge transfer, see equation (74).[268] It also accounted very well for the very small amounts of ^{18}O exchange found in amide hydrolysis,[274] as compared to the large amounts found in ester hydrolysis,[275,276] since this species would break up essentially irreversibly.[268]

(Me)HO, Ph····C=$\overset{+}{N}H_2$, H–O, H, H, O–H, O, H, H ⟶ (Me)HO, Ph····C–$\overset{+}{N}H_3$, HO, H, O, H–O, H, H (74)

This all seemed very reasonable at the time, but subsequent work was not consistent with it. A small but measurable amount of ^{18}O exchange was reported for some amides in reasonably concentrated HCl media,[277,278] and for at least one amide the amount of exchange decreased with increasing acidity,[277] which is the opposite of what would be expected with the Scheme 14 one-water-molecule mechanism taking over from the equation (74) three-water-molecule mechanism as the acidity increased. Also, the solvent deuterium isotope effect was found to be close to unity for at least one amide,[278] a result that has since been confirmed,[279] which is not what would be expected on the basis of either a three- or a one-water-molecule process.[280] Because of this it was decided to re-examine the lactam hydrolysis data; subsequent to the publication of the excess acidity analysis of the ^{1}H NMR results for these,[268] a new study appeared with rate constant data for four of these molecules in aqueous H_2SO_4 media obtained by UV spectroscopy at several temperatures,[281] and this was included too.[282]

The results of this later analysis were disquieting. First, it appeared that different pK_{BH^+} values applied for the UV results and for the NMR results obtained for the same compound; this is possible (or at least conceivable) since the substrate concentrations were much higher for the NMR experiments, and medium effects are large for amides.[101,103] Secondly, when the data analysis computer programs were applied it was found that analysis according to the "three waters followed by one water" mechanistic scenario still worked, but that an analysis according to a "one water followed by unknown mechanism" scenario worked equally well.[282] The data were not of good quality, and it was not possible to discover what the unknown mechanism was; the activation parameters and $m^{\ddagger}$ slopes resulting from the two analyses did not permit a decision to be taken as to which scenario was the more likely.[282]

There the matter rested until I started work on this article, when, upon thinking about the role of water in the excess acidity method following a communication from More O'Ferrall,[283] I suddenly realized that the water activities we had been using for mechanistic work in strong acids up until now were all wrong! As discussed at the beginning of this article, all the species under consideration in strong acid media must be referred to the same standard state (where activity coefficients $f = 1$), the molarity-based one of infinite dilution in the hypothetical ideal 1 M acid solution that is the reference state. But the water activities in use were mole fraction-based ones, which have a pure water standard state (essentially due to Giauque *et al.*[16] in H_2SO_4). I had been using values obtained by dividing these values by the water mole fraction and multiplying by the water molarity, but these still have the wrong standard state. So I calculated correct values, presented here at 25 °C in Tables 3–8, and used these correct values for all the examples given in this chapter; for instance, Fig. 11 displays very good linearity with a two-water-molecule mechanism. This means that the equation (42) mechanism is correct for ester hydrolysis; none of the proposed ester hydrolysis mechanisms[29,267] are changed by this, nor are any of the other mechanisms involving water discussed above changed.

Such is not the case for amide hydrolysis, however! As I write, work on this is still in progress,[284] but I can present here the results of applying a triple linear regression (in X, temperature and water activity) to all of the data obtained for the hydrolysis of methyl benzimidate at several temperatures[253] according to equation (60), see Fig. 16. The coefficients of the triple regression were the free energy of the reaction at 25°C (intercept), and the enthalpy of reaction, the excess acidity slope $(m^{\ddagger} - 1)m^*$, and the number of water molecules in the transition state, which I will continue to call r, as slopes. The latter value is 1.91 ± 0.16, i.e. *two* water molecules; Fig. 16 is drawn as the linear plots resulting from subtracting twice the log water activity from the observed $\log k_{\psi}$ values. The same mechanism seems to apply to benzamides. Thus the current proposed mechanism for amide hydrolysis is that given in Scheme 15.[284]

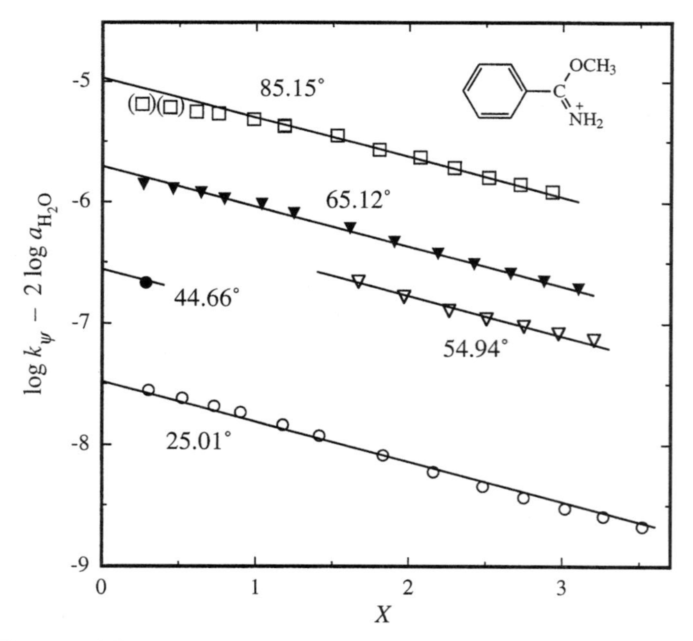

Fig. 16 Excess acidity plot against X according to equation (60) for the hydrolysis of methyl benzimidate at several temperatures, showing the involvement of two water molecules. Data from ref. 253.

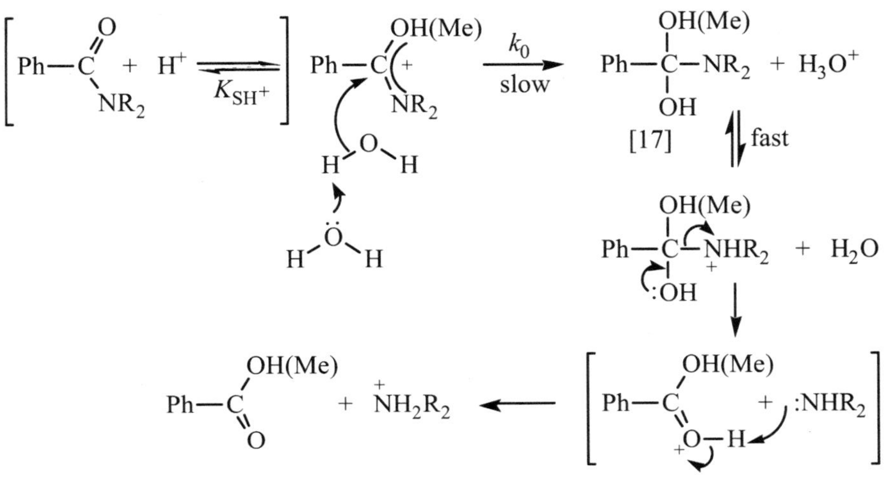

$$\left[\mathrm{Ph{-}C(=O)NR_2} + \mathrm{H^+}\right] \underset{K_{\mathrm{SH^+}}}{\rightleftharpoons} \mathrm{Ph{-}C^+(OH(Me))NR_2} \xrightarrow[\text{slow}]{k_0} \mathrm{Ph{-}C(OH(Me))(OH){-}NR_2} + \mathrm{H_3O^+}$$

$\mathrm{H_2O}$, $\mathrm{H_2O}$

[17] ⇅ fast

$$\mathrm{Ph{-}C(OH(Me))(:OH){-}\overset{+}{N}HR_2} + \mathrm{H_2O}$$

$$\left[\mathrm{Ph{-}C(OH(Me)){=}\overset{+}{O}{-}H} + \mathrm{:NHR_2}\right] \longrightarrow \mathrm{Ph{-}C(OH(Me)){=}O} + \mathrm{\overset{+}{N}H_2R_2}$$

Scheme 15

This is the same mechanism as that given above for esters, in equation (42). The difference between esters and amides is apparent from a comparison of the two tetrahedral intermediates [5] and [17]. The former contains three oxygens, any of which can be protonated, resulting in much ^{18}O exchange being observed when the reaction takes place in ^{18}O-enriched water,[275,276] but [17] contains a much more basic nitrogen, which will be protonated preferentially and lead to much less ^{18}O exchange, as observed.[274,277,278] Also, ammonium ion formation makes the overall reaction irreversible, unlike ester hydrolysis. The calculated solvent isotope effect for the Scheme 15 process is 1.00,[280] exactly in accord with experimental observation.[278,279]

Other amides

One exception to the A2 hydrolysis mechanism given above is the case of β-lactams, which have an A1 hydrolysis mechanism in aqueous sulfuric acid,[273] presumably proceeding from the *N*-protonated form; for this substrate the excess acidity plot of $\log k_\psi - \log C_{\mathrm{H^+}}$ against X is linear.[268]

The hydrolyses of a number of other amides have been examined:[285] *p*-nitroacetanilide and the *p*-nitroanilides of glycine, alanine and α-methylalanine; the derivatives of pyridine with $-CONH_2$ and $-NHCOCH_3$ substituents in the 3- and 4-positions; and some ring-substituted *N*-phenylacetamides. Their basicities were also determined in aqueous H_2SO_4 media, using the excess acidity method among others. These substrates hydrolyze by what would be the Scheme 15 mechanism (*r*-values of ~ 3 being reported), and no evidence was found for the replacement of H by CH_3 on the α-carbon causing any significant rate decrease, either in acid- or in base-catalyzed amide hydrolysis.[285]

Acetamide and a number of *N*-substituted acetamides have been studied in aqueous H_2SO_4 at several temperatures, and the excess acidity and other kinetic analysis techniques led to the proposal of several different hydrolysis mechanisms for them.[286] *N*-Benzylacetamide and its ring-substituted derivatives exhibited *N*-acyl cleavage in an A2 process at low acidities (presumably via Scheme 15), and *N*-alkyl cleavage in an A1 process at higher acidities. Sulfonation was a complicating factor for some substrates at the highest acidities. *N*-*t*-Butylacetamide showed the same behavior (without the sulfonation), with both cleavage modes occurring below ~35 wt% H_2SO_4 but only the A1 process at higher acidities. Acetamide itself showed exclusive A2 *N*-acyl cleavage.[286]

N-*nitro amides*

These all react in aqueous sulfuric acid media, exhibiting a number of different reaction mechanisms, none of which has a close correspondence in the hydrolysis of normal amides. The presence of the *N*-nitro group makes protonation of these molecules quite difficult.[287] Reaction rate constants for the decomposition of several *N*-nitrobenzenesulfonamides at 25°C are given by the Kuznetsov group,[288,289] for some *N*-nitrobenzamides at 25°C by Drozdova *et al.*,[290] and for a few *N*-methyl-*N*-nitrobenzamides at different temperatures by Challis *et al.*[291] At high acidity the less reactive ones lose $^+NO_2$, giving amides or sulfonamides, and the more reactive ones decompose to carboxylic or sulfonic acids and N_2O; see Scheme 16 for the *N*-nitrobenzamides.[287] The *N*-nitrobenzenesulfonamides would give the interesting (but unobserved) species $ArSO_2^+$ rather than $ArCO^+$.

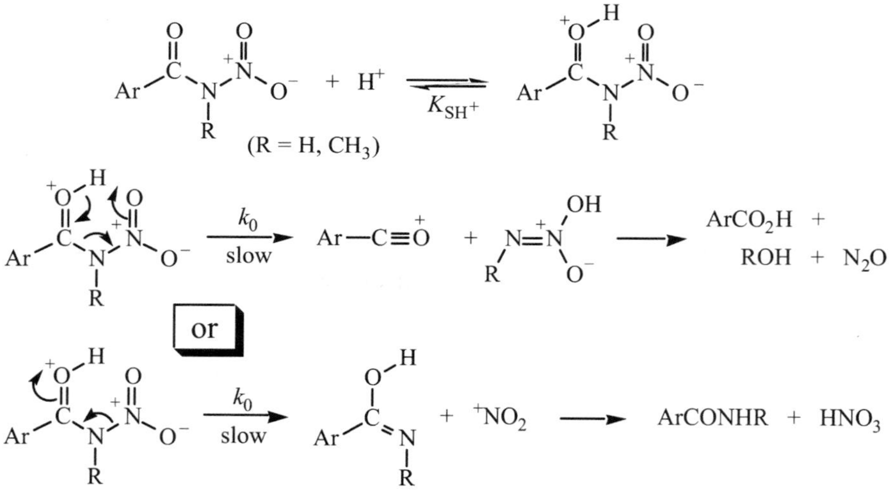

$$\mathrm{ArC(=O)N(R)N^+(=O)O^-} + \mathrm{H^+} \underset{K_{SH^+}}{\rightleftharpoons} \mathrm{ArC(=O^+H)N(R)N^+(=O)O^-} \quad (\mathrm{R = H, CH_3})$$

$$\mathrm{ArC(=O^+H)N(R)N^+(=O)O^-} \xrightarrow[\text{slow}]{k_0} \mathrm{Ar{-}C{\equiv}\overset{+}{O}} + \mathrm{RN{=}\overset{+}{N}(OH)O^-} \longrightarrow \mathrm{ArCO_2H} + \mathrm{ROH} + \mathrm{N_2O}$$

or

$$\mathrm{ArC(=O^+H)N(R)N^+(=O)O^-} \xrightarrow[\text{slow}]{k_0} \mathrm{ArC(OH){=}NR} + {}^+\mathrm{NO_2} \longrightarrow \mathrm{ArCONHR} + \mathrm{HNO_3}$$

Scheme 16

Both of the reactions in Scheme 16 are A1 processes, and all of the substrates give good linear plots of $\log k_\psi - \log C_{H^+}$ against X. The two reactions in

Scheme 16 are easily distinguished by the different products obtained, and by the breaks in the LFER plots of the $\log k_0$ values against σ^+ that occur when the mechanism changes.[287] The *O*-protonation mechanism of Scheme 16 is preferred over the more obvious *N*-protonation one on the basis of the observed $m^\ddagger$ values.[287] The further decomposition of the first-formed benzamide, benzenesulfonamide and *N*-nitroamine reaction products takes place as previously discussed.

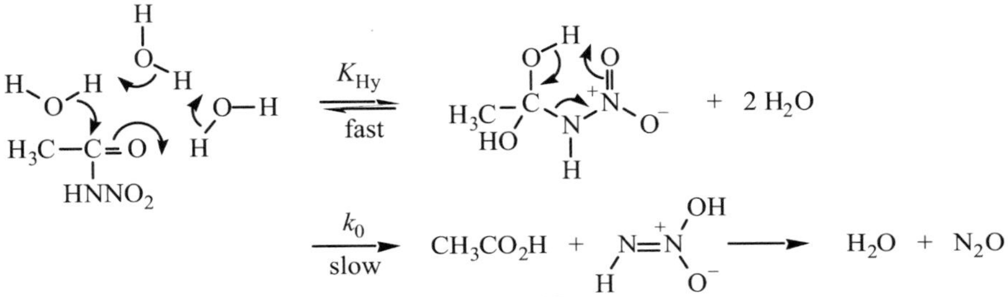

$$H_3C{-}C(=O)HNNO_2 + 3\,H_2O \underset{\text{fast}}{\overset{K_{Hy}}{\rightleftharpoons}} H_3C{-}C(OH)(OH){-}NH{-}N^+(=O)O^- + 2\,H_2O$$

$$\xrightarrow[\text{slow}]{k_0} CH_3CO_2H + HN{=}N^+(OH)O^- \longrightarrow H_2O + N_2O$$

Scheme 17

At lower acidities the *N*-nitrobenzamides and *N*-methyl-*N*-nitrobenzamides have a hydrolysis mechanism that is not acid-catalyzed; for these cases plots of $\log k_\psi - \log a_{H_2O}$ are linear, as for the acylimidazoles discussed above. *N*-Nitroacetamide also hydrolyzes in this way.[291] The proposed mechanism is given in Scheme 17, written for *N*-nitroacetamide; if the hydration shown is a pre-equilibrium (this is a carbonyl compound with a strong electron-withdrawing group attached, so this is likely), only one water molecule will appear in the rate expression (the difference between 3 and 2), as observed.[287] Some evidence for hydroxide-catalyzed processes at the very lowest acidities was also found for some of these compounds.[287]

6 Conclusions

This is a summary of what can be achieved using the excess acidity technique, the method of choice for dealing with aqueous strong acid media. Not everything listed here has been covered in the main body of the text, owing to space limitations, and thus some of the more obvious information obtainable is given in this section, along with that covered earlier.

Reasonably reliable pK_{BH^+} values for the protonation of weak bases or of weakly basic substrates can be obtained via equation (17), together with m^* slope parameters that can be used to classify basic molecules as to type, and for an estimate of the solvation requirements of the protonated base. Measurements at temperatures other than 25°C can be handled using equation (22), and enthalpies and entropies for the protonation can be obtained. Protonation–dehydration processes are covered by equation (26). Medium effects on the spectroscopic measurements used for the determination of basicities can be dealt with relatively easily using equations (27)–(29), and factor analysis if necessary (used carefully).

The kinetics of reactions in these media, when analyzed according to the methods discussed here, are capable of giving a great deal of mechanistic information. Equations (51) and (54) apply to A1 reactions, giving $\log k_0$ values applicable in the same reference state as that used for pH measurements, allowing easy comparison with results obtained in aqueous buffers. Results at several temperatures can be used to evaluate standard-state enthalpies and entropies of activation; using the k_0 values means that the numbers obtained are unencumbered with any variation of the acid system itself with temperature. For the A-S_E2 mechanism equation (56) applies, and the A1 and A-S_E2 mechanisms can be distinguished by the magnitude of the $m^{\ddagger}$ slope parameters, being > 1 in the former case and < 1 in the latter. The A2 mechanism is covered by equations (59) and (60), and the nucleophile or base reacting with the substrates can be uniquely identified, which is difficult to do using any other analysis. Mechanistic information is undoubtedly available from the $m^{\ddagger}$ values for A2 processes, but little work in this area is available yet. The intercept $\log k_0$ values are the best ones to use in LFERs, for evaluating reaction constants ρ^+, ρ^*, etc., because the possible variation of these quantities with acidity is eliminated in the reference state.

Apart from these typical acid-catalyzed reactions, the method is capable of dealing with bimolecular processes, and even reactions that are not acid-catalyzed. For any process that takes place in non-ideal aqueous acid media, it should be the first technique tried. Admixture of the aqueous acid with moderate amounts of inert organic solvent should not be a problem. The broad variety of reaction mechanisms given above should display the versatility and utility of the excess acidity method in physical organic chemistry.

Acknowledgements

It is a pleasure to acknowledge the assistance of, and collaborations with, my many friends and colleagues over the years. In alphabetical order these include Erwin Buncel, Imre Csizmadia, Jack Edward, Ülo Haldna (the latter two both recently deceased, unfortunately), Jerry Kresge, Ed Malinowski, Bob McClelland, Bob McDonald, Mike McKinney, Tom Modro, Ikenna Onyido, Ross Stewart, Mike Thompson, Tom Tidwell, Peter Wan, and, most importantly, Keith Yates; also the members of their research groups, too numerous to mention. Many people have assisted this work by correspondence, talks at meetings, hospitality, etc.: Ned Arnett, Robert Benoit, Stan Brown, Trevor Broxton, Joe Bunnett, Tony Butler, Kallol Ghosh, Peter Guthrie, Don Irish, David Johnson, Sergei Ivanov, Jim Keeffe, Jim King, Leonid Kuznetsov, José-Maria Leal, Don Lee, Howard Maskill, Rory More O'Ferrall, Charlie Perrin,

John Ridd, Gianfranco Scorrano, Ossie Tee, Pietro Traverso, the late Moisei Vinnik, John Warkentin, Lyn Williams, Saul Wolfe, and probably others I have forgotten to mention.

References

1. Bell, R.P. (1973). *The Proton in Chemistry*, 2nd edn. Cornell University Press, Ithaca, New York
2. Bell, R.P. (1959). *The Proton in Chemistry*. Cornell University Press, Ithaca, New York
3. Cox, R.A. and Stewart, R. (1976). *J. Am. Chem. Soc.* **98**, 488
4. Cox, R.A., Stewart, R., Cook, M.J., Katritzky, A.R. and Tack, R.D. (1976). *Can. J. Chem.* **54**, 900
5. Harris, M.G. and Stewart, R. (1977). *Can. J. Chem.* **55**, 3800
6. Stewart, R. and Harris, M.G. (1977). *Can. J. Chem.* **55**, 3807
7. Symons, E.A., Clermont, M.J. and Coderre, L.A. (1981). *J. Am. Chem. Soc.* **103**, 3131
8. Hannigan, T.J. and Spillane, W.J. (1982). *J. Chem. Soc., Perkin Trans. 2*, 851
9. Balón, M., Hidalgo, J., Guardado, P., Muñoz, M.A. and Carmona, M.C. (1993). *J. Chem. Soc., Perkin Trans. 2*, 91, 99; and earlier papers by the Balón group
10. Bagno, A., Scorrano, G. and Terrier, F. (1990). *J. Chem. Soc., Perkin Trans. 2*, 1017
11. More O'Ferrall, R.A. (1972). *J. Chem. Soc., Perkin Trans. 2*, 976
12. Bagno, A., Scorrano, G. and Terrier, F. (1991). *J. Chem. Soc., Perkin Trans. 2*, 651
13. Brønsted, J.N. (1923). *Rec. Trav. Chim.* **42**, 718
14. Lowry, T.M. (1923). *Chem. Ind.* **42**, 43
15. Lewis, G.N. (1923). *Valency and the Structure of Atoms and Molecules*. Reinhold, New York
16. Giauque, W.F., Hornung, E.W., Kunzler, J.E. and Rubin, T.R. (1960). *J. Am. Chem. Soc.* **82**, 62
17. Pearce, J.N. and Nelson, A.F. (1933). *J. Am. Chem. Soc.* **55**, 3075
18. Robinson, R.A. and Baker, O.J. (1946). *Trans. Roy. Soc. New Zealand*, **76**, 250
19. Wai, H. and Yates, K. (1969). *Can. J. Chem.* **47**, 2326
20. Cox, R.A. and Yates, K. (1978). *J. Am. Chem. Soc.* **100**, 3861
21. Hammett, L.P. and Deyrup, A.J. (1932). *J. Am. Chem. Soc.* **54**, 2721
22. Hammett, L.P. (1970). *Physical Organic Chemistry*, 2nd edn. McGraw–Hill, New York
23. Rochester, C.H. (1970). *Acidity Functions*. Academic Press, London
24. Katritzky, A.R., Waring, A.J. and Yates, K. (1963). *Tetrahedron* **19**, 465
25. Arnett, E.M. and Mach, G.W. (1964). *J. Am. Chem. Soc.* **86**, 2671
26. Deno, N.C., Jaruzelski, J.J. and Schriesheim, A. (1955). *J. Am. Chem. Soc.* **77**, 3044
27. Arnett, E.M. and Bushick, R.D. (1964). *J. Am. Chem. Soc.* **86**, 1564
28. Cox, R.A. and Yates, K. (1983). *Can. J. Chem.* **61**, 2225
29. Yates, K. and McClelland, R.A. (1967). *J. Am. Chem. Soc.* **89**, 2686
30. Bunnett, J.F. and Olsen, F.P. (1966). *Can. J. Chem.* **44**, 1899
31. Marziano, N.C., Cimino, G.M. and Passerini, R.C. (1973). *J. Chem. Soc., Perkin Trans. 2*, 1915
32. Passerini, R., Marziano, N.C. and Traverso, P. (1975). *Gazz. Chim. Ital.* **105**, 901
33. Marziano, N.C., Traverso, P.G., Tomasin, A. and Passerini, R.C. (1977). *J. Chem. Soc., Perkin Trans. 2*, 309

34. Marziano, N.C., Traverso, P.G. and Passerini, R.C. (1977). *J. Chem. Soc., Perkin Trans. 2*, 306
35. Perrin, C.L. (1964). *J. Am. Chem. Soc.* **86**, 256
36. Cox, R.A. and Yates, K. (1981). *Can. J. Chem.* **59**, 2116
37. Gillespie, R.J., Robinson, E.A. and Solomons, C. (1960). *J. Chem. Soc.* 4320
38. Robertson, E.B. and Dunford, H.B. (1964). *J. Am. Chem. Soc.* **86**, 5080
39. Bagno, A., Scorrano, G. and More O'Ferrall, R.A. (1987). *Rev. Chem. Intermed.* **7**, 313
40. Brown, K.L. and Yang, T.-F. (1987). *Inorg. Chem.* **26**, 3007; and earlier papers by the Brown group
41. Cox, R.A., Goldman, M.F. and Yates, K. (1979). *Can. J. Chem.* **57**, 2960
42. Cox, R.A. and Yates, K. (1984). *Can. J. Chem.* **62**, 2155
43. Marziano, N.C., Tomasin, A., Tortato, C. and Zaldivar, J.M. (1998). *J. Chem. Soc., Perkin Trans. 2*, 1973
44. Tomikawa, K. and Kanno, H. (1998). *J. Phys. Chem. A*, **102**, 6082
45. Zeleznik, F.J. (1991). *J. Phys. Chem. Ref. Data*, **20**, 1157
46. Robinson, R.A. and Stokes, R.H. (1959). *Electrolyte Solutions*, 2nd edn. Butterworths, London
47. Bidinosti, D.R. and Biermann, W.J. (1956). *Can. J. Chem.* **34**, 1591
48. Randall, M. and Young, L.E. (1928). *J. Am. Chem. Soc.* **50**, 989
49. Åkerlöf, G. and Teare, L.E. (1937). *J. Am. Chem. Soc.* **59**, 1855
50. Bunnett, J.F. (1961). *J. Am. Chem. Soc.* **83**, 4956, 4968, 4973, 4978
51. Liu, Y. and Grén, U. (1991). *Fluid Phase Equilibria*, **63**, 49
52. Cox, R.A., Lam, S.-O., McClelland, R.A. and Tidwell, T.T. (1979). *J. Chem. Soc., Perkin Trans. 2*, 272
53. Noyce, D.S., Avarbock, H.S. and Reed, W.L. (1962). *J. Am. Chem. Soc.* **84**, 1647
54. Marziano, N.C., Tomasin, A. and Traverso, P.G. (1981). *J. Chem. Soc., Perkin Trans. 2*, 1070
55. Traverso, P.G. (1984). *Can. J. Chem.* **62**, 153
56. Marziano, N.C., Sampoli, M. and Gonizzi, M. (1986). *J. Phys. Chem.* **90**, 4347
57. Marziano, N.C., Tomasin, A. and Tortato, C. (1996). *Organic Reactivity (Tartu)*, **30**, 29
58. Marziano, N.C., Tomasin, A. and Tortato, C. (1996). *Organic Reactivity (Tartu)*, **30**, 39
59. Marziano, N.C., Tomasin, A., Tortato, C. and Isandelli, P. (1998). *J. Chem. Soc., Perkin Trans. 2*, 2535
60. Carpentier, J.-M. and Lemetais, P. (1978). *Can. J. Chem.* **56**, 954
61. Kresge, A.J. and Chen, H.-J. (1969). *Anal. Chem.* **41**, 74
62. Bagno, A., Lucchini, V. and Scorrano, G. (1990). *Can. J. Chem.* **68**, 1746
63. Bagno, A., Lovato, G. and Scorrano, G. (1993). *J. Chem. Soc., Perkin Trans. 2*, 1091
64. Attiga, S.A. and Rochester, C.H. (1975). *J. Chem. Soc., Perkin Trans. 2*, 1411
65. Yates, K., Wai, H., Welch, G. and McClelland, R.A. (1973). *J. Am. Chem. Soc.* **95**, 418
66. Edward, J.T. and Wong, S.C. (1977). *Can. J. Chem.* **55**, 2492
67. Mach, G.W. (1964). Ph.D. Thesis, University of Pittsburgh
68. Arnett, E.M. and Mach, G.W. (1966). *J. Am. Chem. Soc.* **88**, 1177
69. Mathivanan, N., McClelland, R.A. and Steenken, S. (1990). *J. Am. Chem. Soc.* **112**, 8454
70. Chandler, W.D. and Lee, D.G. (1990). *Can. J. Chem.* **68**, 1757; and earlier papers by the Lee group
71. Freiberg, W. (1994) *J. Prakt. Chem./Chem.-Ztg.* **336**, 565; and earlier papers by the Freiberg group

72. Kaiser, M., Freiberg, W. and Michalik, M. (1996). *J. Prakt. Chem./Chem.-Ztg.* **338**, 182
73. Géribaldi, S., Grec-Luciano, A., Maria, P.-C. and Azzaro, M. (1982). *J. Chim. Phys.* **79**, 103
74. Haldna, Ü.L., Grebenkova, M. and Ebber, A. (1991). *Comput. Chem.* **15**, 99
75. Haldna, Ü.L., Linask, K., Kuura, H.J. and Kudryashova, M. (1994). *Mendeleev Commun.*, 22
76. Lee, D.G. and Sadar, M.H. (1976). *Can. J. Chem.* **54**, 3464
77. Alberghina, G., Amato, M.E., Fisichella, S. and Occhipinti, S. (1980). *J. Chem. Soc., Perkin Trans. 2*, 1721; and earlier papers by the Fisichella group
78. Consiglio, G., Frenna, V., Mugnoli, A., Noto, R., Pani, M. and Spinelli, D. (1997) *J. Chem. Soc., Perkin Trans. 2*, 309; and earlier papers by the Spinelli group
79. De Maria, P., Barbieri, C.L., Spinelli, D., Dell'Erba, C., Novi, M., Petrillo, G. and Sancassan, F. (1991). *J. Chem. Soc., Perkin Trans. 2*, 373
80. De Maria, P., Fontana, A., Spinelli, D., Dell'Erba, C., Novi, M., Petrillo, G. and Sancassan, F. (1993). *J. Chem. Soc., Perkin Trans. 2*, 649
81. García, B., Casado, R.M., Castillo, J., Ibeas, S., Domingo, I. and Leal, J.M. (1993) *J. Phys. Org. Chem.* **6**, 101; and earlier papers by the García/Leal group
82. Perisic-Janjic, N.U., Arman, L. and Lazarevic, M. (1997). *Spectrosc. Lett.* **30**, 1037; and earlier papers by the Perisic-Janjic group
83. Ghosh, K.K. and Ghosh, S. (1996). *J. Indian Chem. Soc.* **73**, 79
84. Sharma, K., Harjit, J. and Pande, R. (1998). *Asian J. Chem.* **10**, 117; *Chem. Abs.* **128**:120123*a*
85. Edward, J.T. and Wong, S.C. (1979). *Can. J. Chem.* **57**, 1980; and earlier papers by the Edward group
86. Battye, P.J., Cassidy, J.F. and Moodie, R.B. (1981). *J. Chem. Soc., Chem. Commun.*, 68
87. Marziano, N.C., Tortato, C. and Bertani, R. (1992). *J. Chem. Soc., Perkin Trans. 2*, 955
88. Bagno, A. and Scorrano, G. (1996). *Gazz. Chim. Ital.* **126**, 365; and earlier papers by the Bagno/Scorrano group
89. Dorion, F. and Gaboriaud, R. (1980). *J. Chim. Phys.* **77**, 1057
90. Andonovski, B., Spirevska, I. and Nikolovski, A. (1996). *Croat. Chem. Acta*, **69**, 1201
91. Carmona, M.C., Muñoz, M.A., Guardado, P. and Balón, M. (1995) *J. Phys. Org. Chem.* **8**, 559; and earlier papers by the Balón group
92. Gut, I.G. and Wirz, J. (1994). *Angew. Chem. Int. Ed. Engl.* **33**, 1153
93. García, B. and Palacios, J.C. (1988). *Ber. Bunsen-Ges. Phys. Chem.* **92**, 696
94. Benoit, R.L. and Fréchette, M. (1986). *Can. J. Chem.* **64**, 2348; and earlier papers by the Benoit group
95. Pincock, J.A. and Redden, P.R. (1989). *Can. J. Chem.* **67**, 227, 710
96. Brown, K.L, Hakimi, J.M. and Jacobsen, D.W. (1984). *J. Am. Chem. Soc.* **106**, 7894; and other papers by the Brown group
97. Domingo, P.L., García, B. and Leal, J.M. (1987). *Can. J. Chem.* **65**, 583
98. Domingo, P.L., García, B. and Leal, J.M. (1990). *Can. J. Chem.* **68**, 228
99. Davis, C.T. and Geissman, T.A. (1954). *J. Am. Chem. Soc.* **76**, 3507
100. Reeves, R.L. (1966). *J. Am. Chem. Soc.* **88**, 2240
101. Cox, R.A. and Yates, K. (1981). *Can. J. Chem.* **59**, 1560
102. Bevington, P.R. (1969). *Data Reduction and Error Analysis for the Physical Sciences*, McGraw–Hill, New York
103. Cox, R.A., Druet, L.M., Klausner, A.E., Modro, T.A., Wan, P. and Yates, K. (1981). *Can. J. Chem.* **59**, 1568

104. Malinowski, E.R. and Howery, D.G. (1980). *Factor Analysis in Chemistry*. Wiley-Interscience, New York
105. Edward, J.T. and Wong, S.C. (1977). *J. Am. Chem. Soc.* **99**, 4229
106. Haldna, Ü.L. (1990). *Prog. Phys. Org. Chem.* **18**, 65
107. Haldna, Ü.L., Murshak, A. and Kuura, H.J. (1980). *Org. React. (Tartu)*, **17**, 384; *Chem. Abs.* **96**, 68047*h*
108. Haldna, Ü.L., Grebenkova, M. and Ebber, A. (1993). *Org. React. (Tartu)*, **28**, 75; *Chem. Abs.* **121**, 56891*a*.
109. Haldna, Ü.L. and Murshak, A. (1984). *Comput. Chem.* **8**, 201
110. Haldna, Ü.L. and Grebenkova, M. (1993). *Comput. Chem.* **17**, 241
111. Grebenkova, M. and Haldna, Ü.L. (1993). *Eesti Tead. Toim., Keem.* **42**, 138; *Chem. Abs.* **120**, 190985*s*
112. Kudryashova, M. and Haldna, Ü.L. (1994). *Eesti Tead. Toim., Keem.* **43**, 77; *Chem. Abs.* **121**, 300400*c*
113. Ebber, A., Haldna, Ü.L. and Murshak, A. (1986). *Org. React. (Tartu)*, **23**, 40; *Chem. Abs.* **107**, 6686*c*
114. Modro, T.A., Yates, K. and Janata, J. (1975). *J. Am. Chem. Soc.* **97**, 1492
115. Sampoli, M., de Santis, A. and Marziano, N.C. (1985). *J. Chem. Soc., Chem. Commun.*, 110
116. Cox, R.A. and Yates, K. (1986). *J. Org. Chem.* **51**, 3619
117. Cox, R.A. (1991). *J. Phys. Org. Chem.* **4**, 233
118. Cox, R.A., Grant, E., Whitaker, T. and Tidwell, T.T. (1990). *Can. J. Chem.* **68**, 1876
119. Cox, R.A. (1996). *Can. J. Chem.* **74**, 1774
120. Cox, R.A. (1999). *Can. J. Chem.* **77**, 709
121. Lucchini, V., Modena, G., Scorrano, G., Cox, R.A. and Yates, K. (1982). *J. Am. Chem. Soc.* **104**, 1958
122. Kresge, A.J., Chen, H.-J., Capen, G.L. and Powell, M.F. (1983). *Can. J. Chem.* **61**, 249
123. García, B. and Leal, J. M. (1991). *J. Phys. Org. Chem.* **4**, 413
124. Ortiz, M.I., de Campos, D. and Irabien Guilas, J.A. (1987). *Chem. Eng. Sci.* **42**, 2467
125. Ortiz, M.I., Rey, M.I. and Irabien Guilas, J.A. (1987). *J. Mol. Catal.* **43**, 51
126. Ghosh, K.K. and Sar, S.K. (1997). *J. Indian Chem. Soc.* **74**, 187
127. Ghosh, K.K. and Ghosh, S. (1994). *J. Org. Chem.* **59**, 1369
128. Satchell, D.P.N., Wassef, W.N. and Bhatti, Z.A. (1993). *J. Chem. Soc., Perkin Trans. 2*, 2373
129. Motie, R.E., Satchell, D.P.N. and Wassef, W.N. (1993). *J. Chem. Soc., Perkin Trans. 2*, 1087
130. Ali, M., Satchell, D.P.N. and Le, V.T. (1993). *J. Chem. Soc., Perkin Trans. 2*, 917
131. Grant, H.M., McTigue, P.T. and Ward, D.G. (1983). *Aust. J. Chem.* **36**, 2211
132. Farrell, J.R. and McTigue, P.T. (1985). *Aust. J. Chem.* **38**, 985
133. Johnson, C.D. and Stratton, B. (1986). *J. Org. Chem.* **51**, 4100
134. Johnson, C.D. and Stratton, B. (1987). *J. Org. Chem.* **52**, 4798
135. Noto, R., Gruttadauria, M., Spinelli, D. and Consiglio, G. (1990). *J. Chem. Soc., Perkin Trans. 2*, 1975
136. Haldna, Ü.L. and Kuus, H.J. (1979) *Org. React. (Tartu)*, **16**, 5; *Chem. Abs.* **92**, 58014*y*
137. Haldna, Ü.L. (1979) *Org. React. (Tartu)*, **16**, 129; *Chem. Abs.* **92**, 58015*z*
138. Haldna, Ü.L. (1980). *Usp. Khim.* **49**, 1174 (623 in Eng. edn.)
139. Haldna, Ü.L. (1980). *Org. React. (Tartu)*, **17**, 233; *Chem. Abs.* **95**, 96765*p*
140. Traverso, P.G. (1986). *Gazz. Chim. Ital.* **116**, 137
141. Wojcik, J.F. (1982). *J. Phys. Chem.* **86**, 145

142. Wojcik, J.F. (1985). *J. Phys. Chem.* **89**, 1748
143. Edward, J.T., Sjöström, M. and Wold, S. (1981). *Can. J. Chem.* **59**, 2350
144. Bell, R.P., Bascombe, K.N. and McCoubrey, J.C. (1956). *J. Chem. Soc.*, 1286
145. Cox, R.A. (1987). *Acc. Chem. Res.* **20**, 27
146. Zucker, L. and Hammett, L.P. (1939). *J. Am. Chem. Soc.* **61**, 2791
147. Banger, J., Johnson, C.D., Katritzky, A.R. and O'Neill, B.R. (1974). *J. Chem. Soc., Perkin Trans. 2*, 394
148. Kresge, A.J., Hakka, L.E., Mylonakis, S. and Sato, Y. (1965). *Disc. Faraday Soc.*, 75
149. Kresge, A.J., Mylonakis, S.G. and Hakka, L.E. (1972). *J. Am. Chem. Soc.* **94**, 4197
150. Cox, R.A. (1974). *J. Am. Chem. Soc.* **96**, 1059
151. Buncel, E. (1975). *Acc. Chem. Res.* **8**, 132
152. Cox, R.A. and Buncel, E. (1975). *J. Am. Chem. Soc.* **97**, 1871
153. Cox, R.A., Dolenko, A.J. and Buncel, E. (1975). *J. Chem. Soc., Perkin Trans. 2*, 471
154. Buncel, E. and Keum, S.-R. (1983). *J. Chem. Soc., Chem. Commun.*, 578
155. Bunnett, J.F. and Olsen, F.P. (1966). *Can. J. Chem.* **44**, 1917
156. Lucchini, V., Modena, G., Scorrano, G. and Tonellato, U. (1977). *J. Am. Chem. Soc.* **99**, 3387
157. Roth, B. and Bunnett, J.F. (1965). *J. Am. Chem. Soc.* **87**, 340
158. Bunnett, J.F. and Olsen, F.P. (1974). *J. Org. Chem.* **39**, 1156
159. Bunnett, J.F., McDonald, R.L. and Olsen, F.P. (1974). *J. Am. Chem. Soc.* **96**, 2855
160. Modena, G., Rivetti, F., Scorrano, G. and Tonellato, U. (1977). *J. Am. Chem. Soc.* **99**, 3392
161. Cox, R.A. and Yates, K. (1979). *Can. J. Chem.* **57**, 2944
162. Kresge, A.J., More O'Ferrall, R.A., Hakka, L.E. and Vitullo, V.P. (1965). *J. Chem. Soc., Chem. Commun.*, 46
163. Kresge, A.J., Mylonakis, S.G., Sato, Y. and Vitullo, V.P. (1971). *J. Am. Chem. Soc.* **93**, 6181
164. Bell, R.P. and Brown, A.H. (1954). *J. Chem. Soc.*, 774
165. Ali, M. and Satchell, D.P.N. (1991). *J. Chem. Soc., Perkin Trans. 2*, 575
166. Ali, M. and Satchell, D.P.N. (1993). *J. Chem. Soc., Perkin Trans. 2*, 1825
167. Ali, M. and Satchell, D.P.N. (1995). *J. Chem. Soc., Perkin Trans. 2*, 167
168. Satchell, D.P.N. and Wassef, W.N. (1992). *J. Chem. Soc., Perkin Trans. 2*, 1855
169. Motie, R.E., Satchell, D.P.N. and Wassef, W.N. (1992). *J. Chem. Soc., Perkin Trans. 2*, 859
170. Gnedin, B.G., Ivanov, S.N. and Shchukina, M.V. (1988). *Zh. Org. Khim.* **24**, 810 (731 in Eng. edn.)
171. Ivanov, S.N. and Gnedin, B.G. (1989). *Zh. Org. Khim.* **25**, 831 (746 in Eng. edn.)
172. Ivanov, S.N., Kislov, V.V. and Gnedin, B.G. (1998). *Zh. Obshch. Khim.* **68**, 1177 (1123 in Eng. edn.)
173. Lajunen, M., Laine, R. and Aaltonen, M. (1997). *Acta Chem. Scand.* **51**, 1155; and earlier papers by the Lajunen group
174. Lajunen, M., Katainen, E. and Dahlqvist, M. (1998). *Acta Chem. Scand.* **52**, 816
175. Erden, I, Keeffe, J.R., Xu, F.-P. and Zheng, J.-B. (1993). *J. Am. Chem. Soc.* **115**, 9834
176. Alvarez Macho, M.P. and Montequi Martin, M.I. (1990). *An. Asoc. Quím. Argent.* **78**, 91; *Chem. Abs.* **114**, 228194z
177. Alvarez Macho, M.P. and Montequi Martin, M.I. (1992). *An. Quim.* **88**, 149
178. Donaldson, D.J., Ravishankara, A.R. and Hanson, D.R. (1997). *J. Phys. Chem. A*, **101**, 4717
179. Cox, R.A. and Yates, K. (1982). *Can. J. Chem.* **60**, 3061
180. Cox, R.A., McAllister, M.A., Roberts, K.A., Stang, P.J. and Tidwell, T.T. (1989). *J. Org. Chem.* **54**, 4899

181. Bégué, J.-P., Benayoud, F., Bonnet-Delpon, D., Allen, A.D., Cox, R.A. and Tidwell, T.T. (1995). *Gazz. Chim. Ital.* **125**, 399
182. Deno, N.C., Kish, F.A. and Peterson, H.J. (1965). *J. Am. Chem. Soc.* **87**, 2157
183. Ellis, G.W.L. and Johnson, C.D. (1982). *J. Chem. Soc., Perkin Trans. 2*, 1025
184. Ellis, G.W.L. (1982). Ph.D. Thesis, University of East Anglia
185. Noyce, D.S., Hartter, D.R. and Miles, F.B. (1968). *J. Am. Chem. Soc.* **90**, 4633
186. Eaborn, C. and Taylor, R. (1960). *J. Chem. Soc.*, 3301
187. Olsson, S. and Russell, M. (1969). *Ark. Kemi*, **31**, 439
188. Krylov, E.N. (1988). *Zh. Org. Khim.* **24**, 786 (709 in Eng. edn.)
189. Krylov, E.N. (1992). *Zh. Obshch. Khim.* **62**, 147 (121 in Eng. edn.)
190. Guthrie, R.D. and Jencks, W.P. (1989). *Acc. Chem. Res.* **22**, 343
191. Lucchini, V. and Modena, G. (1990). *J. Am. Chem. Soc.* **112**, 6291
192. Kresge, A.J. and Paine, S.W. (1996). *Can. J. Chem.* **74**, 1373
193. Banait, N., Hojatti, M., Findlay, P. and Kresge, A.J. (1987). *Can. J. Chem.* **65**, 441
194. Kresge, A.J. and Tobin, J.B. (1993). *Angew. Chem., Int. Ed. Engl.* **32**, 721
195. Gabelica, V. and Kresge, A.J. (1996). *J. Am. Chem. Soc.* **118**, 3838
196. Chiang, Y. and Kresge, A.J. (1985). *J. Am. Chem. Soc.* **107**, 6363
197. Kresge, A.J. and Leibovitch, M. (1990). *J. Org. Chem.* **55**, 5234
198. Chiang, Y., Chwang, W.K., Kresge, A.J., Powell, M.F. and Szilagyi, S. (1984). *J. Org. Chem.* **49**, 5218
199. Chiang, Y., Kresge, A.J., Obraztsov, P.A. and Tobin, J.B. (1992). *Croat. Chem. Acta*, **65**, 615
200. Burt, R.A., Chiang, Y., Kresge, A.J. and Szilagyi, S. (1984). *Can. J. Chem.* **62**, 74
201. Kresge, A.J. and Yin, Y. (1989). *J. Phys. Org. Chem.* **2**, 43
202. Kresge, A.J., Leibovitch, M. and Sikorski, J.A. (1992). *J. Am. Chem. Soc.* **114**, 2618
203. Kresge, A.J. and Ubysz, D. (1994). *J. Phys. Org. Chem.* **7**, 316
204. Lajunen, M. (1991). *Acta Chem. Scand.* **45**, 377
205. Lajunen, M., Virta, M. and Kylliäinen, O. (1994). *Acta Chem. Scand.* **48**, 122
206. Ali, M. and Satchell, D.P.N. (1991). *J. Chem. Soc., Chem. Commun.*, 866
207. Ali, M. and Satchell, D.P.N. (1992). *J. Chem. Soc., Perkin Trans. 2*, 219
208. Satchell, D.P.N, Satchell, R.S. and Wassef, W.N. (1992). *J. Chem. Res., Synop.*, 46
209. Joseph, V.B., Satchell, D.P.N., Satchell, R.S. and Wassef, W.N. (1992). *J. Chem. Soc., Perkin Trans. 2*, 339
210. Edward, J.T. and Wong, S.C. (1977). *J. Am. Chem. Soc.* **99**, 7224
211. Ghosh, K.K., Rajput, S.K. and Sar, S.K. (1996). *J. Indian Chem. Soc.* **73**, 684
212. Bekdemir, Y., Tillett, J.G. and Zalewski, R.I. (1993). *J. Chem. Soc., Perkin Trans. 2*, 1643
213. Alvarez Macho, M.P. and Montequi Martin, M.I. (1990). *Can. J. Chem.* **68**, 29
214. Ali, M. and Satchell, D.P.N. (1993). *J. Chem. Res. (Synop.)*, 114
215. Gerards, L.E.H. and Balt, S. (1994). *Recl. Trav. Chim. Pays-Bas*, **113**, 137
216. Ghosh, K.K. (1997). *Indian J. Chem., Sect. B*, **36**, 1089
217. Al-Shalchi, W., Selwood, T. and Tillett, J.G. (1985). *J. Chem. Res. (Synop.)*, 10
218. Kutuk, H. and Tillett, J.G. (1993). *Phosphorus, Sulfur, and Silicon*, **85**, 217
219. Saleh, A. and Tillett, J.G. (1981). *J. Chem. Soc., Perkin Trans. 2*, 132
220. Said, Z. and Tillett, J.G. (1982). *J. Chem. Soc., Perkin Trans. 2*, 701
221. Cox, R.A. and Yates, K. (1984). *Can. J. Chem.* **62**, 1613
222. Astrat'ev, A.A., Vorob'eva, E.N., Kuznetsov, L.L. and Gidaspov, B.V. (1979). *Zh. Org. Khim.* **15**, 1136 (1015 in Eng. edn.)
223. Vorob'eva, E.N., Kuznetsov, L.L. and Gidaspov, B.V. (1983). *Zh. Org. Khim.* **19**, 698 (615 in Eng. edn.)
224. Glukhov, A.A., Kuznetsov, L.L. and Gidaspov, B.V. (1983). *Zh. Org. Khim.* **19**, 704 (620 in Eng. edn.)

225. Hughes, M.N., Lusty, J.R. and Wallis, H.L. (1983). *J. Chem. Soc., Dalton Trans.*, 261
226. Cox, R.A. (1996). *Can. J. Chem.* **74**, 1779
227. Marziano, N.C., Traverso, P.G., de Santis, A. and Sampoli, M. (1978). *J. Chem. Soc., Chem. Commun.*, 873
228. Sampoli, M., de Santis, A., Marziano, N.C., Pinna, F. and Zingales, A. (1985). *J. Phys. Chem.* **89**, 2864
229. Marziano, N.C., Tomasin, A. and Sampoli, M. (1991). *J. Chem. Soc., Perkin Trans. 2*, 1995
230. Traverso, P.G., Marziano, N.C. and Passerini, R.C. (1977). *J. Chem. Soc., Perkin Trans. 2*, 845
231. Marziano, N.C. and Sampoli, M. (1983). *J. Chem. Soc., Chem. Commun.*, 523
232. Marziano, N.C., Sampoli, M., Pinna, F. and Passerini, A. (1984). *J. Chem. Soc., Perkin Trans. 2*, 1163
233. Marziano, N.C., Sampoli, M. and Gonizzi, M. (1988). *Bull. Soc. Chim. Fr.*, 401
234. Marziano, N.C., Traverso, P.G. and Cimino, G.M. (1980). *J. Chem. Soc., Perkin Trans. 2*, 574
235. Moodie, R.B., Schofield, K., Taylor, P.G. and Baillie, P.J. (1981). *J. Chem. Soc., Perkin Trans. 2*, 842
236. Marziano, N.C., Tortato, C. and Sampoli, M. (1991). *J. Chem. Soc., Perkin Trans. 2*, 423
237. Marziano, N.C., Tomasin, A. and Tortato, C. (1991). *J. Chem. Soc., Perkin Trans. 2*, 1575
238. Buncel, E. and Onyido, I. (1986). *Can. J. Chem.* **64**, 2115
239. Onyido, I. and Opara, L.U. (1989). *J. Chem. Soc., Perkin Trans. 2*, 1817
240. Cox, R.A., Onyido, I. and Buncel, E. (1992). *J. Am. Chem. Soc.* **114**, 1358
241. Ogunjobi, K.A. (1994). *J. Org. Chem.* **59**, 3386
242. Cox, R.A., Smith, C.R. and Yates, K. (1979). *Can. J. Chem.* **57**, 2952
243. Smith, C.R. (1968). M.Sc. Thesis, University of Toronto
244. Keeffe, J.R., Kresge, A.J. and Toullec, J. (1986). *Can. J. Chem.* **64**, 1224
245. Haldna, Ü.L., Kuura, H.J., Erreline, L. and Palm, V. (1965) *Reakts. Sposobn. Org. Soedin.* **2**, 194; *Chem. Abs.* **64**, 3304*c*
246. Baigrie, L.M., Cox, R.A., Slebocka-Tilk, H., Tencer, M. and Tidwell, T.T. (1985). *J. Am. Chem. Soc.* **107**, 3640
247. Chiang, Y., Kresge, A.J., More O'Ferrall, R.A., Murray, B.A., Schepp, N.P. and Wirz, J. (1990). *Can. J. Chem.* **68**, 1653
248. Cox, R.A., Moore, D.B. and McDonald, R.S. (1994). *Can. J. Chem.* **72**, 1910
249. Blagoeva, I.B., Pojarlieff, I.G. and Rachina, V.T. (1987) *Izv. Khim.* **20**, 119; *Chem. Abs.* **108**, 130785*u*
250. Rachina, V.T., Blagoeva, I.B., Pojarlieff, I.G. and Yates, K. (1990). *Can. J. Chem.* **68**, 1676
251. Marburg, S. and Jencks, W.P. (1962). *J. Am. Chem. Soc.* **84**, 232
252. Jencks, W.P. (1994). Private communication
253. Smith, C.R. and Yates, K. (1972). *J. Am. Chem. Soc.* **94**, 8811
254. Cox, R.A. (1997). *Can. J. Chem.* **75**, 1093
255. Cox, R.A., McKinney, M.A, Pinto, M. and Shiroor, S. (1988). Unpublished results
256. Buncel, E. and Cheon, K.-S. (1998). *J. Chem. Soc., Perkin Trans. 2*, 1241
257. Cheon, K.-S., Cox, R.A., Keum, S.-R. and Buncel, E. (1998). *J. Chem. Soc., Perkin Trans. 2*, 1231
258. Cox, R.A., Cheon, K.-S., Keum, S.-R. and Buncel, E. (1998). *Can. J. Chem.* **76**, 896
259. Yeh, S.-J. and Jaffé, H.H. (1959). *J. Am. Chem. Soc.* **81**, 3274
260. Shine, H.J., Habdas, J., Kwart, H., Brechbiel, M., Horgan, A.G. and San Filippo, J. (1983). *J. Am. Chem. Soc.* **105**, 2823

261. Cox, R.A. and Buncel, E. (1997). In *The Chemistry of the Hydrazo, Azo and Azoxy Groups*, vol. 2 (ed. S. Patai), chapter 15. Wiley, London
262. Liler, M. (1971). *Reaction Mechanisms in Sulphuric Acid.* Academic Press, London
263. O'Connor, C.J. (1970). *Q. Rev. Chem. Soc.* **24**, 553
264. Challis, B.C. and Challis, J.A. (1970). In *The Chemistry of Amides* (ed. J. Zabicky). Interscience, London
265. Yates, K. and Stevens, J.B. (1965). *Can. J. Chem.* **43**, 529
266. Yates, K and Riordan, J.C. (1965). *Can. J. Chem.* **43**, 2328
267. Yates, K. (1971). *Acc. Chem. Res.* **4**, 136
268. Cox, R.A. and Yates, K. (1981). *Can. J. Chem.* **59**, 2853
269. Smith, C.R. (1971). Ph.D. Thesis, University of Toronto
270. Smith, C.R. and Yates, K. (1971). *J. Am. Chem. Soc.* **93**, 6578
271. Bunton, C.A., Farber, S.J., Milbank, A.J.C., O'Connor, C.J. and Turney, T.A. (1972). *J. Chem. Soc., Perkin Trans. 2*, 1869
272. McClelland, R.A. (1975). *J. Am. Chem. Soc.* **97**, 3177
273. Wan, P., Modro, T.A. and Yates, K. (1980). *Can. J. Chem.* **58**, 2423
274. McClelland, R.A. (1975). *J. Am. Chem. Soc.* **97**, 5281
275. Bender, M.L. (1953). *J. Am. Chem. Soc.* **75**, 5986
276. Lane, C.A., Cheung, M.F. and Dorsey, G.F. (1968). *J. Am. Chem. Soc.* **90**, 6492
277. Slebocka-Tilk, H., Brown, R.S. and Olekszyk, J. (1987). *J. Am. Chem. Soc.* **109**, 4620
278. Bennet, A.J., Slebocka-Tilk, H., Brown, R.S., Guthrie, J.P. and Jodhan, A. (1990). *J. Am. Chem. Soc.* **112**, 8497
279. Brown, R.S. (1998). Private communication
280. Kresge, A.J. (1999). Private communication
281. Vinnik, M.I. and Moiseev, Y.V. (1983). *Izv. Akad. Nauk SSSR, Ser. Khim.*, 777 (708 in Eng. edn.)
282. Cox, R.A. (1998). *Can. J. Chem.* **76**, 649
283. More O'Ferrall, R.A. (1998). Private communication
284. Cox, R.A. (2000). Work in progress
285. Denton, P., Horwell, D., Johnson, C.D. and Williamson, D. (1992). *J. Chem. Res., Synop.*, 410
286. Druet, L.M. and Yates, K. (1984). *Can. J. Chem.* **62**, 2401
287. Cox, R.A. (1997). *J. Chem. Soc., Perkin Trans. 2*, 1743
288. Leiman, S.N., Astrat'ev, A.A., Kuznetsov, L.L. and Selivanov, V.F. (1979). *Zh. Org. Khim.* **15**, 2259 (2047 in Eng. edn.)
289. Drozdova, O.A., Astrat'ev, A.A., Kuznetsov, L.L. and Selivanov, V.F. (1982). *Zh. Org. Khim.* **18**, 2335 (2063 in Eng. edn.)
290. Drozdova, O.A., Astrat'ev, A.A., Kuznetsov, L.L. and Selivanov, V.F. (1983). *Zh. Org. Khim.* **19**, 766 (675 in Eng. edn.)
291. Challis, B.C., Rosa, E., Iley, J. and Noberto, F. (1990). *J. Chem. Soc., Perkin Trans. 2*, 179

How Does Structure Determine Organic Reactivity? Partitioning of Carbocations between Addition of Nucleophiles and Deprotonation

JOHN P. RICHARD, TINA L. AMYES, SHRONG-SHI LIN, ANNMARIE C. O'DONOGHUE, MARIA M. TOTEVA, YUTAKA TSUJI and KATHLEEN B. WILLIAMS

Department of Chemistry, University at Buffalo, SUNY, Buffalo, New York, USA

1 Introduction

STEPWISE SUBSTITUTION AND ELIMINATION REACTIONS

Aliphatic substrates that contain a good leaving group and an α-CH group may undergo reactions by competing nucleophilic substitution and elimination pathways (Scheme 1).[1] A stepwise mechanism is favored for these reactions

ADVANCES IN PHYSICAL ORGANIC CHEMISTRY
VOLUME 35 ISBN 0-12-033535-2

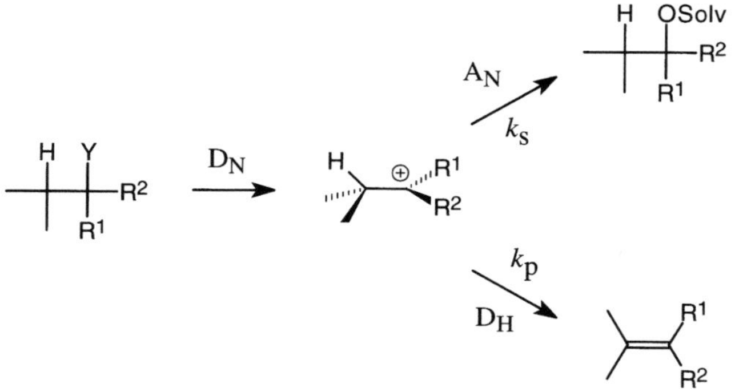

Scheme 1

when heterolytic bond cleavage at carbon yields a carbocation intermediate that is sufficiently stable to exist in a potential energy well for the time of a bond vibration ($t_{1/2} > 10^{-13}$ s).[2,3]

Stepwise nucleophilic substitution and elimination reaction mechanisms through carbocation intermediates are named S_N1 and E1, respectively, by the nomenclature developed by Ingold and coworkers; and, $D_N + A_N$ and $D_N + D_H$, respectively, by the systematic nomenclature for organic reaction mechanisms developed by IUPAC.[4,5] The simplicity and appeal of the IUPAC nomenclature is apparent here, because it clearly delineates that which is common to the two stepwise reactions (dissociation, D_N) from the distinctive pathways for partitioning of the carbocation intermediate to form the products of nucleophilic substitution (A_N) and elimination (D_H) reactions. This focuses attention on developing an understanding of the intrinsic reactivity of carbocations toward Lewis and Brønsted bases, since this is the underlying origin of the tendency of some alkyl derivatives to undergo stepwise nucleophilic substitution and of others to undergo stepwise elimination.

These tendencies reflect the bias of the carbocation intermediate to partition to form either the nucleophile adduct (k_s) or an alkene (k_p) (Scheme 1). An examination of the chemical literature shows examples of determinations of the yields of the products of competing stepwise nucleophilic substitution and elimination reactions, which allow the determination of rate constant ratios k_s/k_p. However, there has been little consideration of how and why this rate constant ratio changes with varying carbocation structure, and no systematic attempt to partition changes in this ratio into changes in the absolute rate constants k_s and k_p. These questions are central to the development of an understanding of the reactivity of aliphatic derivatives in stepwise substitution and elimination reactions. In addition, while proton transfer at saturated carbon and nucleophilic addition to trivalent carbon are reactions ubiquitous in organic chemistry and biochemistry, the details of their mechanisms are still not well understood. A consideration of the effect of changing structure on the relative barriers to these competing reactions of extended series of structurally homologous substrates should provide insight into the factors that control the absolute barriers to the individual reactions.

In this chapter we review published results of studies of the kinetics and products of stepwise nucleophilic substitution and elimination reactions of alkyl derivatives, and we present a small amount of unpublished data from our laboratory. Our review of the literature is selective rather than comprehensive, and focuses on work that provides interesting insight into the factors that control the rate constant ratio k_s/k_p for partitioning of carbocations, and that provides an understanding of how the absolute rate constants k_s and k_p that constitute this ratio change with changing carbocation structure.

INSIGHT FROM RECENT WORK

We are not aware of earlier systematic analyses of substituent effects on the rate constant ratio k_s/k_p for partitioning of carbocations between nucleophilic addition of solvent and deprotonation. This is because the determination of values of k_s/k_p from product analyses has been limited by difficulties in detecting the alkene product of carbocation deprotonation in the presence of the much larger yield of the solvent adduct ($k_s \gg k_p$, Scheme 1). The low yields of the products of $D_N + D_H$ (E1) reactions has severely limited discussion of these stepwise elimination reactions in monographs and texts on organic reaction mechanisms. For example, a 640-page monograph[6] on elimination reactions published in 1973 contains only a 16-page chapter on the $D_N + D_H$ (E1) mechanism and a discussion of the dehydration reactions of alcohols catalyzed by strong mineral acids, where conversion of the alcohol to the alkene is a thermodynamically favorable reaction. The analyses presented in this chapter have been made possible, in part, by the following recent experimental results.

(1) *The determinations of absolute rate constants with values up to* $k_s = 10^{10}\ s^{-1}$ *for the reaction of carbocations with water and other nucleophilic solvents using either the direct method of laser flash photolysis*[7] *or the indirect azide ion clock method.*[8] These values of k_s (s^{-1}) have been combined with rate constants for carbocation formation in the microscopic reverse direction to give values of K_R (M) for the equilibrium addition of water to a wide range of benzylic carbocations.[9–13]

(2) *The determination of large values of the rate constant ratio* k_s/k_p *from the low yields of alkene product that forms by partitioning of carbocations in nucleophilic solvents.* These rate constant ratios may then be combined with absolute rate constants for the overall decay of the carbocation to give absolute values of k_p (s^{-1}).[14–16] For example, the reaction of the 1-(4-methylphenyl)ethyl carbocation in 50/50 (v/v) trifluoroethanol/water gives mainly the solvent adducts and a 0.07% yield of 4-methylstyrene from proton transfer to solvent, which corresponds to $k_s/k_p = 1400$. This can be combined with $k_s = 6 \times 10^9\ s^{-1}$,[14] to give $k_p = 4.2 \times 10^6\ s^{-1}$ (Table 1).

Table 1 Rate and equilibrium constants for partitioning of substituted α-methyl carbocations $R^1(R^2)CCH_3^+$ between nucleophilic addition of solvent (k_s) and deprotonation (k_p) (Scheme 7)[a]

R^1	R^2	Precursor to carbocation[b]	Solvent	k_s/k_p[c]	k_s (s^{-1})	k_p (s^{-1})	pK_R[d]	K_{add}[e]
CH_3	CH_3	Alcohol + H^+	H_2O	30[f]	1×10^{11}[g]	3×10^9	−16.4[h]	7500[i]
$4\text{-}MeOC_6H_4CH_2CH_2$	CH_3	Alkene [2] + H^+	20/80 TFE/H_2O	80[j]	1×10^{11}[g]	1.3×10^9[j]		
$4\text{-}MeOC_6H_4CH_2CH_2$	CH_3	Alkene [2] + H^+	50/50 TFE/H_2O	60[h]	1×10^{11}[g]	1.6×10^9	−16.7[h]	900[h]
$4\text{-}MeC_6H_4$	H	Alcohol + H^+	50/50 TFE/H_2O	1400[k]	6×10^9[k]	4.2×10^6	−12.6[l]	41[m]
$4\text{-}MeOC_6H_4$	H	Phenyl ether + H^+	50/50 TFE/H_2O	7000[n]	5×10^7[l]	7000	−8.6[l]	36[o]
C_6H_5	CH_3	Methyl ether + H^+	25% $MeCN/H_2O$	1000[p]	~5×10^9[p]	~5×10^6		
			50/50 TFE/H_2O	250[q]	1.5×10^{10}[r]	6.0×10^7	−12.3[q]	9[s]
$4\text{-}MeC_6H_4$	CH_3	Methyl Ether + H^+	50/50 TFE/H_2O	700[t]	1.0×10^9[t]	1.4×10^6[t]	−9.6[u]	11[s]
$4\text{-}MeOC_6H_4$	CH_3	Methyl ether + H^+	50/50 TFE/H_2O	3500[v]	1.0×10^7[w]	2900	−6.3[u]	8[s]
Me, Me, Me	CH_3	4-Methoxybenzoate 4-Nitrobenzoate	50/50 TFE/H_2O	0.17[t]	1.4×10^7[t]	8.3×10^7[t]		<0.002[x]

MeO–C$_6$H$_2$(Me)$_2$	CH_3	4-Methoxybenzoate	50/50 TFE/H_2O	0.49^{t}	$\leqslant 7.6 \times 10^{5t}$	$\leqslant 1.6 \times 10^{6t}$		$<8 \times 10^{-4x}$
C_6H_5	C_6H_5	Methyl ether + H^+	20/80 DMSO/H_2O	70^{y}	3×10^{8z}	4×10^{6}		
			50/50 TFE/H_2O	70^{q}	6.8×10^{7q}	9.7×10^{5}	-9.3^{q}	
4-$Me_2NC_6H_4$	4-$Me_2NC_6H_4$	Alkene	80/20 MeOH/H_2O	50	2.5^{A}	0.048^{A}	3.6^{A}	0.13^{A}
C_6H_5	MeO	Dimethyl acetal + H^+	H_2O	1.7×10^{4B}	5×10^{7C}	3000		
4-$MeOC_6H_4$	EtOC(O)	Pentafluorobenzoate	50/50 MeOH/H_2O	$\sim 50^{D}$	6.0×10^{7D}	$\sim 1.2 \times 10^{6D}$		
4-$MeOC_6H_4$	$Me_2NC(O)$	Pentafluorobenzoate	50/50 MeOH/H_2O	1.2^{D}	3.6×10^{6D}	2.9×10^{6D}		
4-$MeOC_6H_4$	$Me_2NC(S)$	4-Methoxybenzoate 4-Nitrobenzoate	50/50 MeOH/H_2O	$<0.01^{D}$	<2	210^{E}		$<0.01^{D}$

[a]At 25°C, unless noted otherwise. Abbreviations: DMSO, dimethyl sulfoxide; MeCN, acetonitrile; MeOH, methanol; TFE, 2,2,2-trifluoroethanol. [b]Substrate used to determine the value of k_s/k_p. [c]Rate constant ratio for partitioning of the carbocation between nucleophilic addition of solvent (k_s) and deprotonation (k_p). [d]K_R is the equilibrium constant for addition of water to the carbocation to give the alcohol, as defined in Scheme 7. [e]Equilibrium constant for addition of water to the alkene to give the alcohol, as defined in Scheme 7. [f]At 55°C. Calculated as the ratio of second-order rate constants for acid-catalyzed exchange of ^{18}O between water and *t*-butyl alcohol and dehydration of *t*-butyl alcohol to give 2-methylpropene,[18] see text. [g]The estimated rate constant for addition of solvent to tertiary carbocations, which is limited by the rate constant for solvent reorganization, $k_{reorg} \sim 10^{11}\ s^{-1}$.[19,20] [h]Ref. 21. [i]Ref. 22. [j]Ref. 23. [k]Ref. 14. [l]Ref. 13. [m]Average of values determined in 3.5–5.5 M $HClO_4$.[24] [n]Tsuji, Y., unpublished results. [o]Average of values determined in 1.3–3.5 M $HClO_4$.[24] [p]Ref. 25. [q]Ref. 9. [r]Extrapolated from a Yukawa–Tsuno correlation of data for the addition of solvent to more stable ring-substituted cumyl carbocations.[26] [s]Ref. 12. [t]Ref. 27. [u]Obtained from a small correction of the previously published data.[12] [v]Calculated with the assumption that the change from a 4-Me to a 4-MeO ring substituent at cumyl carbocations results in the same 5-fold increase in k_s/k_p as that observed for ring-substituted 1-phenylethyl carbocations (data from this table). [w]Calculated from $k_{az}/k_s = 380\ M^{-1}$[28] and $k_{az} = 3.9 \times 10^9\ M^{-1}\ s^{-1}$.[29] [x]Amyes, T.L., unpublished results. [y]Ref. 30. [z]Calculated from $k_{az}/k_{HOH} = 660$,[30] $[H_2O] = 44.4$ M in 20/80 DMSO/H_2O and $k_{az} = 5 \times 10^9\ M^{-1}\ s^{-1}$. [A]Ref. 31. [B]Ref. 15. [C]Ref. 32. [D]Ref. 33. [E]At 20°C.[34]

(3) *Kinetic studies of stepwise hydration reactions of alkenes.* This work has shown that carbocations with labile β-CH bond(s) that are stabilized by an α-amino group,[35–37] or by two α-thiol groups[38–40] undergo preferential deprotonation to form the products of an elimination reaction ($k_p > k_s$, Scheme 1).

(4) *The recent characterization of carbocations that partition preferentially to form an alkene* ($k_p > k_s$, Scheme 1). The reaction of the 1-(4-methylphenyl)ethyl carbocation exemplifies the low yields of alkene usually obtained from partitioning of α-methylbenzyl carbocations. However, there have been several reports that modifications in the structure of benzylic carbocations may result in a sharp change in the preferred pathway for their reaction, from addition of solvent ($k_s > k_p$, Scheme 1) to deprotonation to form the alkene ($k_p > k_s$): (a) The addition of an α-amide[41] or α-thioamide group[42] to α-methylbenzyl carbocations strongly favors their partitioning to form the corresponding α-amide- or α-thioamide-substituted alkene. (b) The addition of two *ortho*-methyl groups to the cumyl carbocation strongly favors the partitioning of the carbocation to form the corresponding α-methylstyrene.[41]

In summary, there now exists a body of data for the reactions of carbocations where the values of k_s/k_p span a range of $>10^6$-fold (Table 1). This requires that variations in the substituents at a cationic center result in a >8 kcal mol^{-1} differential stabilization of the transition states for nucleophile addition and proton transfer which have not yet been fully rationalized. We discuss in this review the explanations for the large changes in the rate constant ratio for partitioning of carbocations between reaction with Brønsted and Lewis bases that sometimes result from apparently small changes in carbocation structure.

2 Partition rate constant ratios from product analyses

HPLC ANALYSIS OF PRODUCTS

The values of k_s/k_p for partitioning of carbocations are most conveniently determined as the ratio of the yields of products from the competing nucleophile addition and proton transfer reactions (equation 1 derived for Scheme 2). The determination of these product yields has been simplified in recent years by the application of high-pressure liquid chromatography (HPLC). Typically, the product peaks from an HPLC analysis are detected and quantified by UV-vis spectroscopy. In cases where the absorbance of reactants and products is small, substrates may be prepared with a chromophore placed at a sufficient distance so that its effects on the intrinsic reactivity of the carbocationic center are negligible. For example, the aliphatic substrates [1]-Y have proved to be very useful in studies of the reactions of the model tertiary carbocation [1^+].[21,23]

The limit on the value of k_s/k_p that can be determined by HPLC analysis of product yields depends, to some extent, on the effort that an investigator is

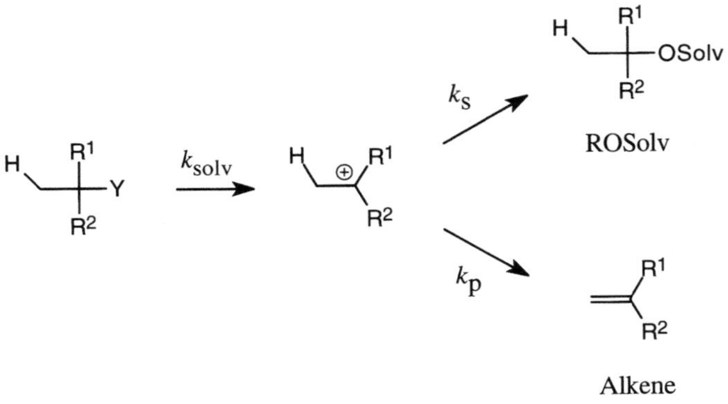

Scheme 2

$$\frac{k_s}{k_p} = \frac{[\text{ROSolv}]}{[\text{Alkene}]} \tag{1}$$

willing to expend on the determination of these yields. Product yields as low as 1% may be determined relatively easily by HPLC analyses when the extinction coefficients of the solvent adduct and alkene are similar. A yield of 0.07% 4-methylstyrene was determined for partitioning of the 1-(4-methylphenyl)ethyl carbocation.[14] These experiments were facilitated by the large extinction coefficient of 4-methylstyrene relative to that of the solvent adducts.[14] Considerable effort was required to determine the yield of 0.006% α-methoxystyrene formed by partitioning of the oxocarbenium ion intermediate of the acid-catalyzed cleavage of acetophenone dimethyl acetal in water.[15,16] Similarly, the

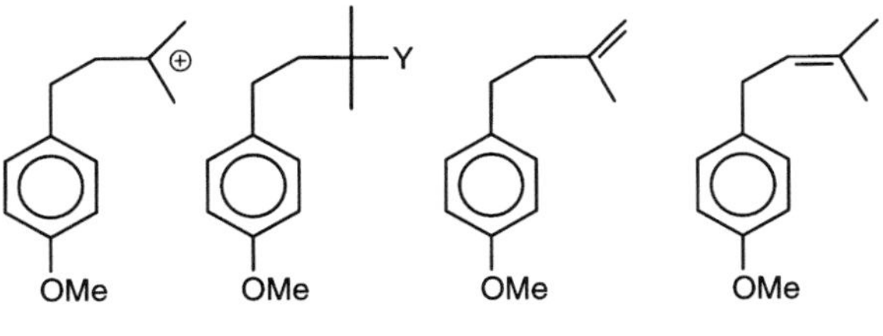

[1+] [1]-Y [2] [3]

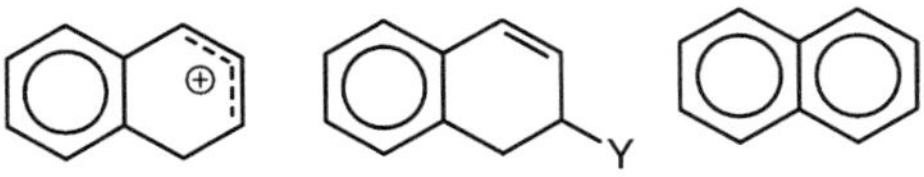

[4+] [4]-Y [5]

reaction of $[4^+]$ in 50/50 (v/v) methanol/water gives mainly naphthalene [5] from proton transfer to solvent, and only 0.09% of the methyl ether [4]-OMe.[44]

The quantitation of products that form in low yields requires special care with HPLC analyses. In cases where the product yield is $<1\%$, it is generally not feasible to obtain sufficient material for a detailed physical characterization of the product. Therefore, the product identification is restricted to a comparison of the UV-vis spectrum and HPLC retention time with those for an authentic standard. However, if a minor reaction product forms with a UV spectrum and HPLC chromatographic properties similar to those for the putative substitution or elimination reaction, this may lead to errors in structural assignments. Our practice is to treat rate constant ratios determined from very low product yields as limits, until additional evidence can be obtained that our experimental value for this ratio provides a chemically reasonable description of the partitioning of the carbocation intermediate. For example, verification of the structure of an alkene that is proposed to form in low yields by deprotonation of the carbocation by solvent can be obtained from a detailed analysis of the increase in the yield of this product due to general base catalysis of carbocation deprotonation.[14,16]

REACTION MECHANISMS

The interpretation of product data for competitive solvolysis and elimination reactions requires that the mechanism for these reactions be known. Two experiments are sufficient to show that the formation of solvolysis and elimination products occurs by partitioning of a common carbocation intermediate (Scheme 3a) rather than by competing bimolecular reactions of the substrate (Scheme 3b).[3]

(1) The demonstration that the rate of formation of a nucleophile adduct R-Nu is kinetically zero-order in the concentration of the nucleophilic reagent Nu^-.

(2) The demonstration that formation of the nucleophile adduct R-Nu results in the same proportional decrease in the yields of the alkene and solvent adducts, so that the ratio of the yields of these reaction products is independent of $[Nu^-]$. If the solvolysis and elimination reactions proceed by competing stepwise and concerted pathways, respectively, then the yield of R-OSolv will decrease with increasing trapping of the carbocation intermediate by added nucleophile, while the yield of alkene from elimination will remain constant, so that the ratio [R-OSolv]/[Alkene] will decrease as $[Nu^-]$ is increased.

Simple tertiary carbocations represent a benchmark against which to compare the reactivity of other α-methyl carbocations. Therefore, it is necessary to deal with complex questions about the mechanism for substitution and elimination reactions at tertiary aliphatic carbon in order to evaluate the rate constant

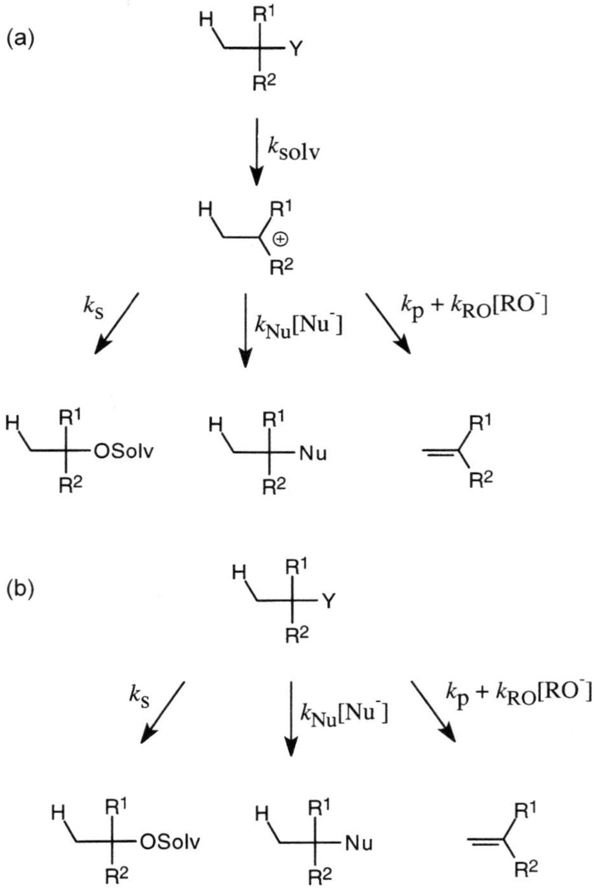

(a)

$$\text{H-CH}_2\text{-C(R}^1\text{)(R}^2\text{)-Y} \xrightarrow{k_{solv}} \text{H-CH}_2\text{-C}^{\oplus}\text{(R}^1\text{)(R}^2\text{)}$$

k_s → H-CH2-C(R1)(R2)-OSolv; $k_{Nu}[Nu^-]$ → H-CH2-C(R1)(R2)-Nu; $k_p + k_{RO}[RO^-]$ → $CH_2=C(R^1)(R^2)$

(b)

H-CH2-C(R1)(R2)-Y

k_s → H-CH2-C(R1)(R2)-OSolv; $k_{Nu}[Nu^-]$ → H-CH2-C(R1)(R2)-Nu; $k_p + k_{RO}[RO^-]$ → $CH_2=C(R^1)(R^2)$

Scheme 3

ratios for partitioning of the putative tertiary carbocation reaction intermediates.

Only low yields of the azide ion adduct are obtained from the reaction of simple tertiary derivatives in the presence of azide ion;[21,45,46] and it is not possible to rigorously determine the kinetic order of the reaction of azide ion, owing to uncertainties in the magnitude of specific salt effects on the rate constants for the solvolysis and elimination reactions. Therefore, these experiments do not distinguish between stepwise and concerted mechanisms for substitution reactions at tertiary carbon.

There are two lines of evidence that the reactions of simple tertiary derivatives such as [1]-Cl to give solvent adducts proceed through the corresponding tertiary carbocation intermediates: (1) There is a large steric barrier to the concerted nucleophilic substitution reaction of solvent;[1] (2) the best estimate for the lifetime of simple tertiary carbocations in mixed aqueous/organic solvents is $t_{1/2} \approx 10^{-12}$ s,[21] which is the time for several bond vibrations. Therefore, it is unlikely that a concerted mechanism for solvolysis of tertiary derivatives is enforced by the absence of a chemical barrier for combination of the putative tertiary carbocation intermediate with solvent.[2,3] Relatively large yields of the alkenes [2] and [3] are obtained from the reaction of [1]-Cl and [1]-pentafluorobenzoate in 50/50 (v/v) trifluoroethanol/water.[21] The reaction of [1]-Cl in the presence of increasing concentrations of azide ion gives [1]-N_3, a corresponding decrease in the yield of the solvent adduct, but no change in the yields of the alkenes [2] and [3].[21] This is consistent with concurrent stepwise and concerted pathways, respectively, for the nucleophilic substitution and alkene-forming elimination reactions of [1]-Cl and probably [1]-pentafluorobenzoate.

The cleavage of [1]-Y gives, initially, an ion pair or ion–dipole complex that contains [1^+] and the leaving group. There is good evidence that the lifetime of [1^+] in aqueous/organic solvents is so short that this carbocation undergoes essentially quantitative reaction with solvent or anions present in the surrounding solvation shell faster than competing diffusional processes such as separation of the ion pair to free ions, or encounter with added anions.[21] The fleeting lifetime of tertiary carbocations such as [1^+] is manifested in the following results of experiments in nucleophilic solvents.

(1) It was expected that values of k_s/k_p for partitioning of [1^+] could be obtained from the yields of the products of acid-catalyzed reactions of [1]-OH and [2]. However, significantly different relative yields of these products are obtained from the perchloric acid-catalyzed reactions of [1]-OH and [2] in several mixed alcohol/water solvents.[21] This demonstrates that the nucleophilic substitution and elimination reactions of these two substrates do not proceed through identical tertiary carbocation intermediates (Scheme 4). The observed

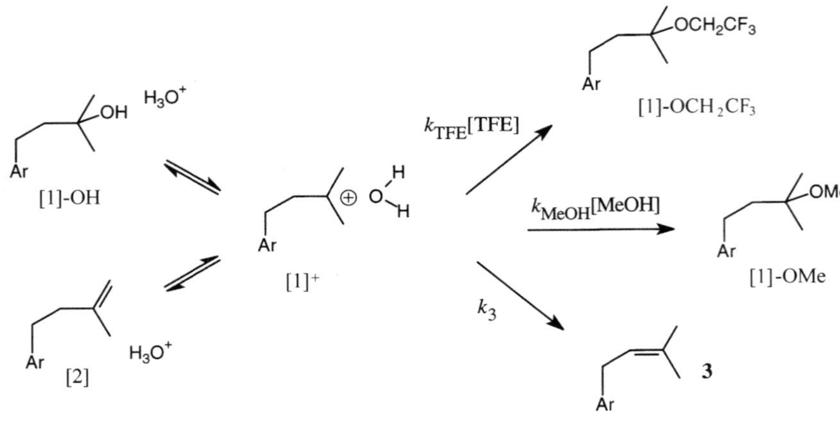

[1]-OH $\xrightarrow{H_3O^+}$ [1]$^+$ ⊕ OH_2; [2] $\xrightarrow{H_3O^+}$ [1]$^+$

[1]$^+$ $\xrightarrow{k_{TFE}[TFE]}$ [1]-OCH_2CF_3

[1]$^+$ $\xrightarrow{k_{MeOH}[MeOH]}$ [1]-OMe

[1]$^+$ $\xrightarrow{k_3}$ **3**

Scheme 4

differences in the relative yields of products from the acid-catalyzed reactions of [1]-OH and [2] might be explained by small differences in the composition and reactivity of the solvation shell that surrounds $[1^+]$ at the time of its formation from the two different substrates (Scheme 4). Alternatively, [1]-OH and/or [2] may undergo competing stepwise reactions through the unselective carbocation $[1^+]$ and concerted reactions that avoid the formation $[1^+]$.

(2) If any concerted reactions of [1]-OH and [2] are assumed to be negligible, then the yield of the disubstituted alkene [2] formed by partitioning of $[1^+]$ generated from the perchloric acid-catalyzed reaction of [1]-OH is ~1.3%.[23] This is 19-fold larger than the yield of 0.07% 4-methylstyrene formed by partitioning of the 1-(4-methylphenyl)ethyl carbocation generated from the perchloric acid-catalyzed cleavage of 1-(4-methylphenyl)ethanol.[14] This is a surprising result, because the larger yield of the alkene [2] from the reaction of $[1^+]$ obtains despite the ~1.8 kcal mol^{-1} larger thermodynamic driving force for formation of 4-methylstyrene from 1-(4-methylphenyl)ethanol ($1/K_{add} = 0.024$) than for formation of [2] from [1]-OH ($1/K_{add} = 0.001$) (Table 1). These results are consistent with the formation of [1]-OSolv (Figure 1) by rate-determining solvent reorganization that places a molecule of solvent in a position to add to $[1^+]$, $k_{reorg} \approx 10^{11}$ s^{-1},[19,20] followed by fast solvent addition, $(k_s)_{chem} \approx 10^{12}$ s^{-1}.[21] This results in a decrease in the *observed* product rate constant ratio $k_s/k_p = k_{reorg}/k_p$ for $[1^+]$ relative to that for partitioning of the 1-(4-methylphenyl)ethyl carbocation, because the *full chemical reactivity* of solvent is expressed in the rate-determining step for solvent addition to the latter but not to the former carbocation. Alternatively, these results may result from the failure of our assumption that the concerted reactions of [1]-OH and [2] are negligible.

3 Partition rate constant ratios from kinetic analyses of alkene hydration

A knowledge of the rate-determining step for the stepwise acid-catalyzed addition of solvent to an alkene (Scheme 5) provides a qualitative measure of k_s/k_p for partitioning of the carbocation intermediate. The observed rate constant for the overall solvent addition reaction will be limited by $k_H[H^+]$ (first step rate-determining) when partitioning of the carbocation favors formation of the solvent adduct ($k_s/k_p > 1$, Figure 2a), but it will be limited by k_s (second step rate-determining) when partitioning favors the regeneration of substrate ($k_p > k_s$, Figure 2b). There are several types of experiments that distinguish between alkene protonation and addition of solvent to the carbocation intermediate as rate-determining steps for the addition of solvent to alkenes (Scheme 5).

(1) When the reaction of the alkene proceeds by rate-determining addition of solvent to the carbocation intermediate (k_s, Figure 2b), proton transfer in the

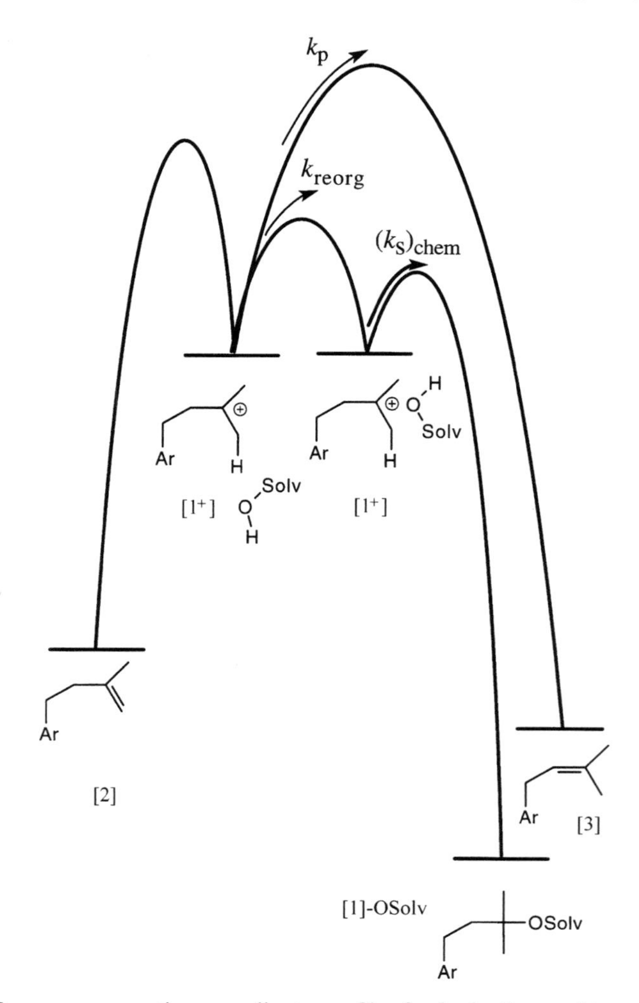

Fig. 1 Free energy reaction coordinate profiles for hydration and isomerization of the alkene [2] through the simple tertiary carbocation [1^+]. The rate constants for partitioning of [1^+] to form [1]-OSolv and [3] are limited by solvent reorganization ($k_s = k_{reorg}$) and proton transfer (k_p), respectively. For simplicity, the solvent reorganization step is not shown for the conversion of [1^+] to [3], but the barrier for this step is smaller than the chemical barrier to deprotonation of [1^+] ($k_{reorg} > k_p$).

first step is reversible and may be detected by monitoring the reaction of the alkene in D_2O. Provided the initial protonation gives a carbocation with two or more β-hydrons, this will result in exchange of the olefinic hydrogen of the substrate for deuterium from solvent. The failure to observe such deuterium exchange during the hydrolysis of vinyl ethers[47,48] or the hydration of

$k_H[H^+] + k_{HA}[HA]$ ⇌ $k_p + k_A[A^-]$ → k_s → SolvO

$k_{Nu}[Nu^-]$ → Nu

Scheme 5

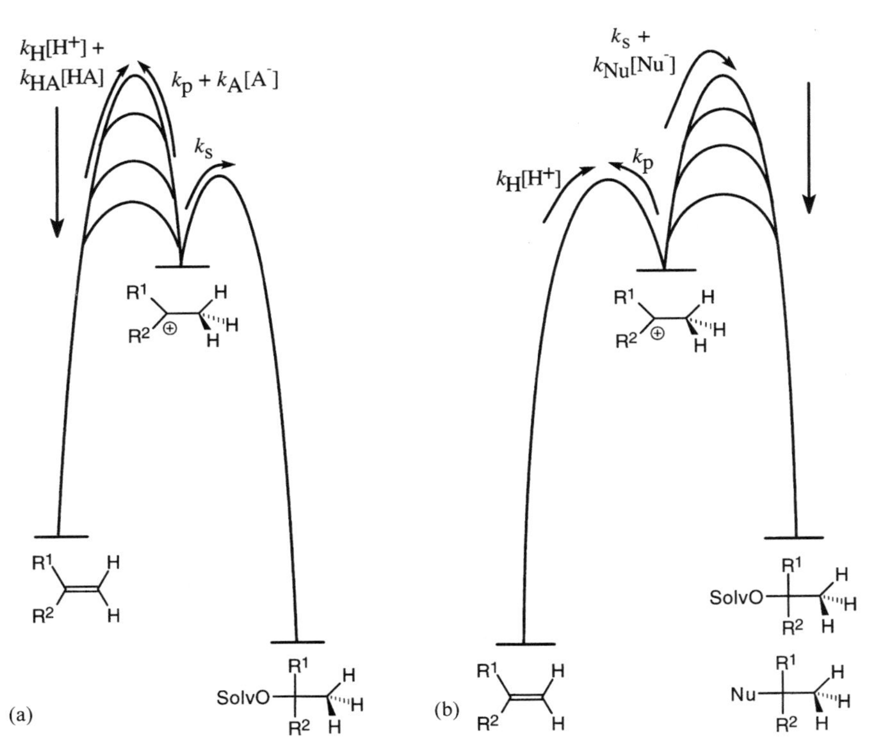

Fig. 2 Free energy reaction coordinate profiles for the stepwise acid-catalyzed hydration of an alkene through a carbocation intermediate (Scheme 5). (a) Reaction profile for the case where alkene protonation is rate determining ($k_s \gg k_p$). This profile shows a change in rate-determining step as a result of Brønsted catalysis of protonation of the alkene. (b) Reaction profile for the case where addition of solvent to the carbocation is rate determining ($k_s \ll k_p$). This profile shows a change in rate-determining step as a result of trapping of the carbocation by an added nucleophilic reagent.

2-phenylpropene in D_2O[49] provides convincing evidence that these reactions proceed by rate-determining protonation of the alkene reactant (Figure 2a).

(2) The observation of Brønsted general acid catalysis of alkene hydration provides evidence for reaction by rate-determining protonation of carbon. This is because the fractional rate acceleration from general Brønsted acid–base catalysis of proton transfer at carbon (first step, Scheme 5) is much larger than the small or negligible rate accelerations from general acid–base catalysis of the addition of water to carbocations or oxocarbenium ions. For example, the hydrolysis of a variety of vinyl ethers[47,48,50–62] and ring-substituted styrenes[24,63] by rate-determining protonation of the alkene to form a carbocation intermediate is subject to strong catalysis by Brønsted acids. By contrast, the cleavage of ketals,[64] acetals[64,65] or alcohols[66] to form carbocations, or the addition of hydroxylic solvents to oxocarbenium ions[67] or benzylic carbocations[17,68] in the reverse direction are either not subject to Brønsted acid–base catalysis or this catalysis is weak.

(3) The observation of a primary solvent deuterium isotope effect $(k_H/k_D) =$ 2–4 on the specific acid-catalyzed hydrolysis of vinyl ethers provides evidence for reaction by rate-determining protonation of the alkene.[69] Values of $k_H/k_D \approx 1$ are expected if alkene hydration proceeds by rate-determining addition of solvent to an oxocarbenium ion intermediate, since there is no "motion" of a solvent hydron at the transition state for this step. However, in the latter case, determination of the solvent isotope effect on the reaction of the fully protonated substrate is complicated by the competing exchange of deuterium from solvent into substrate (see above).

(4) When nucleophilic addition of solvent to a carbocation intermediate is rate-determining for hydration of an alkene ($k_p > k_s$, Scheme 5), trapping of the intermediate by added nucleophilic reagents will lower the barrier to the rate-determining step, resulting in an increase in the observed rate constant for reaction of the alkene (Figure 2b).[38,40,70] Thiols are particularly effective as trapping reagents because they are more strongly nucleophilic than the aqueous solvent; and, unlike many nucleophilic anions, thiols exist in the nucleophilic form in the acidic solutions required for protonation of the alkene.

In cases where Brønsted general acid catalysis results in a change in rate-determining step for hydration of the alkene (Figure 2a) it is possible to determine the value of k_s/k_p for partitioning of carbocation intermediate (Scheme 5) by analysis of curvature in plots of the observed first-order rate constants for alkene hydration against the concentration of the added buffer catalyst. Reports of buffer-induced changes in rate-determining step for alkene hydration are not common,[40,70] because the difference in the heights of the activation barriers for addition of solvent and deprotonation is generally too large to allow for their reversal by the addition of buffer catalysts. In cases where the addition of nucleophilic thiols results in a change in rate-determining step for addition to the alkene it is possible to determine values of k_s/k_p by analysis of curvature in plots of the observed first-order rate constants for reaction of the alkene against the concentration of added nucleophilic thiol.[40,70]

4 Kinetic and thermodynamic considerations

It is often difficult to understand at an intuitive level the explanation for the effect of changing substituents on the rate constant ratio k_s/k_p for partitioning of carbocations between nucleophilic addition of solvent and deprotonation. In these cases, insight into the origins of the changes in this rate constant ratio requires a systematic evaluation of substituent effects on the following:

(1) The absolute values of the individual rate constants k_s and k_p for the nucleophile addition and proton transfer reactions.
(2) The extent to which the effect of changing substituents on the values of k_s and k_p is the result of a change in the thermodynamic driving force for the reaction (ΔG°), a change in the relative intrinsic activation barriers Λ for k_s and k_p, or whether changes in both of these quantities contribute to the overall substituent effect. This requires at least a crude Marcus analysis of the substituent effect on the rate and equilibrium constants for the nucleophile addition and proton transfer reactions (equation 2).[71,72]

$$\Delta G^\ddagger = \Lambda(1 + \Delta G^\circ/4\Lambda)^2 \qquad (2)$$

Changes in the height of the activation barrier to chemical reactions with changing substrate structure may be modeled by construction of reaction coordinate profiles from intersecting potential energy surfaces, which for simplicity are often drawn as simple parabolas. The Marcus equation (2)[71,72] shows the derived dependence for this model parabolic reaction coordinate of the activation barrier $\Delta G^\ddagger$ on the thermodynamic driving force, ΔG°, and the intrinsic barrier, Λ, to the thermoneutral reaction (Fig. 3). The values of Λ for chemical reactions will depend upon the steepness of the approach to the reaction transition state, and these barriers will change with changing curvature along this energy surface as illustrated by the barriers Λ_1 and Λ_2 in Fig. 3.[8] In the same manner, a difference in the observed barriers for deprotonation and nucleophile addition to a simple carbocation may reflect differences in the curvature of the energy surface that connects the ground and transition states for proton transfer and nucleophilic addition (Fig. 4),[15,73] which would correspond to different intrinsic barriers for deprotonation (Λ_p) and for nucleophilic addition of solvent (Λ_s) in hypothetical thermoneutral reactions.

Rate and equilibrium data for groups of chemically related reactions such as proton transfer,[74] nucleophilic addition to the carbonyl group,[75] and aldol condensation[76,77] show fair to good fits to the Marcus equation using a constant intrinsic barrier across a series of structurally related reactions. However, it is also well recognized that there are significant changes in the intrinsic barrier for

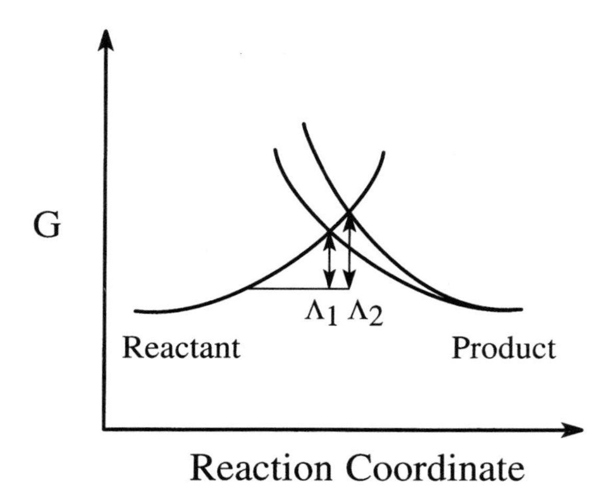

Fig. 3 Free energy profiles for reactions in which there is no change in thermodynamic driving force ΔG° but a change in the intrinsic barrier Λ.

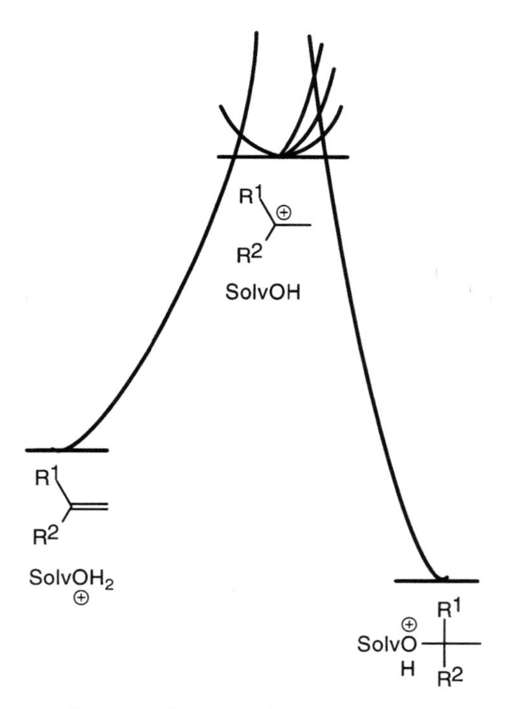

Fig. 4 Free energy reaction coordinate profiles that illustrate a change in the relative kinetic barriers for partitioning of carbocations between nucleophilic addition of solvent and deprotonation resulting from a change in the curvature of the potential energy surface for the nucleophile addition reaction. This would correspond to an increase in the intrinsic barrier for the thermoneutral carbocation–nucleophile addition reaction.

proton transfer at carbon with changing delocalization of negative charge at the carbanion product of these reactions.[78,79] In addition, the observation that an extensive linear logarithmic rate-equilibrium correlation of rate constants for the hydroxide ion-catalyzed deprotonation of simple α-carbonyl substrates fails to show Marcus curvature requires that there be a systematic decrease in the intrinsic barrier with increasing thermodynamic barrier to deprotonation of the carbon acid to form a carbanion.[80] Systematic changes in intrinsic barriers have been observed for other proton transfer reactions at carbon.[81] Similarly, there is good qualitative evidence that the intrinsic barriers for nucleophilic addition to carbocations increase with increasing resonance stabilization of the carbocation.[8,15]

Our analysis of literature data will focus on two closely related questions about the influence of changes in the relative thermodynamic driving force and Marcus intrinsic barrier for the reaction of simple carbocations with Brønsted bases (alkene formation) and Lewis bases (nucleophile addition) on the values of k_s/k_p determined by experiment.

(1) To what extent is the partitioning of simple aliphatic and benzylic α-CH-substituted carbocations in nucleophilic solvents controlled by the relative thermodynamic driving force for proton transfer and nucleophile addition reactions? It is known that the partitioning of simple aliphatic carbocations favors the formation of nucleophile adducts ($k_s/k_p > 1$, Scheme 2); and there is good evidence that this reflects, at least in part, the larger thermodynamic driving force for the nucleophilic addition compared with the proton transfer reaction of solvent ($K_{add} \gg 1$, Scheme 6).[12,21,22,24]

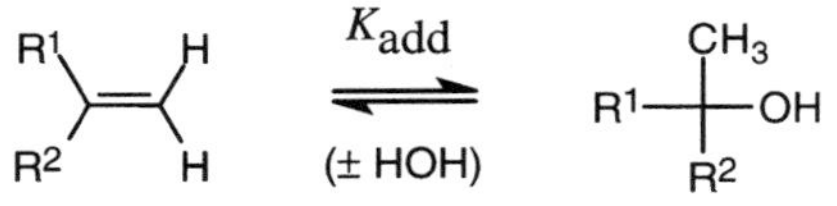

$$R^1R^2C{=}CH_2 \underset{(\pm\,HOH)}{\overset{K_{add}}{\rightleftharpoons}} R^1R^2C(CH_3)OH$$

Scheme 6

(2) To what extent are the variations in the rate constant ratio k_s/k_pobserved for changing structure of aliphatic and benzylic carbocations the result of changes in the Marcus intrinsic barriers Λ_p and Λ_s for the deprotonation and solvent addition reactions? It is not generally known whether there are significant differences in the intrinsic barriers for the nucleophile addition and proton transfer reactions of carbocations.

5 Reactions of aliphatic and benzylic carbocations

Table 1 summarizes experimentally determined values of the following rate and equilibrium constants for the reactions of aliphatic and benzylic α-methyl carbocations (Scheme 7).

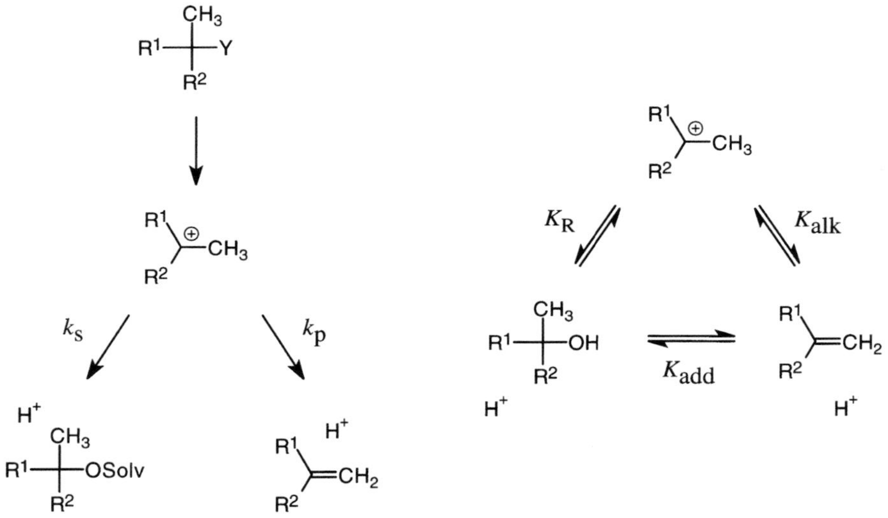

Scheme 7

(1) Values of the rate constant ratios k_s/k_p for partitioning of carbocations between nucleophilic addition of solvent and deprotonation. These rate constant ratios were determined by analysis of the yields of the solvent adduct and alkene that form by partitioning of the carbocation (equation 1).

(2) Absolute values of the rate constants k_s (s^{-1}) and k_p (s^{-1}). In most cases these rate constants were determined from the values of k_{az}/k_s (M^{-1}) or k_{az}/k_p (M^{-1}) for partitioning of the carbocation between reaction with azide ion and solvent, by using the diffusion-limited reaction of azide ion, $k_{az} = 5 \times 10^9\ M^{-1}\ s^{-1}$, as a "clock" for the slower reactions of solvent.[7,8,13,32,82]

(3) Values of pK_R for the addition of water to carbocations to give the corresponding alcohols. The equilibrium constants K_R (M) were determined as the ratio k_{HOH}/k_H, where k_{HOH} (s^{-1}) is the first-order rate constant for reaction of the carbocation with water and k_H ($M^{-1}\ s^{-1}$) is the second-order rate constant for specific acid-catalyzed cleavage of the alcohol to give the carbocation.[9,12,13]

(4) Values of K_{add} for the addition of water (hydration) of alkenes to give the corresponding alcohols. These equilibrium constants were obtained directly by determining the relative concentrations of the alcohol and alkene at chemical equilibrium. The acidity constants pK_{alk} for deprotonation of the carbocations by solvent are not reported in Table 1. However, these may be calculated from data in Table 1 using the relationship $pK_{alk} = pK_R + \log K_{add}$ (Scheme 7).

Wherever possible, values of k_s/k_p are reported that have been determined for partitioning of carbocations generated by acid-catalyzed cleavage of an alcohol or ether precursor, because these reactions give, initially, an ion–dipole pair that contains neutral water or an alcohol leaving group whose chemical reactivity is similar to that of the bulk solvent. This minimizes the effect of the leaving group on the yield of products from partitioning of the carbocation. In the cases where data are reported for reactions of carbocations generated by heterolytic cleavage of neutral substrates to form (initially) an ion pair intermediate, the rate constant for reaction of the carbocation with solvent (k_s) is sufficiently small to allow for essentially complete diffusional separation of the ion pair to the free ions ($k_{-d} \approx 1.6 \times 10^{10}\ s^{-1}$)[14] before its capture by solvent.

Table 1 includes partitioning data only for carbocations that are sufficiently stable to form in the nucleophilic aqueous/organic solvents used in these experiments. For example, it is not possible to obtain values of k_s for reaction of secondary aliphatic carbocations in water and other nucleophilic solvents, because the *chemical* barrier to k_s is smaller than that for a bond vibration.[83] The vanishingly small barriers for reaction of secondary carbocations with nucleophilic solvents results in "enforced" concerted mechanisms[2,3] for the nucleophilic substitution and elimination reactions of secondary derivatives in largely aqueous solvents.[83,84]

TERTIARY ALIPHATIC CARBOCATIONS

If it is assumed that the acid-catalyzed exchange of oxygen-18 between *t*-butyl alcohol and water, k_{ex}, and the acid-catalyzed dehydration of *t*-butyl alcohol to give 2-methylpropene, k_{dehyd}, proceed by the common first step of alcohol cleavage to form the *t*-butyl carbocation, then a value of $k_s/k_p = 30$ for partitioning of the *t*-butyl cation between addition of water and deprotonation can be calculated as the ratio $k_{ex}/k_{dehyd} = 30$ (Table 1).[18] This is in fair agreement with the values of $k_s/k_p = 80$ and 60 for partitioning of [1^+] determined by analysis of the products of reaction of [1]-OH and [2] in 20/80 (v/v) and 50/50 (v/v) trifluoroethanol/water, respectively[21,23] (Table 1). However, there is a significant difference between the reported values of $K_{add} = 7500$ for the conversion of isobutylene to *t*-butyl alcohol[22] and $K_{add} = 900$ for the conversion of [2] to [1]-OH[21] (Table 1). We are uncertain of the explanation for this difference but believe that the later value is more reliable.

The partitioning of [1^+] between addition of solvent and deprotonation was discussed in Section 2, where it was suggested that the observed ratio of $k_s/k_p = 60$ in 50/50 (v/v) trifluoroethanol/water underestimates the relative chemical barriers for k_s and k_p. This is because the addition of solvent is limited by rotational reorganization of solvent, $k_s = k_{reorg} \sim 10^{11}\ s^{-1}$.[21] Use of the larger value of $(k_s)_{chem} = 10^{12}\ s^{-1}$ estimated for the addition of solvent that is prepositioned to react with [1^+][21] gives $k_s/k_p = 600$. The latter ratio is close to $k_s/k_p = 1400$ for partitioning of the 1-(4-methylphenyl)ethyl carbocation (Table 1).

The more favorable partitioning of [1^+] to form [1]-OH than to form [2] must be due, at least in part, to the 4.0 kcal mol^{-1} larger thermodynamic driving force for the former reaction ($K_{add} = 900$ for conversion of [2] to [1]-OH, Table 1). However, thermodynamics alone cannot account for the relative values of k_s and k_p for reactions of [1^+] that are limited by the rate of chemical bond formation, which may be as large as 600. A ratio of $k_s/k_p = 600$ would correspond to a 3.8 kcal mol^{-1} difference in the activation barriers for k_s and k_p, which is almost as large as the 4.0 kcal mol^{-1} difference in the stability of [1]-OH and [2]. However, only a small fraction of this difference should be expressed at the relatively early transition states for the reactions of [1^+], because these reactions are strongly favored thermodynamically. These results are consistent with the conclusion that nucleophile addition to [1^+] is an inherently easier reaction than deprotonation of this carbocation, and therefore that nucleophile addition has a smaller Marcus intrinsic barrier. However, they do not allow for a rigorous estimate of the relative intrinsic barriers $\Lambda_s - \Lambda_p$ for these reactions.

RING-SUBSTITUTED 1-PHENYLETHYL CARBOCATIONS

The rate and equilibrium constants for the reactions of ring-substituted 1-phenylethyl carbocations (X-[6^+]) in 50/50 (v/v) trifluoroethanol/water (Table 2 and Scheme 8),[13,14,17,43], and for interconversion of ring-substituted 1-phenyl-

X-[6^+]

$(k_H)_{solv}$ k_{HOH} k_p $(k_H)_{alk}$

H^+ K_{add} H^+

X-[6]-OH X-[7]

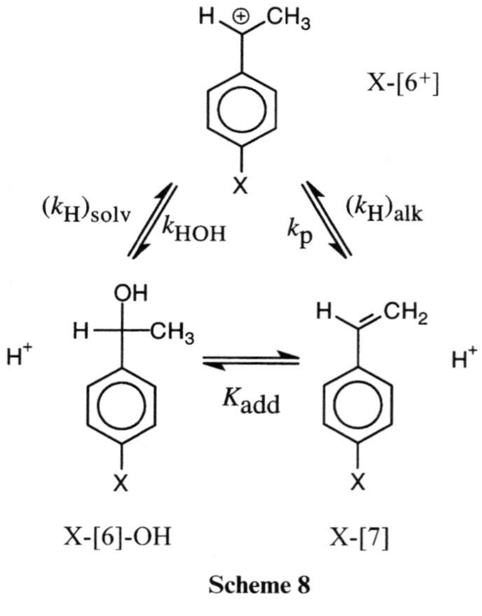

Scheme 8

Table 2 Rate constants, equilibrium constants, and estimated Marcus intrinsic barriers for the formation and reaction of ring-substituted 1-phenylethyl carbocations X-[6^+] (Scheme 8)[a]

Ring substituent	$\log k_{HOH}$ (s^{-1})[b]	$\log (k_H)_{solv}$ ($M^{-1}\,s^{-1}$)[c]	$\log K_R$ (M)[d]	Λ_s[e] (kcal mol^{-1})	$\log k_p$ (s^{-1})[f]	$\log (k_H)_{alk}$ ($M^{-1}\,s^{-1}$)[g]	$\log K_{alk}$ (M)[h]	Λ_p[i] (kcal mol^{-1})	$\Lambda_s - \Lambda_p$ (kcal mol^{-1})
4-Me	9.70[j]	−3.08	12.8	11.3	6.62	−4.89[k]	11.5	15.3	−4.0
4-MeO	7.62	−0.96	8.6	12.2	3.85	−3.41[l]	7.3	16.8	−4.6
4-Me_2N	1.57[m]	2.83	−1.3	14.4					

[a]At 25°C in 50/50 (v/v) trifluoroethanol/water ($I = 0.50$, $NaClO_4$). Unless noted otherwise, data are taken from Ref. 13. [b]First-order rate constant for reaction of carbocation with water. [c]Second-order rate constant for acid-catalyzed cleavage of the ring-substituted 1-phenylethyl alcohol to give the carbocation. [d]Equilibrium constant for addition of water to the carbocation calculated as $K_R = k_{HOH}/(k_H)_{solv}$. [e]Marcus intrinsic barrier for nucleophilic addition of water to the carbocation calculated from the values of $\log k_{HOH}$ and $\log K_R$ using equation 3. [f]First-order rate constant for deprotonation of the carbocation by solvent taken from Table 1. [g]Second-order rate constant for specific acid-catalyzed protonation of the ring-substituted styrene to give the carbocation. [h]Equilibrium constant for deprotonation of the carbocation by solvent calculated as $K_{alk} = k_p/(k_H)_{alk}$. [i]Marcus intrinsic barrier for deprotonation of the carbocation by solvent calculated from the values of $\log k_p$ and $\log K_{alk}$ using equation 3. [j]Calculated from $k_s = 6 \times 10^9\ s^{-1}$ [14] and the dimensionless product rate constant ratio $k_{HOH}/k_{TFE} = 1.3$ determined for reaction of 1-(4-methylphenyl)ethyl chloride.[14] [k]Data from ref. 14. [l]Calculated from the value of $(k_H)_{alk}$ for 4-methylstyrene and a ratio of 30 for the values of $(k_H)_{alk}$ for 4-methoxystyrene and 4-methylstyrene in 3.5 M $HClO_4$ in water.[24] [m]Calculated from $k_s = 40\ s^{-1}$ [10] and the dimensionless product rate constant ratio $k_{HOH}/k_{TFE} = 3.0$ determined for the acid-catalyzed cleavage of 1-(4-dimethylaminophenyl)-ethyl methyl ether.[13]

ethyl alcohols ([X-[6]-OH) and styrenes (X-[7]) (Table 1)[24] provide a detailed description of how the barriers for partitioning of these carbocations are controlled by the relative thermodynamic driving forces and intrinsic barriers for the nucleophile addition and proton transfer reactions of X-[6^+].

The difference in the values of $K_{add} = 900$ for hydration of [2] and $K_{add} \approx 40$ for hydration of X-[7] (Table 1) shows that an α-aryl substituent provides substantial stabilization of an alkene relative to the alcohol. The value of $k_s/k_p = 1400$ for partitioning of Me-[6^+][14] is slightly larger than $(k_s)_{chem}/k_p = 600$ for partitioning of [1^+] that can be calculated by correcting the observed ratio of $k_s/k_p = 60$ (Table 1) for the difference in the values of $k_s = k_{reorg} = 10^{11}\ s^{-1}$ for solvent addition that is limited by solvent reorganization and $(k_s)_{chem} = 10^{12}\ s^{-1}$ estimated for chemical bond formation between solvent and [1^+] (see previous section).

The ratios $k_s/k_p = 1400$ for partitioning of Me-[6^+][14] and $k_s/k_p = 7000$ for partitioning MeO-[6^+] (Tsuji, Y., unpublished results) are much larger than the values of $K_{add} = 41$ (X = 4-Me) and 36 (X = 4-MeO) for hydration of X-[7] to give X-[6]-OH[24] (Table 1). Therefore, the difference in the relative thermodynamic driving force for the solvent addition and deprotonation reactions of X-[6^+] can account for only a small part of the observed difference in the activation barriers to these reactions. The remaining difference between the values of K_{add} and k_s/k_p must reflect a smaller Marcus intrinsic barrier for k_s than for k_p. The following related quantitative analyses provide evidence that the intrinsic barrier for nucleophilic addition of water to X-[6^+] to give X-[6]-OH is $\sim 4\ kcal\ mol^{-1}$ smaller than that for proton transfer to give X-[7].

(1) Table 2 gives rate and equilibrium constants for the deprotonation of and nucleophilic addition of water to X-[6^+]. These data are plotted as logarithmic rate-equilibrium correlations in Fig. 5, which shows: (a) correlations of $\log k_p$ for deprotonation of X-[6^+] and $\log k_{HOH}$ for addition of water to X-[6^+] with $\log K_{alk}$ and $\log K_R$, respectively; (b) correlations of $\log(k_H)_{solv}$ for specific-acid-catalyzed cleavage of X-[6]-OH (the microscopic reverse of nucleophilic addition of water to X-[6^+]) and $\log(k_H)_{alk}$ for protonation of X-[7] (the microscopic reverse of deprotonation of X-[6^+]) with $\log K_{alk}$ and $\log K_R$, respectively.

The intersection of the correlation lines for the plots of $\log k_{HOH}$ and $\log(k_H)_{solv}$ occurs at $\log K_R = 0$ and $\log k_{HOH} = \log(k_H)_{solv} = 2.4$, which corresponds to an intrinsic rate constant of $240\ s^{-1}$ for the interconversion of X-[6^+] and X-[6]-OH. By contrast, the intersection of the correlation lines for the plots of $\log k_p$ and $\log(k_H)_{alk}$ occurs at $\log K_{alk} = 0$ and $\log k_p = \log(k_H)_{alk} = -0.90$, which corresponds to an intrinsic rate constant of $0.13\ s^{-1}$ for the interconversion of X-[6^+] and X-[7]. These intrinsic rate constants can be combined to give $(k_{HOH}/k_p)_o = 1850$ for partitioning of X-[6^+] between hypothetical thermoneutral nucleophile addition and proton transfer reactions.

$$\log k_{obsd} = \frac{1}{1.36}\left[17.44 - \Lambda\left(1 - \frac{1.36 \log K}{4\Lambda}\right)^2\right] \tag{3}$$

$$\log k_o = 12.8 - \frac{\Lambda}{1.36} \tag{4}$$

(2) Values of the intrinsic barriers Λ_s for the addition of water to individual X-[6^+] and Λ_p for the deprotonation of individual X-[6^+] can be calculated from the corresponding rate and equilibrium constants using equation (3) (Table 2). The resulting values of Λ_s increase from 11.3 to 14.4 kcal mol^{-1} as the ring substituent is changed from 4-Me to 4-Me_2N and there is a similar increase in the values of Λ_p. The result of these increases in Λ_s and Λ_p is that the difference $\Lambda_s - \Lambda_p = -4.3 \pm 0.3$ kcal mol^{-1} is nearly constant for the reactions of X-[6^+]. This difference determined for reactions of individual X-[6^+] is in good agreement with $\Lambda_s - \Lambda_p = -4.4$ kcal mol^{-1} that can be calculated (equation 4) from the "intrinsic ratio" $(k_{HOH}/k_p)_o = 1850$ obtained from extrapolation of the rate-equilibrium correlations in Fig. 5.

The data in Table 2 suggest that the intrinsic barriers for the reactions of X-[6^+] increase with increasing stabilization of these carbocations by resonance electron donation from the ring substituent. This complements the indirect evidence for changes in these intrinsic barriers that has been described in recent reviews.[8,15] The data in Table 2 and Fig. 5 provide support for the following generalizations about the reactions of X-[6^+]:

(1) A defining feature of data that follow the Marcus equation is curvature in logarithmic rate-equilibrium correlations;[15,123] the failure to observe such curvature for extended sets of structure–reactivity data has been used to illustrate the limitations of a simple Marcus treatment with a constant intrinsic barrier.[80] Curvature in direct logarithmic plots of rate against equilibrium data is extremely difficult to detect experimentally, particularly for small sets of data such as those shown in Fig. 5. The analysis given in Table 2 illustrates an alternative approach to assess whether small sets of data are adequately fitted by the Marcus equation. Here the values of Λ for the reaction of individual members of the series X-[6^+], which were calculated using equation (3) with the assumption that the rate and equilibrium constants are described by the Marcus equation, increase systematically with increasing stabilization of X-[6^+] by resonance electron donation from the ring substituent. These changes in Λ are not normally accounted for in Marcus-type treatments of rate-equilibrium data.[74,75,85]

(2) The similarity between the increase in the intrinsic barrier for nucleophile addition to carbocations that occurs with increasing delocalization of positive charge on to electron-donating substituents[8,15] and the increase in the intrinsic barrier for protonation of carbanions that occurs with increasing delocalization of negative charge on to electron-withdrawing substituents (equation 5)[78] suggests that there are similar explanations for these changes.[8,15,78,86–88]

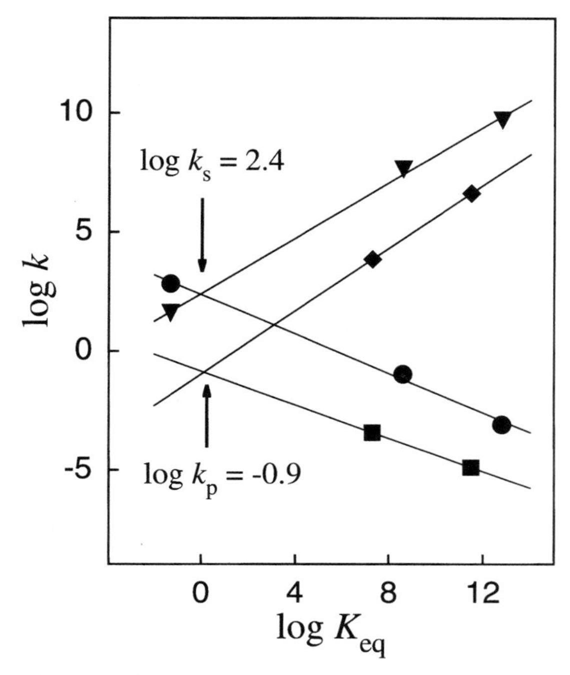

Fig. 5 Logarithmic plots of rate-equilibrium data for the formation and reaction of ring-substituted 1-phenylethyl carbocations X-[6^+] in 50/50 (v/v) trifluoroethanol/water at 25°C (data from Table 2). Correlation of first-order rate constants k_{HOH} for the addition of water to X-[6^+] (▼) and second-order rate constants $(k_H)_{solv}$ for the microscopic reverse specific-acid-catalyzed cleavage of X-[6]-OH to form X-[6^+] (●) with the equilibrium constants K_R for nucleophilic addition of water to X-[6^+]. Correlation of first-order rate constants k_p for deprotonation of X-[6^+] (◆) and second-order rate constants $(k_H)_{alk}$ for the microscopic reverse protonation of X-[7] by hydronium ion (■) with the equilibrium constants K_{alk} for deprotonation of X-[6^+]. The points at which equal rate constants are observed for reaction in the forward and reverse directions (log K_{eq} = 0) are indicated by arrows.

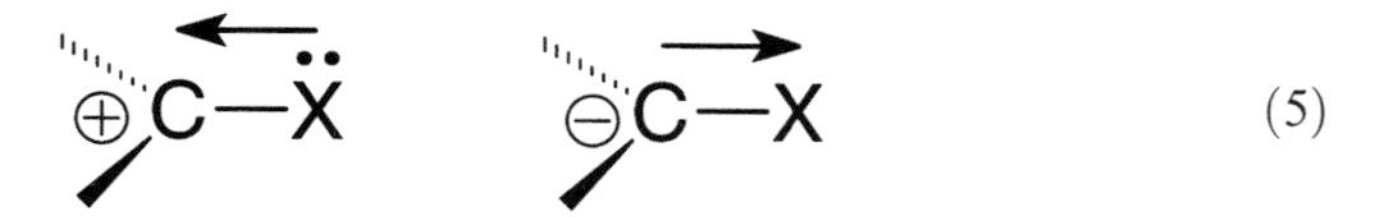

(5)

(3) There is a ~4 kcal mol^{-1} smaller intrinsic barrier Λ_s for nucleophilic addition of water to the benzylic carbocations X-[6^+] than for deprotonation of X-[6^+] by solvent. This difference reflects the greater ease of direct addition of solvent to the charged benzylic carbon of X-[6^+] than of proton transfer at the adjacent α-methyl carbon. This may result in some way from the greater number of bonds formed and cleaved in the proton transfer than in the nucleophile addition reaction. However, it is our impression that there is little or no theoretical justification for generalizations of this type.

Bunting and Kanter have developed a modified form of the Marcus equation to treat the changes in intrinsic barrier Λ observed for deprotonation of β-keto esters and amides.[81] It would be useful to consider similar modifications of the Marcus equation to model the variable intrinsic barriers observed for carbocation–nucleophile addition reactions.

CUMYL AND DI-*ortho*-METHYLCUMYL CARBOCATIONS

The addition of an α-methyl group to 1-phenylethyl carbocations X-[6^+] to give cumyl carbocations X-[8^+] has the following effects on the rate and equilibrium constants for reaction of the carbocations: (1) The α-methyl group results in a ~4-fold decrease in k_s for nucleophilic addition of a solvent of 50/50 (v/v) trifluoroethanol/water,[26,28] which is consistent with stabilization of the carbocation by the α-methyl group. (2) The addition of an α-methyl group to X-[7] to give X-[9] results in a 2-fold decrease in the equilibrium constant K_{add} for alkene hydration (Tables 1 and 5), which shows that the α-methyl group provides a small stabilization of the alkene relative to the alcohol.[12] However, the addition of an α-methyl group to X-[6^+] to give X-[8^+] results in only a small change in k_s/k_p for partitioning of the carbocation. For example, the value of $k_s/k_p = 700$

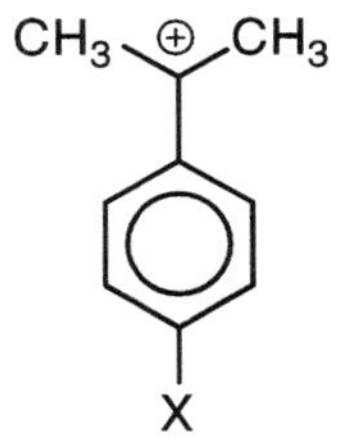

X-[8^+] 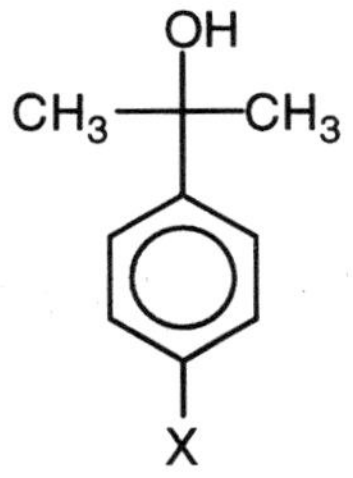 X-[8]-OH 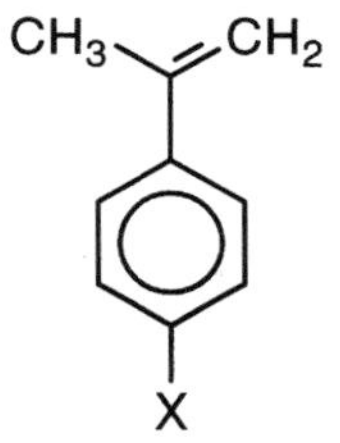 X-[9]

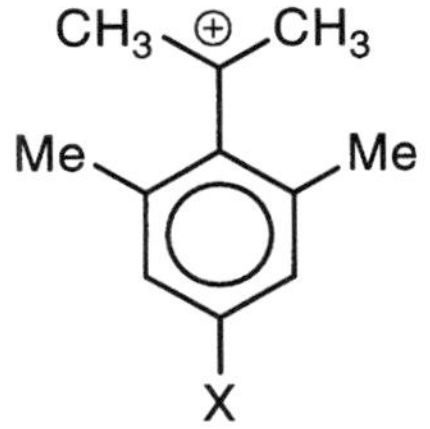

X-[10^+] 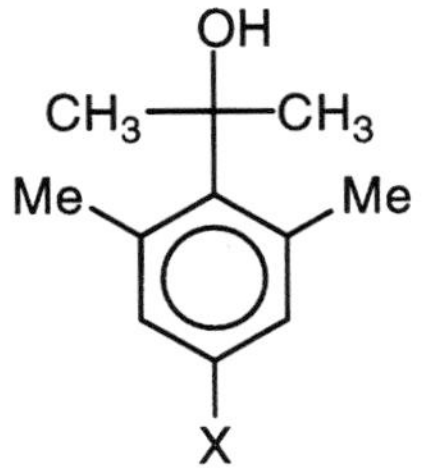 X-[10]-OH 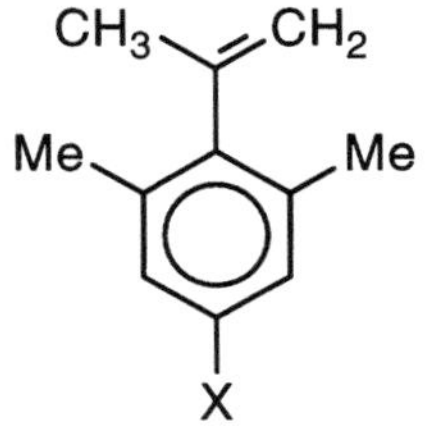 X-[11]

for partitioning of Me-[8^+] between nucleophilic addition of a solvent of 50/50 (v/v) trifluoroethanol/water and deprotonation[27] is very similar to $k_s/k_p = 1400$ for partitioning of Me-[6^+] in the same solvent[14] (Table 1).

By contrast, the addition of a pair of *ortho*-methyl groups to the aromatic ring of X-[8^+] to give X-[10^+] has the following dramatic effects on partitioning of the carbocations between nucleophilic addition of solvent and proton transfer:[27,41]

(1) The addition of two *ortho*-methyl groups to Me-[8^+] to give Me-[10^+] results in a 70-fold decrease in the rate constant k_s for nucleophilic addition of 50/50 (v/v) trifluoroethanol/water.[27]
(2) The addition of two *ortho*-methyl groups to Me-[8^+] to give Me-[10^+] results in a 60-fold *increase* in the rate constant k_p for proton transfer to solvent.[27]
(3) The net effect of the two *ortho*-methyl groups at Me-[10^+] is a 4200-fold decrease in the rate constant ratio k_s/k_p for partitioning of the carbocation between nucleophilic addition of solvent and proton transfer.[27]
(4) The addition of a pair of *ortho*-methyl groups to the alkene Me-[9] to give Me-[11] results in a greater than 5000-fold decrease in the equilibrium constant K_{add} for hydration of the alkene to give the corresponding cumyl alcohol (Amyes, T.L., unpublished results).

There are two important interactions – steric and electronic – between the *ortho*-methyl groups and the cationic center that contribute to the observed effects of these groups on the rate and equilibrium constants for partitioning of ring-substituted cumyl carbocations. Our analysis of these substituent effects assumes that the electronic effects of the *ortho*-methyl groups are similar to the well-characterized electronic effects of *para*-methyl groups, and that changes in rate and equilibrium constants that cannot be accounted for by these electronic substituent effects are the result of steric effects. This analysis is complicated because it requires separate consideration of steric interactions between the *ortho*-methyl groups and the benzylic center at the neutral alcohol X-[10]-OH, the charged carbocation X-[10^+], and the neutral alkene X-[11].

Neutral alcohol (A, Figure 6). The pair of *ortho*-methyl groups result in a decrease in the equilibrium constant for addition of water to the alkene from $K_{add} = 11$ for hydration of Me-[9] to give Me-[8]-OH to $K_{add} < 0.002$ for hydration of Me-[11] to give Me-[10]-OH (Table 1). The large destabilization of X-[10]-OH relative to X-[11] by the *ortho*-methyl groups probably reflects the tendency of these substituents to introduce greater steric strain at the more crowded sp^3-hybridized benzylic carbon of X-[10]-OH than at the corresponding sp^2-hybridized carbon of the alkene X-[9].

Benzylic carbocation (B, Figure 6). A crude analysis predicts that the addition of two *ortho*-methyl groups to Me-[8]-Y to give Me-[10]-Y will result

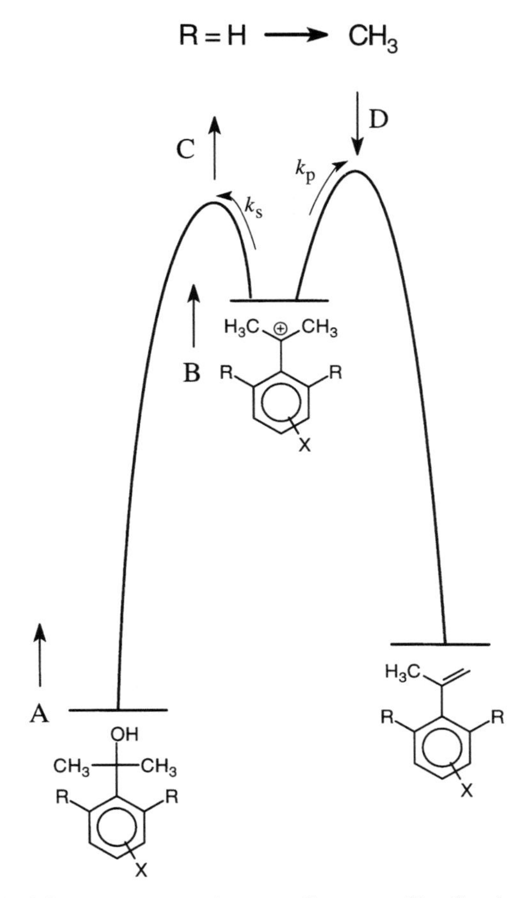

Fig. 6 Hypothetical free energy reaction coordinate profiles for the interconversion of X-[8]-OH and X-[9] (R = H) and X-[10]-OH and X-[11] (R = CH_3) through the corresponding carbocations. The arrows indicate the proposed effects of the addition of a pair of *ortho*-methyl groups to X-[8]-OH, X-[8^+] and X-[9] to give X-[10]-OH, X-[10^+] and X-[11]. A: Effect of a pair of *ortho*-methyl groups on the stability of cumyl alcohols. B: Effect of a pair of *ortho*-methyl groups on the stability of cumyl carbocations. C: Effect of a pair of *ortho*-methyl groups on the stability of the transition state for nucleophilic addition of water to cumyl carbocations. D: Effect of a pair of *ortho*-methyl groups on the stability of the transition state for deprotonation of cumyl carbocations.

in a 400-fold increase in the rate constant for cleavage to give the corresponding carbocations owing to stabilization of the carbocation-like transition state by the electron-donating methyl groups.[27] Therefore, the observed smaller 1.7-fold difference in the rate constants for cleavage of Me-[8]-(4-nitrobenzoate) and Me-[10]-(4-nitrobenzoate) requires that the two *ortho*-methyl substituents also result in an offsetting 240-fold *decrease* in the rate constant for carbocation

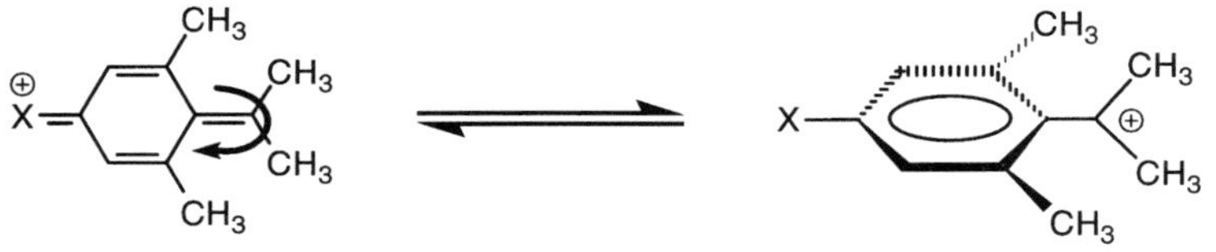

Scheme 9

formation.[27] This probably reflects the tendency of steric interactions between the *ortho*-methyl groups and the methyl groups at the benzylic carbon to induce twisting about the C_{Ar}—C_{α} bond away from the conformation in which the aromatic ring and the isopropyl group lie coplanar (Scheme 9) . Such twisting will reduce the destabilizing steric interactions between the methyl groups, but at the same time it will reduce carbocation stabilization by electron donation from the aromatic ring, so that Me-[10$^+$] is more weakly resonance-stabilized, and thus inherently more reactive, than Me-[8$^+$].

The addition of two *ortho*-methyl groups to cumyl derivatives also results in a decrease in the Hammett reaction constant from $\rho^+ = -4.3$ for cleavage of X-[8]-(4-nitrobenzoate) to $\rho^+ = -3.0$ for cleavage of X-[10]-(4-methoxybenzoate).[27] This provides additional support for the conclusion that rotation about the C_{Ar}—C_{α} bond, which will minimize destabilizing steric interactions between methyl groups in the transition state, also reduces stabilizing resonance electron donation from the ring substituent to the benzylic carbon (Scheme 9).

Substituent effect on the stability of the transition state for solvent addition (C, Figure 6). The difference in the values of $k_s = 1.4 \times 10^7\ s^{-1}$ for the addition of a solvent of 50/50 (v/v) trifluoroethanol/water to Me-[10$^+$] and $k_s = 1.0 \times 10^9\ s^{-1}$ for addition of the same solvent to Me-[8$^+$] (Table 1) reflects the balance between: (1) the destabilization of X-[10$^+$] resulting from rotation about the C_{Ar}—C_{α} bond, which results in a less stable carbocation and hence an increase in k_s; (2) destabilization of the transition state due to steric interactions between solvent and the *ortho*-methyl groups, which will result in a decrease in k_s. The 70-fold smaller value of k_s for Me-[10$^+$] than for Me-[8$^+$] shows that the net effect of the two *ortho*-methyl groups is dominated by destabilizing steric interactions that develop in the transition state for nucleophilic addition of solvent.[27]

Substituent effect on the stability of the transition state for deprotonation (D, Figure 6). The addition of two *ortho*-methyl groups to Me-[8$^+$] to give Me-[10$^+$] results in an increase in k_p for proton transfer to solvent from $1.4 \times 10^6\ s^{-1}$ to $8.3 \times 10^7\ s^{-1}$ (Table 1). There should be relatively little steric hindrance to the reaction of solvent with the β-hydrogens of Me-[10$^+$] because these are relatively distant from the *ortho*-methyl groups. However, the twisting about the C_{Ar}–C_{α} bond that minimizes steric interactions between the methyl groups (Scheme 9) will destabilize the carbocation by reducing electron donation from the aromatic ring. We propose that this destabilization is reflected in an increase in the thermodynamic driving force for conversion of Me-[10^+] to the alkene Me-[11] and a corresponding increase in the rate constant k_p for deprotonation of Me-[10^+] by solvent.

α,α-DIPHENYL CARBOCATIONS

The value of $k_s/k_p = 70$ for partitioning of the methylbenzhydryl carbocation $Ph_2CCH_3^+$ [9,30] is considerably smaller than rate constant ratios for partitioning of X-[6^+] and X-[8^+] (Table 1). A value of $k_s/k_p = 50$ has been determined for the partitioning of $(4\text{-}Me_2NC_6H_4)_2CCH_3^+$ in 80/20 (v/v) methanol/water.[31] However, a smaller partitioning ratio for this carbocation is expected to obtain in solvents of higher water content for the following reasons: (1) The Brønsted basicities of methanol ($pK_a = 15.5$)[89] and water ($pK_a = 15.7$) are similar, so that the change in solvent should not result in a large change in k_p. (2) Methanol shows a significantly larger Lewis basicity than water: a value of $k_{MeOH}/k_{HOH} = 70$ has been determined for partitioning of the 1-(4-dimethylamino-phenyl)ethyl carbocation,[14] so that k_s will decrease as methanol is replaced by water.[13]

To the best of our knowledge, equilibrium constants for the interconversion of $Ar_2C{=}CH_2$ and $Ar_2C(CH_3)OH$ have not been determined. The limited data presented here are not sufficient for an evaluation of the effect of a second α-aryl group on the rate and equilibrium constants for the partitioning of benzylic carbocations. However, they do show that the addition of a second α-aryl ring to $ArCH(CH_3)^+$ results in a decrease in the value of k_s/k_p. This change reflects, in part, the tendency of the phenyl group to stabilize the alkene relative to the corresponding alcohol. This effect is manifested in a decrease in the equilibrium constant for alkene hydration along the series [2] ($K_{add} = 900$), Me-[7] ($K_{add} = 41$) and Me-[9] ($K_{add} = 11$) (Table 1).

OXOCARBENIUM IONS

A value of $k_s/k_p = 17\,000$ has been determined for partitioning of the acetophenone oxocarbenium ion [12^+] in water.[15,16] It is not possible to estimate an equilibrium constant for the addition of water to [12^+], because of the instability of the hemiketal product of this reaction. However, kinetic and thermodynamic parameters have been determined for the reaction of [12^+] with methanol to form protonated acetophenone dimethyl ketal [12]-$OMeH^+$ and for loss of a proton to form α-methoxystyrene [13] in water (Scheme 10).[15,16] Substitution of these rate and equilibrium constants into equation (3) gives values of $\Lambda_{MeOH} = 6.5$ kcal mol^{-1} and $\Lambda_p = 13.8$ kcal mol^{-1} for the intrinsic

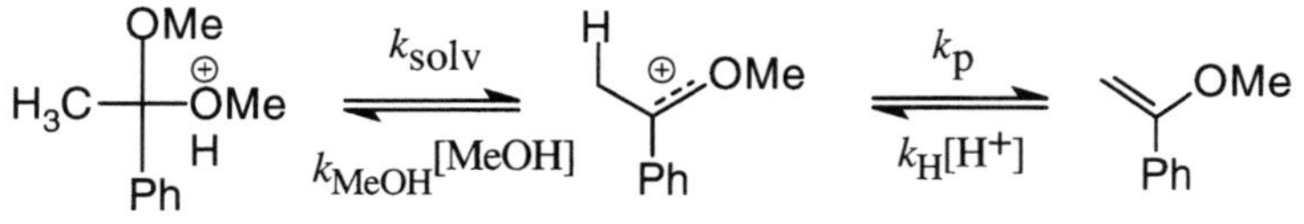

k_{solv} / $k_{MeOH}[MeOH]$; k_p / $k_H[H^+]$

[12]-OMeH$^+$ [12$^+$] [13]

Scheme 10

barriers for the nucleophilic addition of methanol and deprotonation of [12$^+$]. There is no detectable change in the value of Λ_p for deprotonation of [12$^+$] as the ring substituent is changed from hydrogen to 4-MeO.[16]

The intrinsic barriers for the reaction of [12$^+$] correspond to intrinsic rate constants of $(k_{MeOH})_o = 1 \times 10^8\ M^{-1}\ s^{-1}$ and $(k_p)_o = 450\ s^{-1}$ (equation 4). This analysis shows that the thermoneutral addition of methanol to [12$^+$] is an intrinsically *fast* reaction, with a rate constant that is only ~50-fold smaller than that for a diffusion-limited reaction.[16]

The rate constants for nucleophilic addition of a solvent of 50/50 (v/v) trifluoroethanol/water to the α-methoxy 4-methoxybenzyl carbocation ($k_s = 2.2 \times 10^7\ s^{-1}$) and the 1-(4-methoxylphenyl)ethyl carbocation ($k_s = 5 \times 10^7\ s^{-1}$) are similar, despite the ~9 kcal mol^{-1} greater thermodynamic driving force for reaction of the latter carbocation.[28,90] This surprising result requires that any increase in k_s resulting from the more favorable driving force for reaction of the 1-(4-methoxylphenyl)ethyl carbocation be balanced by a decrease in k_s resulting from a larger Marcus intrinsic barrier to this reaction.[8] The ~10^5-fold difference between the intrinsic rate constants of $(k_s)_o = 240\ s^{-1}$ for addition of water to X-[6$^+$] (Figure 5) and $(k_{MeOH})_o = 1 \times 10^8\ M^{-1}\ s^{-1}$ for addition of methanol to [12$^+$] provides additional support for the conclusion that α-alkoxy substituents act to lower the Marcus intrinsic barrier for nucleophile addition to benzylic carbocations.[8,28,91]. A more detailed comparison of these intrinsic rate constants is not warranted because of uncertainties in the effects of a change in the nucleophile (first-order reaction of aqueous solvent compared with second-order reaction of methanol) on the intrinsic reaction barrier.

α-CARBONYL AND α-THIOCARBONYL SUBSTITUTED BENZYL CARBOCATIONS

The changes in the values of k_s/k_p observed for partitioning of the carbocations R-[14$^+$] (Table 3) requires that the addition of α-ester, α-amide and α-thioamide substituents result in different changes in k_s and k_p for partitioning of the parent 1-(4-methoxyphenyl)ethyl carbocation.[33,41,42] The explanation for the changes in these ratios is complex, and is most easily understood by separate considerations of the effects of these substituents on k_s and k_p (Scheme 11).

Table 3 The effects of α-carbonyl and α-thiocarbonyl substituents on the rate and equilibrium constants for the formation and reaction of α-methyl 4-methoxybenzyl carbocations R-[14^+] (Scheme 11)[a]

α-Substituent R	k_{solv} [b] (s^{-1})	k_{az} [c] ($M^{-1}\,s^{-1}$)	K_{az} [d] (M)	$\Delta G - \Delta G_{Me}$ [e] ($kcal\,mol^{-1}$)	k_s [f] (s^{-1})	k_p [f] (s^{-1})	k_s/k_p [f]
CH_3	3.4×10^{-5} [g]	5×10^9	6.8×10^{-15}	0	2.8×10^8	2900 [h]	9×10^4
EtOC(O)	5.6×10^{-10} [i]	5×10^9	1.1×10^{-19}	6.5	6.0×10^7	$\sim 1.2 \times 10^6$	~50
$Me_2NC(O)$	3.2×10^{-10} [j]	5×10^9	6.4×10^{-20}	6.8	3.6×10^6	2.9×10^6	1.2
$Me_2NC(S)$	2.7×10^{-7} [j]	4.9×10^4 [k]	5.5×10^{-12}	−4.0	<2 [l]	210 [k]	<0.01

[a] At 25°C, unless noted otherwise. [b] Rate constant for cleavage of the azide ion adduct R-[14]-N_3 to give the carbocation R-[14^+] in 50/50 (v/v) trifluoroethanol/water. [c] Second-order rate constant for reaction of the carbocation with azide ion. A value of $k_{az} = 5 \times 10^9\ M^{-1}\,s^{-1}$ for the diffusion-limited reaction of azide ion was used, unless noted otherwise.[33] [d] Equilibrium constant for formation of the carbocation from the neutral azide ion adduct in 50/50 (v/v) trifluoroethanol/water, calculated as $K_{az} = k_{solv}/k_{az}$. [e] Difference in Gibbs free energy change for the cleavage of R-[14]-N_3 and CH_3-[14]-N_3 to give the corresponding carbocations. [f] Data for reactions in 50/50 (v/v) methanol/water taken from Ref. 33, unless noted otherwise. [g] Ref. 28. [h] Value in 50/50 (v/v) trifluoroethanol/water taken from Table 1. [i] Calculated from k_{solv} for R-[14]-(pentafluorobenzoate)[33] and the 6.2×10^4-fold larger value of k_{solv} for R-[14]-(pentafluorobenzoate) than of R-[14]-N_3 that was calculated from the data in Ref. 28. [j] Calculated from k_{solv} for R-[14]-(4-nitrobenzoate)[33] and the 2650-fold larger value of k_{solv} for R-[14]-(4-nitrobenzoate) than of R-[14]-N_3 that was calculated from the data in Ref. 28. [k] At 20°C. Determined directly by monitoring decay of the carbocation in 50/50 (v/v) methanol/water.[34] [l] Calculated from the values of k_s/k_p and k_p.

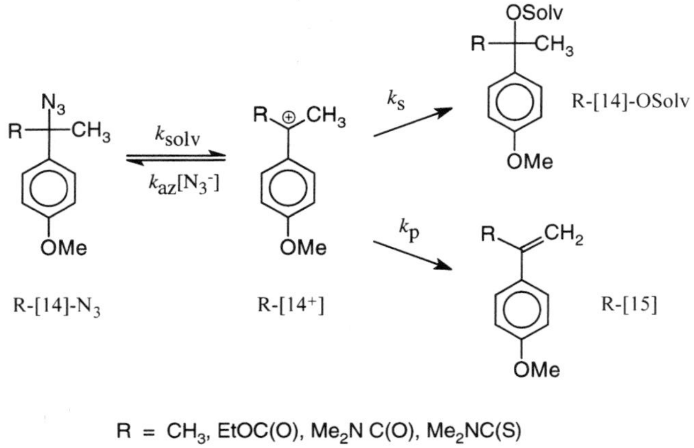

R-[14]-N$_3$ $\underset{k_{az}[N_3^-]}{\overset{k_{solv}}{\rightleftharpoons}}$ R-[14$^+$] $\xrightarrow{k_s}$ R-[14]-OSolv; R-[14$^+$] $\xrightarrow{k_p}$ R-[15]

R = CH_3, EtOC(O), Me_2N C(O), $Me_2NC(S)$

Scheme 11

Substituent effects on k_s. The replacement of an α-methyl group at the 4-methoxycumyl carbocation CH_3-[14$^+$] by an α-ester or α-amide group destabilizes the parent carbocation by ~7 kcal mol^{-1} relative to the neutral azide ion adduct (Scheme 11 and Table 3) and results in 5-fold and 80-fold decreases, respectively, in k_s for nucleophilic addition of a solvent 50/50 (v/v) methanol/water.[33] These results follow the trend that strongly electron-withdrawing substituents, which destabilize α-substituted 4-methoxybenzyl carbocations relative to neutral adducts to nucleophiles, do not lead to the expected large *increases* in the rate constants for addition of solvent.[28,33,92–95]

By contrast, the substitution of an α-thioamide group for the α-methyl group at CH_3-[14$^+$] *stabilizes* this carbocation by 4.0 kcal mol^{-1} relative to the neutral azide ion adduct and results in a more than 10^8-fold decrease in the rate constant for the nucleophilic addition of solvent[34] (Table 3). This substituent effect on carbocation reactivity is extraordinarily large and requires that there be a large stabilization of $Me_2NC(S)$-[14$^+$] by resonance electron donation from the α-thioamide group, together with the loss of a relatively large *fraction* of these stabilizing interactions on proceeding from the ground to the transition state for reaction of the carbocation.[33,86]

The results of *ab initio* calculations provide evidence that $Me_2NC(S)$-[14$^+$] is stabilized by resonance electron donation from the α-thioamide group (A, Scheme 12) and by covalent bridging of sulfur to the benzylic carbon (B, Scheme 12).[96] Direct resonance stabilization of the carbocation will increase the barrier to the nucleophile addition reaction, because of the requirement for the relatively large fractional loss of the stabilizing resonance interaction (A, Scheme 12) at the transition state for nucleophile addition to α-substituted benzyl carbocations.[8,13,28,91–93] If the solvent adds exclusively to an open carbocation that is the minor species in a mixture of open and closed ions, then

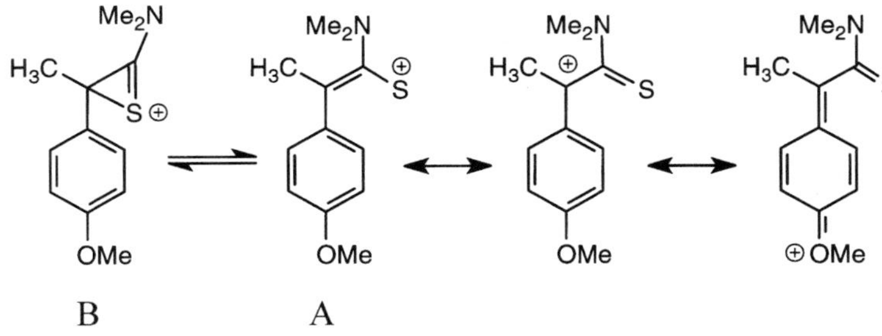

B A

Scheme 12

sulfur bridging will result in close to the expected maximum decrease in k_s. This is because the stabilizing bridging interaction must be *completely* lost in the transition state for nucleophile addition to the open form of $Me_2NC(S)$-[14^+].

Substituent effects on k_p. The substitution of an α-ester or an α-amide group for the α-methyl group at CH_3-[14^+] results in ~400-fold and 1000-fold increases, respectively, in k_p for deprotonation of the carbocation by solvent, and even larger ~1800-fold and 75 000-fold decreases in the rate constant ratio k_s/k_p (Table 3).[33] These changes probably do not arise from strong interactions between the α-ester or α-amide groups and an adjacent carbon–carbon double bond which stabilize the alkene R-[15], because the results of *ab initio* calculations for a related system (Scheme 13) predict that the α-ester group *destabilizes* MeOC(O)-[17] relative to MeOC(O)-[16]-OH, and that the α-amide group stabilizes $Me_2NC(O)$-[17] relative to $Me_2NC(O)$-[16]-OH by only 1.6 kcal mol^{-1} (Table 4).[33] The observed decreases in k_s/k_p for partitioning of EtOC(O)-[14^+] and $Me_2NC(O)$-[14^+] relative to that for CH_3-[14^+] may be the result of the larger effect of resonance electron donation from the α-ester and α-amide groups at EtOC(O)-[14^+] and $Me_2NC(O)$-[14^+] on k_s for nucleophilic addition of solvent than on k_p for deprotonation of these carbocations.

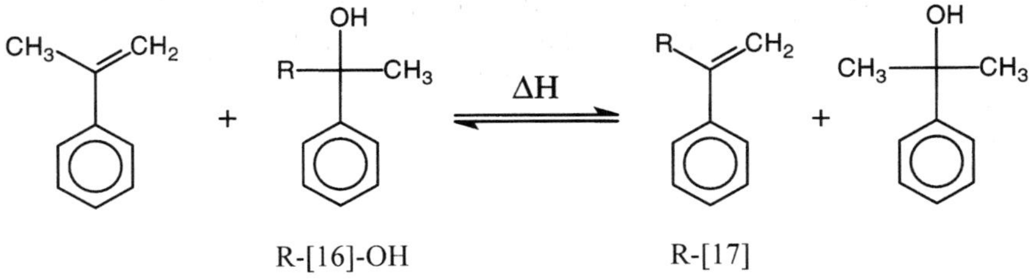

R-[16]-OH R-[17]

R = $Me_2NC(S)$, $Me_2NC(O)$, MeOC(O), CH_3

Scheme 13

Table 4 The effects of α-carbonyl and α-thiocarbonyl substituents on the relative stabilities of α-methyl benzyl alcohols R-[16]-OH and the corresponding styrenes R-[17] (Scheme 13)[a]

α-Substituent R	ΔH (kcal mol^{-1})[b] 3-21G//3-21G	6-31G*//3-21G
CH_3	0	0
MeOC(O)	1.07	0.27
$Me_2NC(O)$	−1.52	−1.57
$Me_2NC(S)$	−1.79	−5.04

[a] Energy changes are for the isodesmic reaction shown in Scheme 13.[33]
[b] Calculated from the *ab initio* energies of R-[16]-OH and R-[17].[33]

The stabilization of $Me_2NC(S)$-[14^+] by resonance electron donation from the α-thioamide group is particularly large (Table 3); and a differential expression of this resonance effect on k_s (large expression) and k_p (small expression) may also contribute to the relatively *small* value of $k_s/k_p < 0.01$ for this carbocation. In addition, *ab initio* calculations show that the α-thioamide group stabilizes the alkene $Me_2NC(S)$-[17] relative to the alcohol $Me_2NC(S)$-[16]-OH by 5.0 kcal mol^{-1}, as compared with the α-methyl group at Me-[17] and Me-[16]-OH (Table 4).[33] Such a preferential stabilization of the product of deprotonation of $Me_2NC(S)$-[14^+] is expected to result in a sizable increase in k_p and probably contributes to the small value of k_s/k_p for partitioning of $Me_2NC(S)$-[14^+] (Table 3).

The calculated values of ΔH for the isodesmic reactions shown in Scheme 13 (Table 4) suggest that the effects of α-carbonyl and α-thiocarbonyl substituents are controlled largely by the balance between two opposing interactions at the alkenes R-[17]:

(1) A *stabilizing* interaction between adjacent sp^2-hybridized centers, which is of roughly equal magnitude for the α-thioamide, α-amide and α-ester groups. This is not likely to be a simple resonance interaction, because the adjacent π-centers are not coplanar in the ground states of R-[17] optimized at the 3-21G level, but are twisted away from coplanarity by 70°, 60°, and 10° for $Me_2NC(S)$-[17], $Me_2NC(O)$-[17] and MeOC(O)-[17], respectively. Rather, the stabilization of the alkenes R-[17] by α-thiocarbonyl and α-carbonyl substituents (Table 4) is likely a consequence of the σ-bond between the sp^2-hybridized carbon of these α-substituents and that of the alkene double bond, which is expected to be stronger than the corresponding σ-bond between these α-substituents and the benzylic carbon at R-[16]-OH.[97]

(2) A *destabilizing* dipole–dipole interaction between the electron-deficient α-thiocarbonyl or α-carbonyl substituent and the double bond at R-[17],

which decreases in magnitude on proceeding from the strongly electron-withdrawing α-ester substituent to the more weakly electron-withdrawing α-amide and α-thioamide substituents.[98]

6 Aromaticity as a driving force for deprotonation of carbocations

The deprotonation of carbocations is favored both kinetically and thermodynamically when the π-bond of the product alkene is formed as part of a stable aromatic ring system. A dramatic example is the reaction of the benzallylic carbocation [4^+] (Scheme 14).[44,99,100] This carbocation has been generated by the acid-catalyzed cleavage of 1-methoxy-1,4-dihydronaphthalene ([4′]-OMe) in 25% acetonitrile in water, where it partitions to give >99.9% naphthalene [5] and only a trace (<0.1%) of the alcohol [4]-OH. Analysis of the products of this reaction in the presence of azide ion gave values of $k_{az}/k_p = 0.32\ \text{M}^{-1}$ and $k_{az}/k_s = 500\ \text{M}^{-1}$,[44] which can be combined to give $k_s/k_p = 0.00063$ (Scheme 14). The value of $k_{az}/k_p = 0.32\ \text{M}^{-1}$ falls within the range of the small, limiting, nucleophile selectivities (0.2–0.75 M^{-1}) observed when a carbocation intermediate undergoes reaction with solvent *faster* than its diffusional encounter with added azide ion.[21,26,101] Therefore, the use of the diffusion-limited azide ion clock ($k_{az} = 5 \times 10^9\ \text{M}^{-1}\ \text{s}^{-1}$) to calculate absolute values of k_s and k_p is not appropriate here, because only a small fraction of the azide adduct is formed by a pathway involving the diffusional encounter of azide ion with the carbocation. However, the azide ion clock does set a lower limit on the reactivity of azide ion

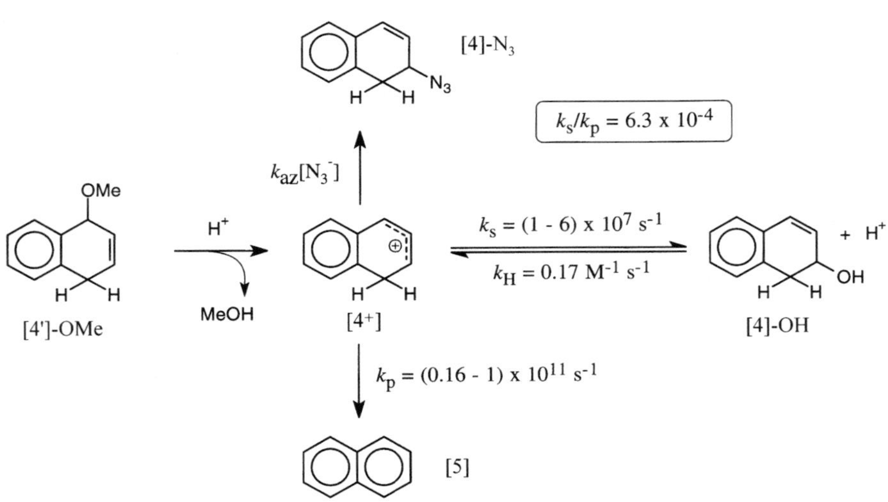

Scheme 14

to give the azide ion adduct, so that the value of $k_{az}/k_p = 0.32\ \text{M}^{-1}$ gives the *lower limit* on the rate constant for deprotonation of $[4^+]$ as $k_p \geqslant 1.6 \times 10^{10}\ \text{s}^{-1}$. The upper limit on this rate constant, $k_p \leqslant 10^{11}\ \text{s}^{-1}$, obtains when the rate-limiting step for proton transfer is the rotation of a molecule of solvent into a position to deprotonate $[4^+]$ ($k_{reorg} \approx 10^{11}\ \text{s}^{-1}$),[19,20] so that $k_p = (0.16 - 1) \times 10^{11}\ \text{s}^{-1}$. Combining this range of values with the product rate constant ratio $k_s/k_p = 0.000\,63$ gives $k_s = (1 - 6) \times 10^7\ \text{s}^{-1}$ (Scheme 14). The second-order rate constant for the acid-catalyzed cleavage of [4]-OH was determined as $k_H = 0.17\ \text{M}^{-1}\ \text{s}^{-1}$.[44] This can be combined with the range of values for k_s to give $pK_R = -7.8$ to -8.5 for $[4^+]$.

The very small value of $k_s/k_p = 0.00063$ for partitioning of $[4^+]$ reflects the strong aromatic stabilization of naphthalene [5]. This will reduce the activation barrier for deprotonation, provided aromaticity is partly developed in the reaction transition state. Thus, this aromatic stabilization results in both a large ~20 kcal mol^{-1} ($K \approx 10^{15}$) driving force for the dehydration of [4]-OH to give [5],[44,99] and a very small barrier for deprotonation of $[4^+]$ ($k_p \geqslant 1.6 \times 10^{10}\ \text{s}^{-1}$).

There is also substantial stabilization of $[4^+]$ by electron delocalization from the cyclic α-vinyl group. This is shown by a comparison of the thermodynamic driving force (pK_R lies between -7.8 and -8.5) and absolute rate constant ($k_s = 1 - 6 \times 10^7\ \text{s}^{-1}$) for the reaction of $[4^+]$ in 25% acetonitrile in water with the corresponding parameters for reaction of the resonance-stabilized 1-(4-methoxyphenyl)ethyl carbocation in water ($pK_R = -9.4$ and $k_s = 1 \times 10^8\ \text{s}^{-1}$, Table 5).

7 Cyclic benzylic carbocations

A comparison of rate and equilibrium constants for partitioning of the cyclic carbocation $[18^+]$ with those for the 1-(4-methylphenyl)ethyl carbocation Me-$[6^+]$ (Table 5) shows that placement of the cationic benzylic carbon in a five-membered ring results in the following complex changes in the reactivity of the carbocation towards deprotonation and nucleophilic addition of solvent (Scheme 15).

(1) The 3.6 kcal mol^{-1} less favorable change in Gibbs free energy for the hydration of [19] in water to give [18]-OH ($K_{add} = 0.10$)[103] than for the hydration of Me-[7] in the same solvent ($K_{add} = 41$, Table 1) provides evidence for an interaction which stabilizes [19] toward hydration. We suggest that this represents, at least in part, stabilization of [19] compared to Me-[7] by enhanced π-overlap between the phenyl ring and the cyclic vinyl group. This may reflect the smaller loss in rotational entropy required to restrict the carbon–carbon double bond of the cyclic vinyl group and the phenyl ring at [19] to the coplanar conformation most favorable to π-overlap, compared with Me-[7], where there is free rotation about the C_{Ar}—C_α bond.

Table 5. Rate and equilibrium constants for the formation and reaction of cyclic benzylic carbocations [18^+] and [20^+] and analogous ring-substituted 1-phenylethyl carbocations (Scheme 15)[a]

Carbocation	Precursor to carbocation[b]	Solvent	k_s/k_p [c]	k_s (s^{-1})	k_p (s^{-1})	pK_R [d]	pK_{alk} [e]	K_{add} [f]
Me, H, ⊕, CH_3	[g] Alcohol + H^+	50/50 TFE/H_2O	1400	6×10^9	4.2×10^6	−12.8	−11.5	20[h]
⊕, H, H	[i] Trifluoroacetate	H_2O	1300	2.6×10^9	2×10^6	−11.9	−12.9	0.10[j]
		50/50 TFE/H_2O		1.9×10^9		−11.3		
MeO, H, ⊕, CH_3	[g] Alcohol + H^+	50/50 TFE/H_2O	7000	5×10^7	7000	−8.6	−7.3	20[h]
		H_2O		1×10^8 [k]		−9.4[l]	−7.8[h]	36
⊕, H, H, O	[i] Chloroacetate	H_2O	1.7	1.6×10^7	9.4×10^6	−9.6		≪0.10[j]

[a]For reactions at 25°C. [b]Substrate used to determine the value of k_s/k_p. [c]Rate constant ratio for partitioning of the carbocation between nucleophilic addition of solvent (k_s) and deprotonation (k_p). [d]K_R is the equilibrium constant for addition of water to the carbocation to give the alcohol (Scheme 7). [e]K_{alk} is the equilibrium constant for deprotonation of the carbocation to give the alkene (Scheme 7). [f]Equilibrium constant for addition of water to the alkene to give the alcohol (Scheme 7). [g]Data from Tables 1 and 2, unless noted otherwise. [h]Calculated from the relationship $pK_{alk} = pK_R + \log K_{add}$. [i]Data from Ref. 104, unless noted otherwise. [j]Ref. 103. [k]Ref. 13. [l]Calculated with the assumption that the 0.8 unit more negative pK_R for the 9-methylfluorenyl carbocation in 100% water than in 50/50 (v/v) trifluoroethanol/water[9] will also be observed for the 1-(4-methoxyphenyl)ethyl carbocation.

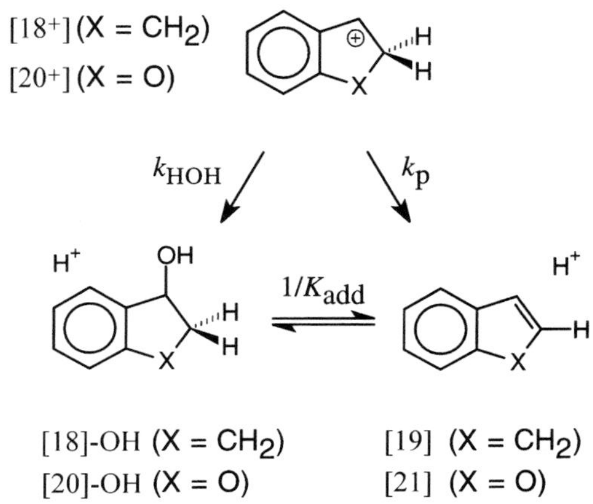

[18^+] (X = CH_2)
[20^+] (X = O)

k_{HOH} k_p

$1/K_{add}$

[18]-OH (X = CH_2) [19] (X = CH_2)
[20]-OH (X = O) [21] (X = O)

Scheme 15

(2) The 1.8 kcal mol^{-1} less favorable change in Gibbs free energy for the addition of water to [18^+] to give [18]-OH in 50/50 (v/v) trifluoroethanol/water ($pK_R = -11.3$)[104] than for addition of water to Me-[6^+] in the same solvent ($pK_R = -12.6$)[13] shows that the former carbocation is stabilized relative to the alcohol. This stabilization may be the result of the smaller entropic price paid to restrict the β—CH bonds in the five-membered ring at [18^+] to conformations that are favorable for hyperconjugation with the cationic carbon.

(3) The greater acidity of [18^+] ($pK_{alk} = -12.9$)[104] than of Me-[6^+] ($pK_{alk} = -11.5$)[14] is consistent with both (a) stronger hyperconjugation between the cationic center and the β-CH bonds of [18^+] relative to that at Me-[6^+], because this should weaken these bonds and favor proton transfer to give [19]; and (b) greater stabilization of [19] relative to Me-[7] by more favorable π-overlap between the phenyl ring and the cyclic vinyl group.

The more favorable change in Gibbs free energy for the dehydration of [18]-OH to give [19] than for the dehydration of Me-[6]-OH to give Me-[7] (Table 5) is not reflected by any significant difference in the rate constant ratios k_s/k_p for formation of the respective alcohols and alkenes by partitioning of [18^+] ($k_s/k_p = 1300$)[104] and Me-[6^+] ($k_s/k_p = 1400$).[14] We do not understand why these values of k_s/k_p are not sensitive to changes in the relative thermodynamic driving force for the nucleophile addition and proton transfer reactions.

The substitution of a 4-methoxy for the 4-methyl group at Me-[6^+] results in a 100-fold decrease in k_s ($\Delta \log k_s = 2$) and a larger 4.2-unit increase in pK_R (Table 5).[13] Therefore, approximately 50% of the 4-methoxy substituent effect on pK_R for the equilibrium addition of water is expressed in the value of k_s, which suggests the loss of about 50% of the stabilizing interactions between the 4-methoxy group and the cationic center at the transition state for solvent addition.[13] By contrast, the substitution of oxygen for carbon at [18^+] to give [20^+] results in a similar 160-fold decrease in k_s ($\Delta \log k_s = 2.2$) but only a 2.3-unit increase in pK_R, so that essentially 100% of the equilibrium oxygen substituent effect is expressed in the rate constant k_s. These results are consistent with: (1) a large stabilization of [20^+] by resonance electron donation from the *ortho* oxygen; (2) a smaller destabilization of [20^+] by electrostatic interactions between the closely spaced partial positive charges at the *ortho* oxygen and the benzylic carbon; and (3) an imbalance between the large fractional expression of the stabilizing resonance substituent effect and the smaller expression of the destabilizing inductive effect on k_s for solvent addition.[28,86–88] The result of the dominance of the resonance substituent effect is that k_s for addition of solvent to [20^+] is smaller than expected for the overall change in pK_R. A similar imbalance between resonance and inductive substituent effects has been proposed to explain the nearly constant values of k_s observed with increasing destabilization of the 4-methoxybenzyl carbocation by electron-withdrawing α-substituents.[28]

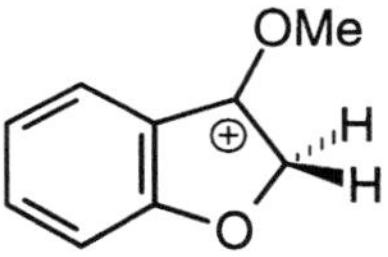

[22] 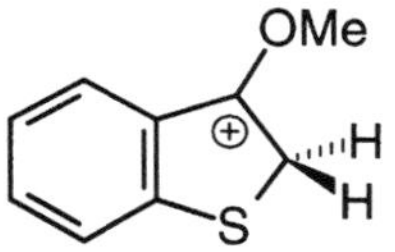 [23]

The substitution of oxygen for carbon at [18^+] to give [20^+] results in a 5-fold increase in k_p and a 760-fold decrease in the ratio k_s/k_p from 1300 to 1.7.[104] These changes reflect the tendency of the 6π-aromatic stabilization of the furan ring of the alkene [21] to increase the thermodynamic driving force for deprotonation of [20^+] relative to that for deprotonation of [18^+].[103,104] The large difference between the values of $k_s/k_p > 20$ for partitioning of [22] and $k_s/k_p = 0.032$ for partitioning of [23] is consistent with a larger stabilization of the 3-methoxybenzothiophene than the 3-methoxybenzofuran reaction product.[105]

8 Kinetic studies of alkene hydration

The results of studies of the acid-catalyzed hydration of oxygen-, sulfur-, seleno- and nitrogen-substituted alkenes and the relevance of this work to partitioning of the corresponding carbocation intermediates (Chart 1) between deprotonation and nucleophile addition was reviewed in 1986.[70]. We present here a brief summary of this earlier review, along with additional discussion of recent literature.

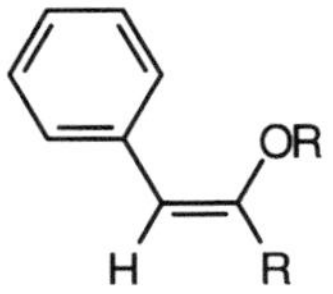

[24] 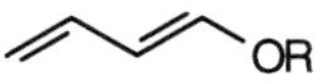 [25] 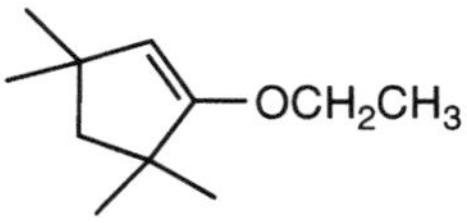 [26] 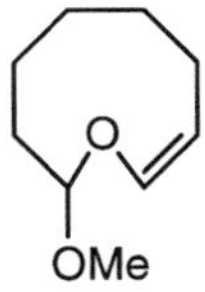 [27]

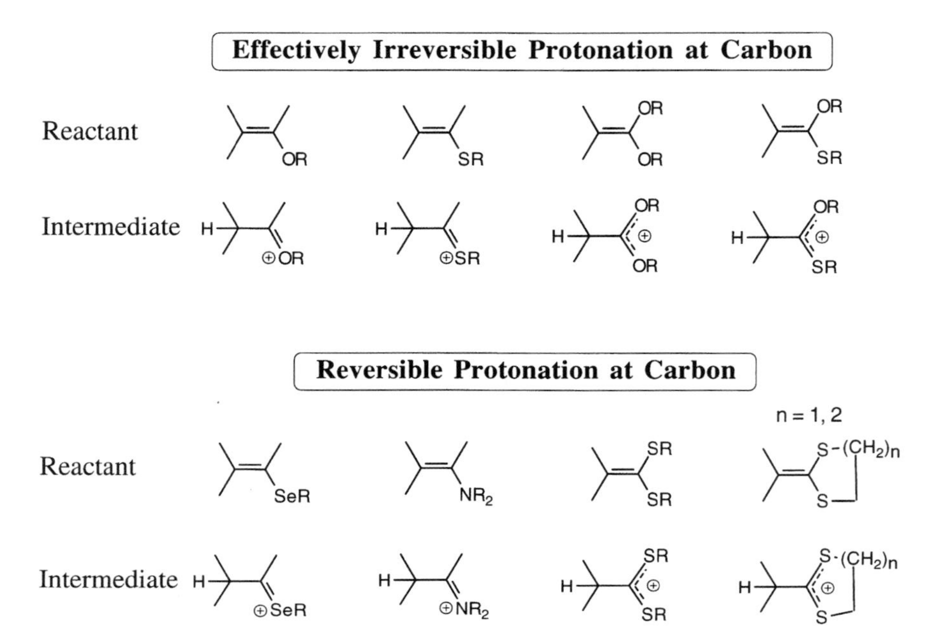

Chart 1

The specific acid-catalyzed reactions of simple vinyl ethers and vinyl thioethers proceed by rate-determining protonation at carbon ($k_s > k_p$, Scheme 5). The following modifications in the structure of simple vinyl ethers, or in the reaction conditions for their hydrolysis, failed to induce a change in rate-determining step to the addition of solvent ($k_s < k_p$):[70] (1) addition of a phenyl substituent β to the ether oxygen in [24], which should stabilize the olefin by conjugation and hence lower the barrier to formation of the olefin (k_p) relative to that for solvent addition (k_s);[56] (2) stabilization of the olefin by a conjugating vinyl group ([25]);[106] (3) the addition of a strong base (and the corresponding weak conjugate acid in a buffered solution), which will catalyze deprotonation of the carbocation and lower the barrier to deprotonation (k_p) relative to that for solvent addition (k_s);[70] (4) the introduction of substituents (e.g. [26]) that are expected to reduce k_s relative to k_p for reaction of the oxocarbenium ion by the introduction of *I* strain into the sp^3-hybridized product of the solvent addition reaction.[51]

The hydration of [27] has been reported to proceed by rate-determining hydration of an oxocarbenium ion intermediate.[107] However, studies of compounds with structural features common to [27],[108–110] and a reinvestigation of the mechanism for hydrolysis of [27] have demonstrated that protonation of the vinylic carbon of [27] is irreversible.[70]

The value of $k_s/k_p = 17\,000$ for partitioning of the acetophenone oxocarbenium ion [12^+] between deprotonation and nucleophilic addition of water[15,16] shows that the common rate-determining step of alkene protonation for hydrolysis of a wide range of vinyl ethers is a consequence of the tendency of simple α-methyl oxocarbenium ions to partition to form the adduct to solvent. Apparently, the structural variations in vinyl ethers examined to date are not sufficient to result in the ~20 000-fold decrease in k_s/k_p for partitioning of the oxocarbenium intermediate that would be required for the observation of a change in rate-determining step.

Chart 1 compares alkenes that undergo hydration by rate-determining proton transfer to carbon with those that react by rate-determining addition of solvent to a carbocation intermediate. The progression down the periodic table from an α-oxygen to an α-sulfur to an α-selenium substituent favors a change in rate-determining step for alkene hydration. For example: (1) Alkene protonation is the rate-determining step for hydration of ketene acetals, but addition of solvent to a α,α-dithio carbocation intermediate is rate-determining for hydration of ketene dithioacetals.[38–40,111–114] (2) Alkene protonation is the rate-determining step for hydration of vinyl ethers and vinyl thioethers, but addition of solvent to a α-seleno carbocation intermediate is rate-determining for hydration of vinyl seleno ethers.[115,116]

We suggest that the substitution of an α-oxygen by a α-sulfur or α-selenium at an oxocarbenium ion has several effects on the reactivity of the carbocation, each of which contributes to the large decrease in value of k_s/k_p for partitioning of the carbocation:

(1) There is good evidence that the sp^2-hybridized alkenyl group is destabilized relative to a sp^3-hybridized alkyl group by electron-withdrawing substituents.[98] This interaction is expected to be most strongly destabilizing for a vinyl ether and to become progressively weaker on changing from an oxygen to sulfur or seleno substituents. This will favor an increase in k_p for deprotonation relative to k_s for solvent addition to the corresponding carbocations (Scheme 16).

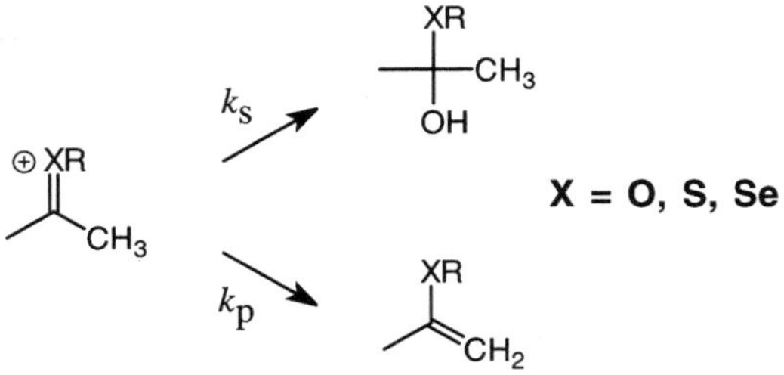

Scheme 16

(2) There is strong stabilization of sp^3-hybridized carbon by geminal interactions between groups that contain electron-withdrawing α-oxygen atoms; this stabilization becomes weaker as a single α-oxygen atom is replaced first by an α-sulfur and then by an α-seleno atom.[117,118] This should result in a progressive destabilization of the product that forms by addition of water to α-oxygen-, α-

sulfur- and α-selenium-stabilized carbocations relative to the corresponding alkenes that form by proton transfer to solvent (Scheme 16). Destabilization of the product of water addition relative to the product of proton transfer from these carbocations will favor an increase in k_p relative to k_s as the heteroatom X at the α-substituent is changed from O to S to Se (Scheme 16).

(3) The α-CF_3CH_2S group at [28] and the α-CF_3CH_2O group at [29] provide nearly equal stabilization of the parent 4-methoxybenzyl carbocation relative to the neutral azide ion adducts.[119] Therefore, the observation of a significantly smaller rate constant for addition of an aqueous solvent to [28] than to [29] is consistent with a larger intrinsic barrier for reaction of the carbocation stabilized by the α-CF_3CH_2S group.[8,119] This difference in the intrinsic barriers for the reaction of α-oxygen and α-sulfur substituted carbocations will result in a smaller value of k_s/k_p for partitioning of an α-sulfur carbocation relative to that for a structurally homologous α-oxygen carbocation, provided the intrinsic barrier for solvent addition (k_s) shows a larger change with the sulfur for oxygen substitution than does the intrinsic barrier for deprotonation (k_p).

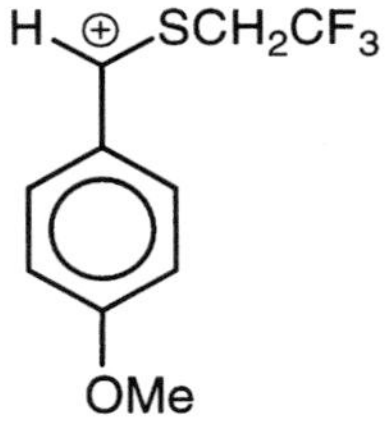

H ⊕ SCH_2CF_3 — OMe

[28]

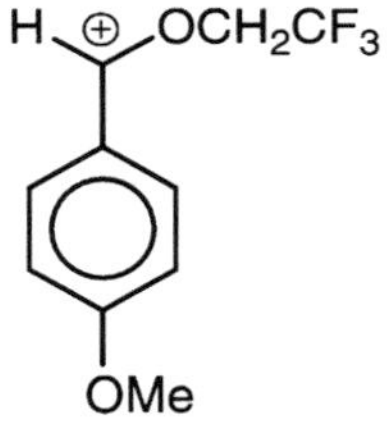

H ⊕ OCH_2CF_3 — OMe

[29]

The rate-determining step for the hydrolysis of enamines (Scheme 17) at neutral pH is the addition of solvent to an iminium ion intermediate ($k_p > k_s$). A change to rate-determining protonation at carbon is observed at high pH, owing to the contribution of the rate constant for direct addition of a hydroxide ion to the iminium ion $[(k_s + k_{HO}[HO^-]) > k_p)]$.[35–37]. The dramatic difference between the partitioning of simple oxocarbenium ions ($k_s/k_p \gg 1$) and iminium ions ($k_s/k_p \ll 1$) suggests that stabilization of the carbocation by electron donation from an α-substituent, which is much stronger for an iminium ion than for an oxocarbenium ion, has a larger effect on k_s for reaction of solvent as a Lewis base than on k_p for its reaction as a Brønsted base.

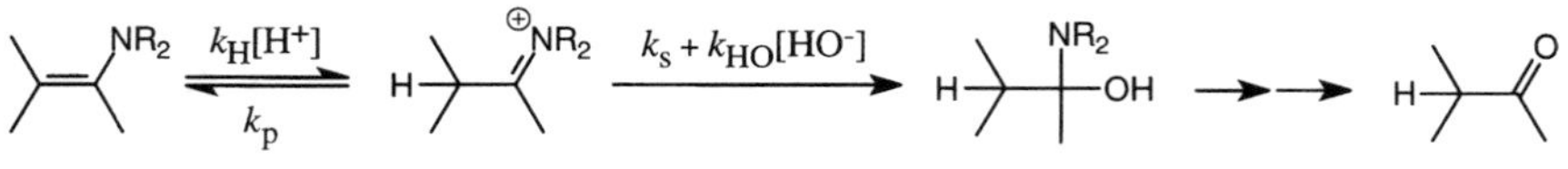

NR_2 $\xrightleftharpoons[k_p]{k_H[H^+]}$ H ⊕NR_2 $\xrightarrow{k_s + k_{HO}[HO^-]}$ H NR_2 OH ⟶⟶ H O

Scheme 17

9 Other systems

There are at least two other studies of competitive reactions to form the products of solvolysis and elimination reactions that may provide insight into the relationships between carbocation structure and reactivity toward nucleophile addition and deprotonation.

(1) The partitioning of ferrocenyl-stabilized carbocations [30] between nucleophile addition and deprotonation (Scheme 18) has been studied by Bunton and coworkers. In some cases the rate constants for deprotonation and nucleophile addition are comparable, but in others they favor formation of the nucleophile adduct. However, the alkene product of deprotonation of [30] is always the thermodynamically favored product.[120]. In other words, the addition of water to [30] gives an alcohol that is *thermodynamically less stable* than the alkene that forms by deprotonation of [30], but the reaction passes over an activation barrier whose height is equal to, or smaller than, the barrier for deprotonation of [30]. These data require that the intrinsic barrier for thermoneutral addition of water to [30] (Λ_s) be smaller than the intrinsic barrier for deprotonation of [30] (Λ_p). It is not known whether the magnitude of ($\Lambda_p - \Lambda_s$) for the reactions of [30] is similar to the values of ($\Lambda_p - \Lambda_s$) = 4–6 kcal mol^{-1} reported here for the partitioning of α-methyl benzyl carbocations.

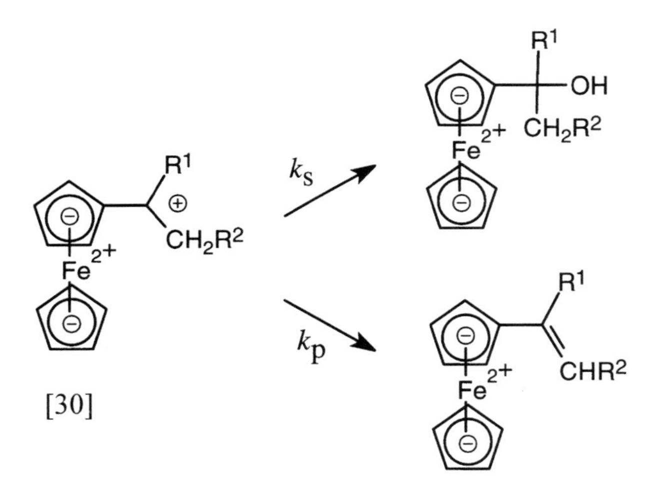

Scheme 18

(2) The results of a thorough study of the kinetics, products and stereochemical course for the nucleophilic substitution and elimination reactions of ring-substituted 9-(1-Y-ethyl)fluorenes ([31]-Y, Y = Br, I, brosylate) have been reported (Scheme 19).[121,122]. The reactions of the halides [31]-Br and [31]-I were proposed to proceed exclusively by "a solvent-promoted E1cB reaction or an E2 reaction with a large component of hydron transfer in the transition state".[122]

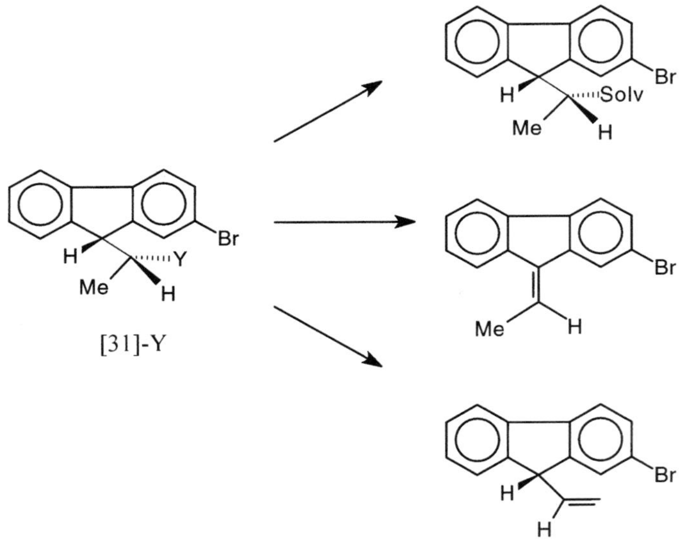

[31]-Y

Scheme 19

By contrast, [31]-(brosylate) was proposed to show a large amount of reaction through a secondary carbocation intermediate,[122] which partitions between nucleophile addition and deprotonation. We are uncertain of how to relate the partitioning of the putative carbocation intermediate of the reaction of [31]-(brosylate) to the partitioning of other carbocations discussed in this review.

10 Summary

The results described in this review provide support for the following generalizations about the influence of thermodynamics and intrinsic kinetic barriers on the partitioning of carbocations between nucleophilic addition of aqueous solvents to form a tetrahedral adduct (k_s) and proton transfer to these solvents to form an alkene (k_p).

(1) The observed value of k_s/k_p for partitioning of the simple tertiary carbocation [1^+] is *smaller* than that expected if the nucleophilic addition of solvent were to occur by rate-determining chemical bond formation. This is probably because solvent addition is limited by the rate constant for reorganization of the solvation shell that surrounds the carbocation.

(2) The partitioning of simple tertiary carbocations, ring-substituted 1-phenylethyl carbocations, and cumyl carbocations between deprotonation and nucleophilic addition of solvent strongly favors formation of the solvent adduct. The more favorable partitioning of these carbocations to form the solvent adduct is due, in part, to the greater thermodynamic stability of the solvent

adduct relative to the corresponding alkene. In two cases, quantitative analyses of the data have shown that the intrinsic barrier for deprotonation of the carbocation to form the alkene, Λ_p, is 4–6 kcal mol^{-1} larger than that for nucleophilic addition of an aqueous solvent, Λ_s. It is not difficult to propose qualitative explanations for this difference in intrinsic barriers. However, we are not aware of any theoretical or computational attempts to model this difference in intrinsic barriers.

(3) There is a ~3 kcal mol^{-1} increase in the intrinsic barrier Λ_s for addition of a largely aqueous solvent of 50/50 (v/v) trifluoroethanol/water to ring-substituted 1-phenylethyl carbocations on changing from a 4-Me to a 4-Me_2N ring substituent. This shows that this nucleophile addition reaction becomes intrinsically more difficult with increasing stabilization of the carbocation by resonance electron donation from the aromatic ring.

(4) The intrinsic barrier for the addition of solvent to an α-alkoxy benzyl carbocation is several kcal mol^{-1} smaller than that for the corresponding reaction of ring-substituted 1-phenylethyl carbocations. This result is consistent with the conclusion that these nucleophile addition reactions become intrinsically easier as stabilizing resonance electron donation from an α-phenyl group to the cationic center is replaced by electron donation from an α-alkoxy group.

(5) The addition of two *ortho*-methyl groups to ring-substituted cumyl carbocations results in greater steric crowding at the tetrahedral nucleophile adducts than at the corresponding α-methyl styrenes. This results in a decrease in k_s for reaction of the carbocation with a nucleophilic solvent relative to k_p for deprotonation of the carbocation.

(6) The addition of two *ortho*-methyl groups to ring-substituted cumyl carbocations results in steric crowding of the fully planar resonance-stabilized carbocation, which can be relieved by rotation about the C_{Ar}—C_α bond. The relief of steric strain that occurs with bond rotation is balanced by a decrease in stabilizing electron donation from the aromatic ring. The latter effect on carbocation *stability* is probably responsible for the larger rate constants k_p for deprotonation of ring-substituted 2,5-dimethyl cumyl carbocations than for the corresponding cumyl carbocation (Table 1).

(7) The change from an α-methyl to an α-ester or α-amide substituent results in 2000-fold and 75 000-fold decreases, respectively, in k_s/k_p for partitioning of the 4-methoxycumyl carbocation, due to both decreases in k_s and increases in k_p. The origins of these changes are complex.

(8) The change from an α-methyl to an α-thioamide substituent results in a $>10^7$-fold decrease in k_s/k_p for partitioning of the 4-methoxycumyl carbocation. This is the result of: (a) A large decrease in k_s for addition of solvent to the α-thioamide substituted carbocation. This decrease in k_s is the result of resonance electron donation from the α-thioamide group to the benzylic carbon, and/or the formation of a covalent sulfur bridge to the benzylic carbon (Scheme 12). (b) The strong stabilization of the alkene relative to the corresponding

alcohol by an α-thioamide group. This results in an increase in k_p for deprotonation of α-thioamide substituted carbocations to form the alkene relative to k_s for solvent addition to form the alcohol.

(9) The incorporation of the alkene product of carbocation deprotonation into an aromatic system results in the expected changes in the absolute rate constant k_p and the product rate constant ratio k_s/k_p for reaction of the carbocation.

(10) The incorporation of a cationic benzylic carbon into a five-membered ring results in complex changes in the reactivity of the carbocation toward deprotonation and nucleophilic addition of solvent (Scheme 15).

(11) The substitution of an α-oxygen by an α-sulfur or an α-selenium results in a decrease in k_s/k_p for partitioning of oxocarbenium ions between addition of solvent and deprotonation. The largest effect is observed for the α-selenium substitution. These changes are probably the result of several different effects on carbocation reactivity, each of which contributes to the observed substituent effect on k_s/k_p.

(12) Different rate-determining steps are observed for the acid-catalyzed hydration of vinyl ethers (alkene protonation, $k_s \gg k_p$) and hydration of enamines (addition of solvent to an iminium ion intermediate, $k_s \ll k_p$). This suggests that increasing stabilization of α-CH substituted carbocations by π-electron donation from an adjacent electronegative atom results in a larger decrease in k_s for nucleophile addition of solvent than in k_p for deprotonation of the carbocation by solvent.

Acknowledgement

We acknowledge the National Institutes of Health Grant GM 39754 for its generous support of the work from our laboratory described in this review.

References

1. Ingold, C.K. (1969). *Structure and Mechanism in Organic Chemistry*, 2nd edn. Cornell University Press, Ithaca
2. Jencks, W.P. (1980). *Acc. Chem. Res.* **13**, 161
3. Jencks, W.P. (1981). *Chem. Soc. Rev.* **10**, 345
4. Commission on Physical Organic Chemistry, IUPAC (1989). *Pure Appl. Chem.* **61**, 23
5. Guthrie, R.D. and Jencks, W.P. (1989). *Acc. Chem. Res.* **22**, 343
6. Saunders, W.H. and Cockerill, A.F. (1973). *Mechanisms of Elimination Reactions*, Wiley, New York
7. McClelland, R.A. (1996). *Tetrahedron* **52**, 6823
8. Richard, J.P. (1995). *Tetrahedron* **51**, 1535
9. Amyes, T.L., Richard, J.P. and Novak, M. (1992). *J. Am. Chem. Soc.* **114**, 8032

10. Cozens, F.L., Mathivanan, N., McClelland, R.A. and Steenken, S. (1992). *J. Chem. Soc., Perkin Trans.* **2**, 2083
11. Mathivanan, N., McClelland, R.A. and Steenken, S. (1990). *J. Am. Chem. Soc.* **112**, 8454
12. Richard, J.P., Jagannadham, V., Amyes, T.L., Mishima, M. and Tsuno, Y. (1994). *J. Am. Chem. Soc.* **116**, 6706
13. Richard, J.P., Rothenberg, M.E. and Jencks, W.P. (1984). *J. Am. Chem. Soc.* **106**, 1361
14. Richard, J.P. and Jencks, W.P. (1984). *J. Am. Chem. Soc.* **106**, 1373
15. Richard, J.P., Amyes, T.L. and Williams, K.B. (1998). *Pure Appl. Chem.* **70**, 2007
16. Williams, K.B. (1998) Ph.D. Thesis, University at Buffalo, Buffalo NY, USA
17. Richard, J.P. and Jencks, W.P. (1984). *J. Am. Chem. Soc.* **106**, 1396
18. Dostrovsky, I. and Klein, F.S. (1955). *J. Chem. Soc.*, 791
19. Giese, K., Kaatze, U. and Pottel, R. (1970). *J. Phys. Chem.* **74**, 3718
20. Kaatze, U., Pottel, R. and Schumacher, A. (1992). *J. Phys. Chem.* **96**, 6017
21. Toteva, M.M. and Richard, J.P. (1996). *J. Am. Chem. Soc.* **118**, 11434
22. Eberz, W.F. and Lucas, H.J. (1934). *J. Am. Chem. Soc.* **56**, 1230
23. Toteva, M.M. and Richard, J.P. (1997). *Bioorg. Chem.* **25**, 239
24. Schubert, W.M. and Keeffe, J.R. (1972). *J. Am. Chem. Soc.* **94**, 559
25. Thibblin, A. (1989). *J. Phys. Org. Chem.* **2**, 15
26. Richard, J.P., Amyes, T.L. and Vontor, T. (1991). *J. Am. Chem. Soc.* **113**, 5871
27. Amyes, T.L., Mizerski, T. and Richard, J.P. (1999). *Can. J. Chem.* **77**, 922
28. Amyes, T.L., Stevens, I.W. and Richard, J.P. (1993). *J. Org. Chem.* **58**, 6057
29. McClelland, R.A., Cozens, F.L., Steenken, S., Amyes, T.L. and Richard, J.P. (1993). *J. Chem. Soc., Perkin Trans.* **2**, 1717
30. Thibblin, A. (1992). *J. Chem. Soc., Perkin Trans.* **2**, 1195
31. Bernasconi, C.F. and Boyle, W.J. (1974). *J. Am. Chem. Soc.* **96**, 6070
32. Amyes, T.L. and Jencks, W.P. (1989). *J. Am. Chem. Soc.* **111**, 7888
33. Richard, J.P., Lin, S.-S., Buccigross, J.M. and Amyes, T.L. (1996). *J. Am. Chem. Soc.* **118**, 12603
34. McClelland, R.A., Licence, V.E., Richard, J.P., Williams, K.B. and Lin, S.-S. (1998). *Can. J. Chem.* **76**, 1910
35. Coward, J.K. and Bruice, T.C. (1969). *J. Am. Chem. Soc.* **91**, 5329
36. Keeffe, J.R. and Kresge, A.J. (1994). In *The Chemistry of Enamines* (ed. Z. Rappoport), p. 1049. Wiley, New York
37. Sollenberger, P.Y. and Martin, R.B. (1970). *J. Am. Chem. Soc.* **92**, 4261
38. Okuyama, T. (1984). *J. Am. Chem. Soc.* **106**, 7134
39. Okuyama, T. and Fueno, T. (1980). *J. Am. Chem. Soc.* **102**, 6590
40. Okuyama, T., Kawao, S. and Fueno, T. (1983). *J. Am. Chem. Soc.* **105**, 3220
41. Creary, X., Casingal, V.P. and Leahy, C.E. (1993). *J. Am. Chem. Soc.* **115**, 1734
42. Creary, X., Hatoum, H.N., Barton, A. and Aldridge, T.E. (1992). *J. Org. Chem.* **57**, 1887
43. Richard, J.P. and Jencks, W.P. (1984). *J. Am. Chem. Soc.* **106**, 1383
44. Pirinicciogu, N. and Thibblin, A. (1998). *J. Am. Chem. Soc.* **120**, 6512
45. Ta-Shma, R. and Rappoport, Z. (1983). *J. Am. Chem. Soc.* **105**, 6082
46. Thibblin, A. (1987). *J. Am. Chem. Soc.* **109**, 2071
47. Kresge, A.J. and Chiang, Y. (1967). *J. Chem. Soc. (B)*, 53
48. Kresge, A.J. and Chwang, W.K. (1978). *J. Am. Chem. Soc.* **100**, 1249
49. Deno, N.C., Kish, F.A. and Peterson, H.J. (1965). *J. Am. Chem. Soc.* **87**, 2157
50. Chiang, Y., Chwang, W.K., Kresge, A.J., Robinson, L.H., Sagatys, D.S. and Young, C.I. (1978). *Can. J. Chem.* **56**, 456
51. Chiang, Y., Chwang, W.K., Kresge, A.J. and Szilagyi, S. (1980). *Can. J. Chem.* **58**, 124

52. Chiang, Y., Kresge, A.J., Seipp, U. and Winter, W. (1988). *J. Org. Chem.* **53**, 2552
53. Chiang, Y., Kresge, A.J., Tidwell, T.T. and Walsh, P.A. (1980). *Can. J. Chem.* **58**, 2203
54. Chwang, W.K., Eliason, R. and Kresge, A.J. (1977). *J. Am. Chem. Soc.* **99**, 805
55. Kresge, A.J., Chen, H.-L., Chiang, Y., Murrill, E., Payne, M.A. and Sagatys, D.S. (1971). *J. Am. Chem. Soc.* **93**, 413
56. Kresge, A.J. and Chen, H.J. (1972). *J. Am. Chem. Soc.* **94**, 2818
57. Kresge, A.J., Chen, H.J. and Chiang, Y. (1977). *J. Am. Chem. Soc.* **99**, 802
58. Kresge, A.J. and Chiang, Y. (1972). *J. Am. Chem. Soc.* **94**, 2814
59. Kresge, A.J. and Chiang, Y. (1973). *J. Am. Chem. Soc.* **95**, 803
60. Kresge, A.J., Leibovitch, M. and Sikorski, J.A. (1992). *J. Am. Chem. Soc.* **114**, 2618
61. Loudon, G.M. and Berke, C. (1974). *J. Am. Chem. Soc.* **96**, 4508
62. Salomaa, P., Kankaanperä, A. and Lajunen, M. (1966). *Acta Chem. Scand.* **20**, 1790
63. Schubert, W.M. and Jensen, J.L. (1972). *J. Am. Chem. Soc.* **94**, 566
64. Cordes, E.H. and Bull, H.G. (1974). *Chem. Rev.* **74**, 581
65. Jensen, J.L., Herold, L.R., Lenz, P.A., Trusty, S., Sergi, V., Bell, K. and Rogers, P. (1979). *J. Am. Chem. Soc.* **101**, 4672
66. Bunton, C.A., Davoudazedeh, F. and Watts, W.E. (1981). *J. Am. Chem. Soc.* **103**, 3855
67. Young, P.R. and Jencks, W.P. (1977). *J. Am. Chem. Soc.* **99**, 8238
68. Ta-Shma, R. and Jencks, W.P. (1986). *J. Am. Chem. Soc.* **108**, 8040
69. Kresge, A.J., Sagatys, D.S. and Chen, H.L. (1977). *J. Am. Chem. Soc.* **99**, 7228
70. Okuyama, T. (1986). *Acc. Chem. Res.* **19**, 370
71. Marcus, R.A. (1956). *J. Chem. Phys.* **24**, 966
72. Marcus, R.A. (1968). *J. Phys. Chem.* **72**, 891
73. Marcus, R.A. (1969). *J. Am. Chem. Soc.* **91**, 7224
74. Guthrie, J.P. (1997). *J. Am. Chem. Soc.* **119**, 1151
75. Guthrie, J.P. (1998). *J. Am. Chem. Soc.* **120**, 1688
76. Guthrie, J.P. (1991). *J. Am. Chem. Soc.* **113**, 7249
77. Guthrie, J.P. and Guo, J. (1996). *J. Am. Chem. Soc.* **118**, 11472
78. Bernasconi, C.F. (1985). *Tetrahedron* **41**, 3219
79. Bernasconi, C.F., Moreira, J.A., Huang, L.L. and Kittredge, K.W. (1999). *J. Am. Chem. Soc.* **121**, 1674
80. Amyes, T.L. and Richard, J.P. (1996). *J. Am. Chem. Soc.* **118**, 3129
81. Bunting, J.W. and Kanter, J.P. (1993). *J. Am. Chem. Soc.* **115**, 11705
82. Richard, J.P. and Jencks, W.P. (1982). *J. Am. Chem. Soc.* **104**, 4689
83. Dietze, P.E. and Jencks, W.P. (1986). *J. Am. Chem. Soc.* **108**, 4549
84. Dietze, P.E. (1987). *J. Am. Chem. Soc.* **109**, 2057
85. Guthrie, J.P. (1996). *Can. J. Chem.* **74**, 1283
86. Bernasconi, C.F. (1987). *Acc. Chem. Res.* **20**, 301
87. Bernasconi, C.F. (1992). *Adv. Phys. Org. Chem.* **27**, 119
88. Bernasconi, C.F. (1992). *Acc. Chem. Res.* **25**, 9
89. Jencks, W.P. and Regenstein, J. (1976). In *Handbook of Biochemistry and Molecular Biology (Physical and Chemical Data)* (ed. G.D. Fasman), p. 305. CRC Press, Cleveland, OH
90. Amyes, T.L. and Richard, J.P. (1991). *J. Chem. Soc., Chem. Commun.*, 200
91. Richard, J.P., Amyes, T.L., Jagannadham, V., Lee, Y.-G. and Rice, D.J. (1995). *J. Am. Chem. Soc.* **117**, 5198
92. Richard, J.P. (1989). *J. Am. Chem. Soc.* **111**, 1455
93. Richard, J.P., Amyes, T.L., Bei, L. and Stubblefield, V. (1990). *J. Am. Chem. Soc.* **112**, 9513
94. Richard, J.P., Amyes, T.L. and Stevens, I.W. (1991). *Tetrahedron Lett.* **32**, 4255
95. Schepp, N.P. and Wirz, J. (1994). *J. Am. Chem. Soc.* **116**, 11749

96. Lien, M.H. and Hopkinson, A.C. (1988). *J. Am. Chem. Soc.* **110**, 3788
97. Dewar, M.J.S. and Schmeising, H.N. (1960) *Tetrahedron* **11**, 96
98. Hine, J. (1975). *Structural Effects on Equilibria in Organic Chemistry*. Wiley-Interscience, New York
99. Boyd, D.R., McMordie, R.A.S., More O'Ferrall, R.A., Sharma, N.D. and Kelly, S.C. (1990). *J. Am. Chem. Soc.* **112**, 7822
100. Rao, S.N., More O'Ferrall, R.A., Kelly, S.C., Boyd, D.R. and Agarwal, R. (1993). *J. Am. Chem. Soc.* **115**, 5458
101. Finneman, J.I. and Fishbein, J.C. (1995). *J. Am. Chem. Soc.* **117**, 4228
103. Kelly, S.C., McDonnell, C.A., More O'Ferrall, R.A., Rao, N.S., Boyd, D.R., Brannigan, I.N. and Sharma, N.D. (1996). *Gazz. Chem. Ital.* **126**, 747
104. O'Donoghue, A.C. (1999) Ph.D. Thesis, University College Dublin, Dublin, Ireland
105. Capon, B. and Kwok, F.-C. (1987). *Tetrahedron* **43**, 69
106. Chiang, Y., Eliason, R., Guo, G.H.X. and Kresge, A.J. (1994). *Can. J. Chem.* **72**, 1632
107. Cooper, J.D., Vitullo, V.P. and Whalen, D.L. (1971). *J. Am. Chem. Soc.* **93**, 6294
108. Burt, R.A., Chiang, Y., Chwang, W.K., Kresge, A.J., Okuyama, T., Tang, Y.S. and Yin, Y. (1987). *J. Am. Chem. Soc.* **109**, 3787
109. Burt, R.A., Chiang, Y. and Kresge, A.J. (1980). *Can. J. Chem.* **58**, 2199
110. Burt, R.A., Chiang, Y., Kresge, A.J. and Szilagyi, S. (1984). *Can. J. Chem.* **62**, 74
111. Okuyama, T., Inaoka, T. and Fueno, T. (1986). *J. Org. Chem.* **51**, 4988
112. Okuyama, T., Kawao, S. and Fueno, T. (1984). *J. Org. Chem.* **49**, 85
113. Okuyama, T., Kawao, S., Fujiwara, W. and Fueno, T. (1984). *J. Org. Chem.* **49**, 89
114. Okuyama, T., Toyoda, M. and Fueno, T. (1986). *Can. J. Chem.* **64**, 1116
115. Hevesi, L., Piquard, J.-L. and Wautier, H. (1981). *J. Am. Chem. Soc.* **103**, 870
116. Wautier, H., Desauvage, S. and Hevesi, L. (1981). *J. Chem. Soc., Chem. Commun.*, 738
117. Salzner, U. and Schleyer, P.v.R. (1993). *J. Am. Chem. Soc.* **115**, 10231
118. Schleyer, P.v.R., Jemmis, E.D. and Spitznagel, G.W. (1985). *J. Am. Chem. Soc.* **107**, 6393
119. Jagannadham, V., Amyes, T.L. and Richard, J.P. (1993). *J. Am. Chem. Soc.* **115**, 8465
120. Bunton, C.A., Carrasco, N., Cully, N. and Watts, W.E. (1980). *J. Chem. Soc., Perkin Trans.* **2**
121. Meng, Q. and Thibblin, A. (1995). *J. Am. Chem. Soc.* **117**, 1839
122. Meng, Q., Gogoll, A. and Thibblin, A. (1997). *J. Am. Chem. Soc.* **119**, 1217
123. Kresge, A.J. (1973). *Chem. Soc. Rev.* **2**, 475

Electron Transfer, Bond Breaking and Bond Formation

JEAN-MICHEL SAVÉANT

Laboratoire d'Electrochimie Moléculaire, Unité Mixte de Recherche Université – CNRS No 7591, Université de Paris 7, Paris, France

ADVANCES IN PHYSICAL ORGANIC CHEMISTRY
VOLUME 35 ISBN 0-12-033535-2

1 Introduction

Single electron transfer to or from molecules is often accompanied by other reactions involving the formation and/or the cleavage of bonds. The high-energy intermediates thus formed may undergo further electron transfer with the same electron source or sink that initiated the reaction. Establishing the reaction mechanism then consists of unraveling the sequence, which might be rather complex, of electron transfers and coupled reactions. The synthetic value of electrochemistry is based on such reaction sequences. This is also true with other modes of electron injection or removal such as homogeneous electron donors or acceptors, solvated electrons, photoinduced electron transfers and radiolysis.

Among the reactions accompanying electron transfer, bond breaking is a common mode by which a free radical and a diamagnetic leaving group may be produced by single electron transfer to a diamagnetic molecule. Such reactions bridge single electron transfer chemistry and radical chemistry.

As depicted in Scheme 1, reductive and oxidative cleavages may follow either a concerted or a stepwise mechanism. How the dynamics of concerted electron transfer/bond breaking reactions (heretofore called dissociative electron transfers) may be modeled, and particularly what the contribution is of bond breaking to the activation barrier, is the first question we will discuss (Section 2). In this area, the most numerous studies have concerned thermal heterogeneous (electrochemical) and homogeneous reactions.

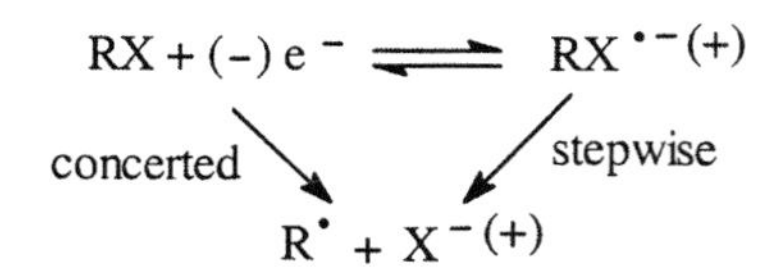

Scheme 1 The same formalism applies to the reduction of a $+n$ charged substrate (with X^{n+} replacing X) and to the oxidation of a $-n$ charged substrate (with X^{n-} replacing X)

The molecular parameters that govern the concerted/stepwise dichotomy will be addressed in Section 3. After reviewing the diagnostic criteria that allow the distinction between the two mechanisms to be made, the molecular factors that control the occurrence of one mechanism or the other will be examined. Particularly revealing is the possibility of passing, in borderline cases, from one mechanism to the other by changing the driving force offered to the reaction. Several illustrative experimental examples in heterogeneous and homogeneous reactions will be described. Competition between the two reaction pathways is discussed after establishing the proper reaction coordinate for each of the three elementary steps involved.

In the case of stepwise processes, the decomposition of the primary radical intermediate (often an ion radical) may be viewed in a number of cases as an intramolecular dissociative electron transfer. A simple extension of the dissociative electron transfer theory gives access to the dynamics of the cleavage of a primary radical into a radical and a charged or neutral leaving group. The theory applies to the reverse reaction, i.e., the coupling of a radical with a nucleophile, the key step of the vast family of nucleophilic substitutions catalyzed by single electron transfer ($S_{RN}1$) reactions. The dynamics of these cleavage and coupling reactions will be discussed in Section 4, including the available evidence for stereoelectronic effects in cases of poor overlap between the orbitals involved in the intramolecular electron transfer process. Although less often investigated, intramolecular dissociative electron transfer may also take place in cleavage of cation radicals. Cation radicals are usually strong acids, much stronger than their parents. They thus form a class of carbon acids where the deprotonation kinetics are accessible as they are in the more traditional case of carbon acids whose acidity is enhanced by the presence of electron-withdrawing groups. As discussed in the same section, their deprotonation is better viewed as an electron transfer concerted with the homolytic cleavage of the carbon–hydrogen bond based on an unexpected application of the intramolecular dissociative electron transfer theory to this class of reactions. As well as intramolecular dissociative electron transfer, homolytic dissociation is another mode by which the primary radical may cleave and this will also be discussed and illustrated by several experimental examples involving reduction and oxidation processes.

Up to this point, the potential energy profile of the fragments resulting from dissociative electron transfer was regarded as purely repulsive. In the gas phase, the fragments often interact within a cluster that corresponds to a shallow minimum. In a number of cases the minimum disappears when the reaction takes place in a polar solvent. However, according to the nature of R, $X^{-(+)}$ and of the solvent, some residual interaction between the two fragments may exist. A model establishing the influence of such an interaction on the dynamics of the electron transfer/bond breaking process will be discussed in Section 5. Several experimental observations that had previously remained unexplained can be rationalized within the framework of this model. It will also be the occasion to refine the definitions of concerted and stepwise processes and of the rules governing the passage from one mechanism to the other.

Although the most numerous investigations of dissociative electron transfer have concerned thermal reactions, photoinduced dissociative electron transfer has also attracted a great deal of recent theoretical and experimental attention. As discussed in Section 6, one of the key questions in the field is whether photoinduced dissociative electron transfers are necessarily endowed with a unity quantum yield as one would predict on purely intuitive grounds. Quantum yield expressions for the concerted and stepwise cases are established and experimental examples are discussed.

Coming back to thermal homogeneous dissociative electron transfer reactions, the question arises whether the electron-donor molecule reacts as a single electron donor or as a nucleophile in an S_N2 reaction. We will review this long-debated question in Section 7, including the most recent developments.

Since the publication of the review on Single Electron Transfer and Nucleophilic Substitution in this same series,[1] reviews or research accounts have appeared concerning several particular points among those addressed here, namely, dynamics of dissociative electron transfer,[2–6] single electron transfer and S_N2 reactions,[2,7–9] and $S_{RN}1$ reactions.[10,11]

2 Dynamics of thermal dissociative electron transfer

THERMODYNAMICS. MICROSCOPIC REVERSIBILITY

The thermodynamics of dissociative electron transfer reactions were first characterized by Hush[12] from thermochemical data in the case of the electrochemical reactions

$$RX + e^- \rightarrow R^\bullet + X^- \quad (X = Cl, Br, I)$$

in water. A similar method was used later to estimate the standard potentials of alkyl halides in non-aqueous solvents.[13,14]

Recently[2] it has been asserted that the very existence of dissociative electron transfer reactions is ruled out by application of the principle of microscopic reversibility. The line of argument was as follows. In the reaction of the cleaving substrate RX, say, with an electron donor D (the same argument could be developed for an oxidative cleavage triggered by an electron acceptor),

$$RX + D \rightleftharpoons R^\bullet + X^- + D^{\bullet +}$$

the reverse reaction must proceed by a termolecular encounter of $R^\bullet$, X^- and $D^{\bullet +}$, in principle a very unlikely event. Thus if D and RX are selected so as to react with a standard free energy of reaction close to nil, the forward reaction could be reasonably fast whereas the reverse reaction would be very slow owing to its termolecular nature. The principle of microscopic reversibility would thus be violated and the only way to avoid this difficulty would be to view the reaction as producing a $R^\bullet \cdots X^-$ complex representing an energy minimum rather than as being truly dissociative. In retrospect, the estimations of the dissociative electron transfer standard potentials quoted above would thus merely reflect the inability to estimate the standard potential for the formation of anion radicals.

In fact, although termolecular collision numbers are certainly much smaller than bimolecular collision numbers , they are sufficient to ensure the reversibility of the reactions. Following Tolman's[15] approach,[16,17] for a reaction

$$\mathrm{AB} + \mathrm{C} \underset{k_{\mathrm{ter}}}{\overset{k_{\mathrm{bi}}}{\rightleftharpoons}} \mathrm{A} + \mathrm{B} + \mathrm{C}$$

involving reactants that are approximated by hard spheres, the forward and reverse rate constants, k_{bi} and k_{ter}, respectively, may be expressed by equations (1) and (2).

$$k_{\mathrm{bi}} = Z_{\mathrm{bi}} \exp\left(-\frac{\Delta E_{+}^{\ddagger}}{RT}\right) \tag{1}$$

with
$$Z_{\mathrm{bi}} = N_{\mathrm{A}}\left(8\pi RT\frac{m_{\mathrm{AB}} + m_{\mathrm{C}}}{m_{\mathrm{AB}} m_{\mathrm{C}}}\right)^{1/2} d^2_{\mathrm{AB}\leftrightarrow\mathrm{C}}$$

$$k_{\mathrm{ter}} = Z_{\mathrm{ter}} \exp\left(-\frac{\Delta E_{-}^{\ddagger}}{RT}\right) \tag{2}$$

with: $$Z_{\mathrm{ter}} = N_{\mathrm{A}}^2 8\pi^2\left(\frac{2RT}{\pi}\right)^{1/2}\left[\left(\frac{m_{\mathrm{A}} + m_{\mathrm{B}}}{m_{\mathrm{A}} m_{\mathrm{B}}}\right)^{1/2} + \left(\frac{m_{\mathrm{B}} + m_{\mathrm{C}}}{m_{\mathrm{B}} m_{\mathrm{C}}}\right)^{1/2}\right] d^2_{\mathrm{A}\leftrightarrow\mathrm{B}} d^2_{\mathrm{B}\leftrightarrow\mathrm{C}}\,\delta$$

where m is the molar masses of the subscript species and d is the distance between the centers of the spheres equivalent to the subscript particles. δ is the distance between the two first spheres when hit by the third; its value is somewhat arbitrary provided it is smaller than the diameter of the spheres. N_A is the Avogadro constant. $\Delta E_{+}^{\ddagger}$ and $\Delta E_{-}^{\ddagger}$ are the activation energies for the forward and reverse reactions respectively, and $\Delta E_{+}^{\ddagger} - \Delta E_{-}^{\ddagger} = \Delta E^0$, ΔE^0 being the standard energy of the reaction. As a typical example, we may take $m_A = m_B = m_C = 100\ \mathrm{g\ mol^{-1}}$, $d_{AB\leftrightarrow C} = 6$ Å, $d_{A\leftrightarrow B} = d_{B\leftrightarrow C} = 4$ Å. If we take δ as small as 0.3 Å, we obtain as the pre-exponential factors $Z_{bi} = 2 \times 10^{11}\ \mathrm{M^{-1}\ L\ s^{-1}}$ and $Z_{ter} = 8 \times 10^{9}\ \mathrm{M^{-2}\ L^2\ s^{-1}}$ for the forward and reverse rate constants, respectively. The pre-exponential factor in the equilibrium constant (equation 3), is equal to 25 corresponding to $T\Delta S^0 = 0.08$ eV.

$$K = \frac{k_{\mathrm{bi}}}{k_{\mathrm{ter}} C^0} = \frac{Z_{\mathrm{bi}}}{Z_{\mathrm{ter}} C^0}\exp\left(-\frac{\Delta E^0}{RT}\right) \quad \text{with} \quad \frac{Z_{\mathrm{bi}}}{Z_{\mathrm{ter}} C^0} = \exp\left(\frac{\Delta S^0}{R}\right) \tag{3}$$

(C^0 = standard state concentration, i.e., 1 mol $\mathrm{L^{-1}}$, ΔS^0 = standard entropy of the reaction).

These values are quite reasonable for the conversion of a molecule AB into two fragments A and B. One may alternatively, more rigorously, and less restrictively (the reactants need not be approximated by hard spheres) analyze the reactive collisions within the framework of transition state theory,[18] leading to the expressions given in equations (4) and (5).

$$k_{\mathrm{bi}} = \frac{1}{C^0}\frac{kT}{h}\frac{\tilde{q}^{\ddagger}}{q^{\mathrm{AB}}q^{\mathrm{C}}}\exp\left(-\frac{\Delta E_{+}^{\ddagger}}{RT}\right) \tag{4}$$

$$k_{\mathrm{ter}} = \frac{1}{C^{0^2}}\frac{kT}{h}\frac{\tilde{q}^{\ddagger}}{q^{\mathrm{A}}q^{\mathrm{B}}q^{\mathrm{C}}}\exp\left(-\frac{\Delta E_{-}^{\ddagger}}{RT}\right) \tag{5}$$

The q terms are the molecular partition functions of the superscript species. For the transition state, ‡, the vibration along which the reaction takes place is omitted in the partition function, $\tilde{q}^{\ddagger}$.

Reversibility falls in line with the fact that the ratio of the expressions of the two rate constants matches the expression of the equilibrium constant, K (equation 6),

$$\frac{k_{\mathrm{bi}}}{k_{\mathrm{ter}}C^0} = \frac{q^{\mathrm{A}}q^{\mathrm{B}}}{q^{\mathrm{AB}}}\exp\left(-\frac{\Delta E^0}{RT}\right) = K \tag{6}$$

The forward and reverse rate constants are thus equal at zero standard free energy. However, this will be difficult to check in practice because both reactions are very slow since a bond breaking/bond forming process endowed with a quite large internal reorganization is involved. The result is that dissociative electron transfer reactions are usually carried out with electron donors that have a standard potential much more negative than the dissociative standard potential. The re-oxidation of the $R^{\bullet}$, X^- system is thus possible only with electron acceptors $D^{\bullet+}$ that are different from the $D^{\bullet+}$ produced in the reduction process (they are more powerful oxidants). There is no requirement then that the oxidation mechanism be the reverse of the reduction mechanism, since the $D/D^{\bullet+}$ couple is not the same in both cases. The same considerations apply for electrochemical reactions as observed, for example, by means of cyclic voltammetry. The mechanistic differences between such reduction and oxidation pathways are not easy to observe experimentally because $R^{\bullet}$ radicals usually undergo fast side-reactions. A notable exception is the reduction of 9-mesityl-fluorene and 9-[α-(9-fluorenylidene)-benzyl]fluorene chlorides leading to stable radicals (Scheme 2). Re-oxidation involves the radicals leading to the

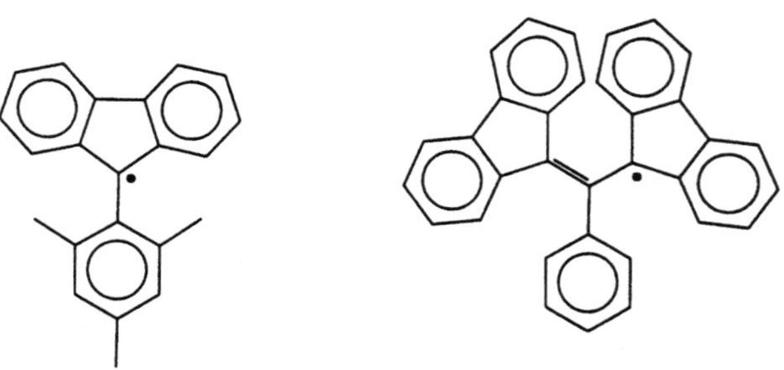

Scheme 2

carbocations, which then combine with chloride ions to regenerate the starting molecules.[19–21] The re-oxidation pathway is thus clearly not the reverse of the forward reaction, with no violation of the principle of microscopic reversibility.

The preceding does not imply that there are no interactions between the fragments formed upon cleavage of the bond. However, the existence of these interactions does not derive from the principle of microscopic reversibility and their magnitude is a function of molecular structure and solvent (see Section 5).

THE MORSE CURVE MODEL

The first attempt to describe the dynamics of dissociative electron transfer started with the derivation from existing thermochemical data of the standard potential for the dissociative electron transfer reaction, $E^0_{RX \rightarrow R^\bullet + X^-}$,[12,14] with application of the Butler–Volmer law for electrochemical reactions[12] and of the Marcus quadratic equation for a series of homogeneous reactions.[13,14] Application of the Marcus–Hush model to dissociative electron transfers had little basis in electron transfer theory (the same is true for applications to proton transfer or S_N2 reactions). Thus, there was no real justification for the application of the Marcus equation and the contribution of bond breaking to the intrinsic barrier was not established.

To overcome these drawbacks, a model for thermal dissociative electron transfers in polar solvents was subsequently proposed,[22] based on a Morse curve approximation of the energy of the cleaving bond in the reactant and on the assumption that the repulsive interaction of the two fragments formed upon electron transfer is the same as the repulsive part of the reactant Morse curve (Fig. 1). Associating this description of bond breaking with a Marcus–Hush modeling of the attending solvent reorganization,[23–27] one obtains the following equations, which summarize the predictions of the model.

$$\Delta G^{\ddagger} = \Delta G_0^{\ddagger}\left(1 + \frac{\Delta G^0}{4\Delta G_0^{\ddagger}}\right)^2 \tag{7}$$

$$\Delta G_0^{\ddagger} = \frac{D + \lambda_0}{4} \tag{8}$$

As with the Marcus–Hush model of outer-sphere electron transfers, the activation free energy, $\Delta G^{\ddagger}$, is a quadratic function of the free energy of the reaction, ΔG^0, as depicted by equation (7), where the intrinsic barrier free energy (equation 8) is the sum of two contributions. One involves the solvent reorganization free energy, λ_0, as in the Marcus–Hush model of outer-sphere electron transfer. The other, which represents the contribution of bond breaking, is one-fourth of the bond dissociation energy (BDE). This approach is

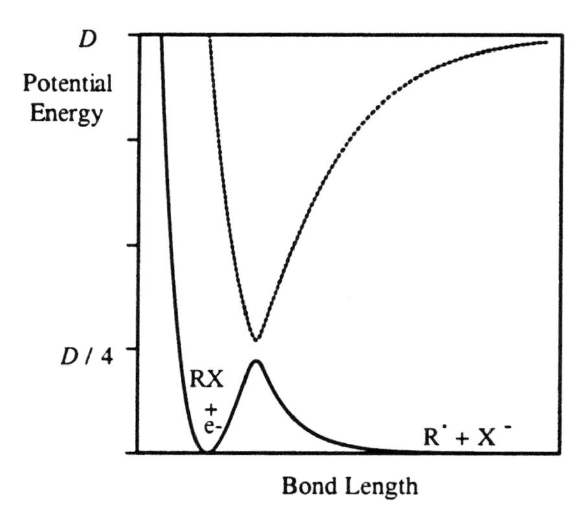

Fig. 1 Morse curve modeling of the contribution of bond-breaking to the dynamics of dissociative electron transfer

intended to apply to adiabatic reactions (see Fig. 1) in which the avoided crossing energy is large enough to ensure adiabaticity but is small as compared to the activation barrier. If necessary, other sources of intramolecular reorganization may be taken into account by including a reorganization energy term, λ_i, in equation (8). *Ab initio* calculations on methyl halides[28] confirmed that the above Morse curve approximation of the diabatic potential energy curves is a reasonable one.

Experimental tests of the theoretical predictions have involved the electrochemical reduction of alkyl and benzyl halides as well as their reduction by homogeneous electron donors.[22,29–31] In the first case, $\pm\Delta G^0 = E - E^0_{RX\to R^\bullet + X^\mp}$ where E is the electrode potential and $E^0_{RX\to R^\bullet + X^\mp}$ is the standard potential of the $RX/R^\bullet + X^\mp$ couple. In the homogeneous case, $\pm\Delta G^0 = E^0_{P/Q} - E^0_{RX\to R^\bullet + X^\mp}$, where $E^0_{P/Q}$ is the standard potential of the outer-sphere electron donor or acceptor couple P/Q, and + stands for a reduction and − for an oxidation.

In the electrochemical case, the predicted and experimental values of the intrinsic barrier for the reduction of butyl and benzyl halides on a glassy carbon electrode were found to agree satisfactorily.[22,31]

Equations (7) and (8) may also be used to obtain a rough estimate of the BDE from cyclic voltammetric peak potentials. The standard potential of the $RX/R^\bullet + X^-$ couple may be expressed, for a reduction, by equation (9).

$$E^0_{RX\to R^\bullet + X^-} = -D + T\Delta S_{RX\to R^\bullet + X^\bullet} + E^0_{X^\bullet/X^-} \tag{9}$$

At the peak potential, the activation free energy may be expressed by equation (10).

$$\Delta G_{\mathrm{p}}^{\ddagger} = \frac{RT}{F}\left\{\ln\left[A\left(\frac{RT}{\alpha_{\mathrm{p}}F\nu D}\right)\right] - 0.78\right\} \tag{10}$$

(A is the pre-exponential factor, D the RX diffusion coefficient, ν the scan rate and α_{p} the transfer coefficient at the peak). Application of a linearized version of equation (7) to the peak potential leads to equation (11).

$$\Delta G_{\mathrm{p}}^{\ddagger} \approx \frac{D + \lambda_0}{4} + \frac{E_{\mathrm{p}} - E^0_{\mathrm{RX}\rightarrow\mathrm{R}^\bullet+\mathrm{X}^-}}{2} \tag{11}$$

In series where $\Delta S_{\mathrm{RX}\rightarrow\mathrm{R}^\bullet+\mathrm{X}^\bullet}$ (formation of two particles out of one), A, D, α_{p}, and λ_0 do not vary much from one compound to another, and equation (12) applies approximately, at a given scan rate and a given temperature.

$$D \approx -\frac{2}{3}(E_{\mathrm{p}} - E^0_{\mathrm{X}^\bullet/\mathrm{X}^-}) + \text{constant} \tag{12}$$

The validity of equation (12) has been checked for several families of alkyl halides for which D and $E^0_{\mathrm{X}^\bullet/\mathrm{X}^-}$ are known (Ref. 32, see particularly figure 6 therein). It was thus found that for $\nu = 0.1\ \mathrm{V\ s^{-1}}$, the constant is equal to 0.3 eV at 20°C (expressing D in eV and the potentials in V). Equation (12) was then applied to the approximate determination of unknown BDEs in several series of compounds undergoing dissociative electron transfer, namely, *N*-halosultams,[32] sulfonium cations,[33] vicinal dihalides,[34] 1,3-dihaloadamantes, 1,4-dihalobicyclo[2.2.2]octanes, and 1,3-dihalobicyclo[1.1.1]pentanes.[35] In the latter case, the mutual influence of the two halogens could be rationalized thanks to the conversion of the peak potential data to bond dissociation energies.

Application of the theory to homogenous dissociative electron transfer has mostly involved the reaction of alkyl halides with outer-sphere electron donors such as aromatic anion radicals,[22,29–31] and, more recently, the reduction of organic peroxides by the same type of electron donors.[36,37] The rate data were gathered by means of cyclic voltammetry, generating the homogeneous electron donor electrochemically from its oxidized form according to a "redox catalysis" approach.[38] The experimental data are conveniently represented as $\log k$ (bimolecular rate constant) versus $E^0_{\mathrm{P/Q}}$ plots that may be analyzed according to equations (7) and (8), taking into account that $\Delta G^0 = E^0_{\mathrm{P/Q}} - E^0$. It is remarkable that among the (*n*-, *s*-, *t*-) butyl halides (I, Br, Cl) and benzyl chlorides, there is a good agreement between experimental data and theoretical predictions with the tertiary derivatives, while the experimental rate constants are larger than predicted with the secondary derivatives and even more with the primary derivatives (see the analysis in Ref. 31 based on the experimental data in Refs 39 and 40). Similar effects, related to steric hindrance at the reacting carbon centers, have been observed when using low-valent iron porphyrins as electron donors as well as the parallel effect of steric congestion in the iron porphyrin donor.[41] It is also remarkable that the entropy of activation of the reaction of aromatic radicals with alkyl halides increases from *n*- to *s*- and to *t*-derivatives.[41–43] These results suggest a mechanism involving a competition between a single electron transfer pathway and a S_N2 pathway. This problem will be analyzed in detail in Section 7.

VARIATION OF THE SYMMETRY FACTOR WITH THE DRIVING FORCE

The theory also predicts that the transfer coefficient, α, should vary with the electrode potential as depicted by equation (13).

$$\alpha = \frac{\partial \Delta G^{\ddagger}}{\partial \Delta G^{0}} = \frac{1}{2}\left(1 \pm \frac{E - E^{0}_{RX \to R^{\bullet} + X^{\mp}}}{4\,\Delta G_{0}^{\ddagger}}\right) \tag{13}$$

The above-mentioned experimental studies revealed that α is significantly smaller than 0.5 and that it is smaller with bromides than with iodides in line with the theoretical predictions. Indeed, at the cyclic voltammetric reduction peak, the standard free energy of the reaction, $E - E^0$, has to be negative enough to overcome (equation 7) the large intrinsic barrier caused by the contribution of the BDE (equation 8). A value of α significantly smaller than 0.5 ensues (equation 13). Since the intrinsic barrier is larger for bromides than for iodides, this effect is larger in the first case than in the second.

A much more precise investigation of the variation of α with the electrode potential has been reported recently.[44] It concerns the electrochemical reductive cleavage of the O—O bond in a series of five organic peroxides:

$$\text{R—O—O—R}' + e^- \rightarrow \text{R—O}^{\bullet} + \text{R}'\text{—O}^-$$

and provides a convincing demonstration of the validity of equations (7) and (13). The possibility of employing mercury as electrode material allowed particularly good reproducibility of the data and the precision was improved by using convolution potential sweep voltammetry[45] instead of plain cyclic voltammetry, a technique successfully applied in the past to detect the variation of α with the electrode potential in the case of outer-sphere electron transfer reactions (see Ref. 46 and references therein).

The variation of α with potential according to equation (13) may also be used to estimate the value of E^0, from which the value of D may be inferred. This approach has been applied to organic peroxides,[44] including endoperoxides of biological interest.[37] Here again, convolution voltammetry proved to be more precise than plain cyclic voltammetry.

The variations of the symmetry factor, α, with the driving force are much more difficult to detect in $\log k$ versus driving force plots derived from homogeneous experiments than in electrochemical experiments because of a lesser precision on the rate and driving force data, because the self-exchange rate constant of the donor couple may vary from one donor to the other and also because the cross-exchange relationship is the result of rather crude approximations.[31] It has been claimed[7,47,48] that the plots are linear instead of quadratic as predicted by the dissociative electron transfer model. However, careful examination of the data pertaining to *t*-BuI and *t*-BuBr does not demonstrate that the slightly curved theoretical plots expected with these two compounds would not fit the experimental data as well as a straight line taking into account the uncertainties indicated above. The parabolic behavior is more apparent in the reaction depicted in Scheme 3[37] because the BDE is small and the variations of α with the driving force are expected to increase when the BDE decreases (see equations 8 and 13).

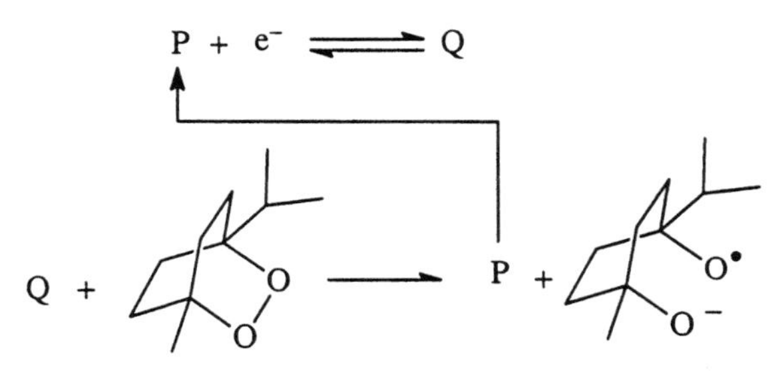

Scheme 3 In DMF. The Ps are aromatic hydrocarbons and the Qs their anion radicals

ENTROPY OF ACTIVATION

The dissociative electron transfer model has been improved since then by a more accurate estimation of the activation entropy, taking into account that $R^{\bullet}$ and X^{-} are formed within a solvent cage from which they successively diffuse out.[49] The free energy and entropy of activation are thus obtained from equations (14) and (15)

$$\Delta G^{\ddagger} = \frac{D + \lambda_0}{4}\left(1 + \frac{\Delta G^0_{\mathrm{C}}}{D + \lambda_0}\right)^2 \tag{14}$$

$$\Delta S^{\ddagger} = \alpha\,\Delta S^0_{\mathrm{C}} = \frac{1}{2}\left(1 + \frac{\Delta G^0_{\mathrm{C}}}{D + \lambda_0}\right)\Delta S^0_{\mathrm{C}} \tag{15}$$

respectively. ΔS^0_{C} is the standard entropy for the formation of the caged products. ΔG^0_{C}, the standard free energy for the formation of the caged products is assumed to differ from the standard free energy for the formation of the separated products only by the entropic term. $\Delta S_C^0 = \Delta S_{F,C}^0 + \Delta S_S^0$ is made of two contributions. ΔS_S^0 corresponds to solvation and is assumed to be the same for the caged and separated product systems. $\Delta S_{F,C}^0$ corresponds to cleavage in the cage. It is a fraction of the cleavage standard entropy for the formation of the separated products, ΔS_F^0. The validity of equations (14) and (15) has been tested for the electrochemical reduction of *t*-BuBr in *N*,*N'*-dimethylformamide (DMF) and for the reaction of the same compound with anthracene anion radical in the same solvent.[49] The results are shown in Fig. 2. In the electrochemical case, the predicted values of the cyclic voltammetric peak potential (at 0.2 V s^{-1}) and the entropy of activation are plotted as functions of the ratio $\Delta S_{F,C}^0/\Delta S_F^0$. That the theory is a correct representation of reality derives from the observation that the agreement between theoretical and experimental values is reached for the same value of $\Delta S_{F,C}^0/\Delta S_F^0$ for the peak potential and the entropy of activation. The same is true for the homogeneous reactions. That this common value of $\Delta S_{F,C}^0/\Delta S_F^0$ is smaller in the latter case than in the former falls in line with the fact that the presence of anthracene renders more difficult the mutual displacement of the R and X moieties within the solvent cage.

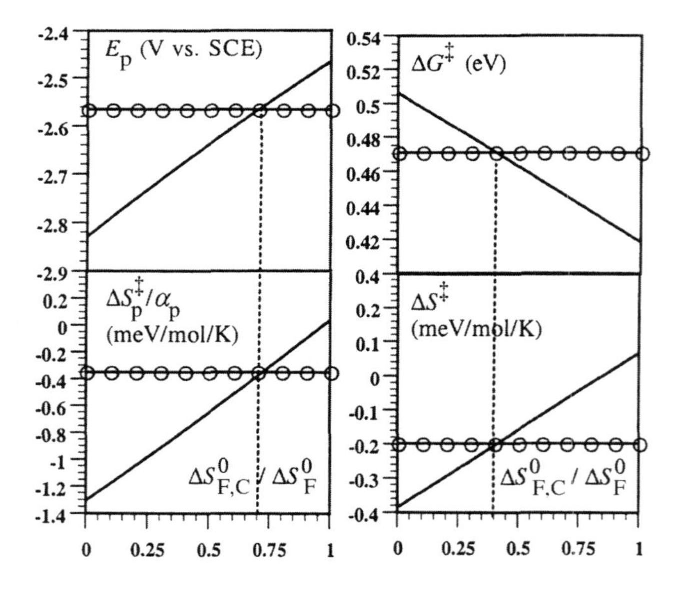

Fig. 2 Reduction of *t*-BuBr in DMF. Left: electrochemical reduction. Right: reduction by anthracene anion radical. ○: experimental data, oblique lines: theoretical predictions.

ADIABATIC AND NON-ADIABATIC REACTIONS

More recent theoretical studies have provided a Hamiltonian formalism for adiabatic dissociative electron transfer at metal electrodes.[169] Large scale numerical simulations of bond breaking and solvent reorganization[200] have confirmed the difficulty of obtaining a precise estimate of the contribution of the latter factor to the intrinsic barrier.[49] The expressions applicable for a non-adiabatic dissociative electron transfer and for the examination of possible quantum effects have been derived.[50] In this work and in related studies,[51–53] the product repulsive potential energy profile, rather than being equated to the repulsive part of the reactant Morse profile, is derived from semi-empirical quantum-chemical calculation (PM3) implemented with a dielectric continuum estimation of the solvation energies. As seen in Section 4 with the example of CCl_4, the drawback of such an approach is the lack of reliability of the semi-empirical quantum-chemical method employed, leading to a result that is worse than the purely empirical Morse curve approach. It is also worth noting that the study concerning *t*-butyl halides[52] is misleading in the sense that the analysis is based on the value of the transfer coefficient, which measure is quite imprecise.

Expressions for the transition between adiabatic and non-adiabatic dissociative electron transfers have also recently been derived.[54]

From an experimental viewpoint, there are strong indications that the reduction of di-tert-butyl peroxide proceeds with a significant degree of non-adiabaticity.[201]

Note also that additional information about the degree of adiabaticity can be derived from the investigation of photoinduced dissociative electron transfers, as discussed in Section 6.

3 Concerted and stepwise reactions. Transition between the two mechanisms

The potential energy curves corresponding to the two mechanisms are sketched in Fig. 3 for the case where the interaction between caged fragments is negligible, i.e., the product potential energy profile is purely repulsive. Figure 3 represents the case where the primary radical is an anion radical deriving from a neutral starting molecule by a reduction process and decomposing into a neutral radical and an anionic leaving group. The same reasoning applies to a neutral primary radical deriving from a positively charged starting molecule by a reduction process and decomposing into a neutral radical and a neutral leaving group. It also applies to the case where the primary radical is a cation radical deriving from a neutral starting molecule by an oxidation process and decomposing into a neutral radical and a cationic leaving group. The same is true for a neutral primary radical deriving from a negatively charged starting molecule by an oxidation process and decomposing into a neutral radical and a neutral leaving group.

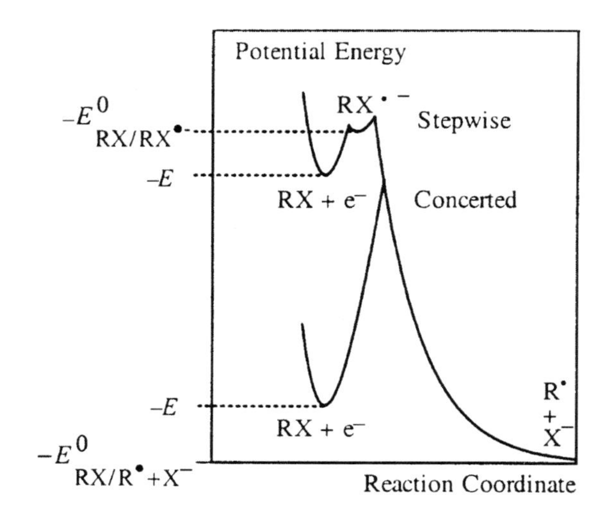

Fig. 3 Potential energy profiles for the concerted and the stepwise mechanism in the case of a thermal reductive process. *E* is the electrode potential for an electrochemical reaction and the standard potential of the electron donor for a homogeneous reaction. For an oxidative process, change – into + and donor into acceptor.

In the stepwise process, the cleavage of the intermediate primary radical is a fast reaction, endowed with a large driving force. Cases where the cleavage of the primary radical is an uphill reaction will be discussed in Sections 4 and 5.

DIAGNOSTIC CRITERIA

How do we know, in practice, whether one or the other mechanism is followed? For electrochemical reactions, the examination of the cyclic voltammetric response is a convenient means to distinguish one mechanism from the other. The stepwise mechanism is the easiest to identify. A simple case is when the cyclic voltammogram becomes reversible upon raising the scan rate. Then, for a reduction, the presence of a reverse anodic trace is the signature of the RX anion radical (and vice versa for an oxidation). Using ultramicroelectrodes, scan rates as high as a few million volts per second may be reached under favorable conditions, corresponding to lifetimes in the sub-microsecond range for the ion radical intermediate.[55] Even if reversibility cannot be reached, either because the splitting of the ion radical is too fast or because such high scan rates cannot be achieved, the characteristics of the irreversible wave may serve to identify the stepwise mechanism. If the cleavage step is not too fast and/or the initial electron transfer step is fast enough, the kinetics of the overall reaction is under the kinetic control of the cleavage step, the initial electron transfer acting as a pre-equilibrium. Under such conditions, the variation of the peak potential with the scan rate and the peak width are given by equations (16).[38] Equations (16) and the following analysis apply for reductions. Transposition to oxidations, with the appropriate changes of sign, is straightforward.

$$\frac{\partial E_p}{\partial \log v} = -\frac{RT}{2F}\ln 10(-29.6 \text{ mV at } 25^\circ\text{C})$$
$$E_{p/2} - E_p = 1.85\frac{RT}{F}(47.5 \text{ mV at } 25^\circ\text{C}) \quad (16)$$

E_p and $E_{p/2}$ are the peak and half-peak potential respectively. v is the scan rate. If the cleavage step is very fast and/or the initial electron transfer step is slow enough, the kinetics of the overall reaction is under the kinetic control of the electron transfer step. The variation of the peak potential with the scan rate and the peak width are given by equations (17).[38,56]

$$\frac{\partial E_p}{\partial \log v} = -\frac{RT}{2\alpha_p F}\ln 10\left(-\frac{29.6}{\alpha_p} \text{ mV at } 25^\circ\text{C}\right)$$
$$E_{p/2} - E_p = 1.85\frac{RT}{\alpha_p F}\left(\frac{47.5}{\alpha_p} \text{ mV at } 25^\circ\text{C}\right) \quad (17)$$

where α_p is the value of the transfer coefficient at the peak. Assuming that the initial electron transfer obeys a Marcus–Hush quadratic kinetic law, the transfer coefficient is predicted to vary linearly with the electrode potential (equations (13) and (7) are still applicable, with the standard potential $E^0_{RX/RX^{\bullet-}}$ replacing the dissociative standard potential $E^0_{RX/R^{\bullet}+X^-}$ and a different expression of the intrinsic barrier, replacing in equation (8) the bond dissociation energy by the intramolecular reorganization energy). However, the cyclic voltammetric waves are narrow enough for the quadratic law to be linearized within the potential range where the wave appears.[30] This is the reason equations (17) are the same as with a Butler–Volmer linear kinetic law, taking for α its value at the peak, α_p. Owing to the effect of the follow-up reaction, the peak potential stands in the vicinity of the standard potential $E^0_{RX/RX^{\bullet-}}$, unless the initial electron transfer is very slow.[56] It follows that α_p is close to 0.5. Thus, passing from a situation where the follow-up reaction is the rate-determining step to a situation where the initial electron transfer is the rate-determining step results, at 25°C, in an increase of $-\partial E_p/\partial \log v$ from 29.6 to 59.2 mV and an increase of $E_{p/2} - E_p$ from 47.5 to 95 mV. Experimental data in these ranges of values are thus an indication that the reaction follows a stepwise mechanism.

Table 1 Molecular factors governing the dichotomy between concerted and stepwise mechanisms

Stepwise	Concerted	Reference
Examples of the prevailing role of $E^0_{RX/RX^{\bullet-}}$		
All aryl halides	All aliphatic halides	1
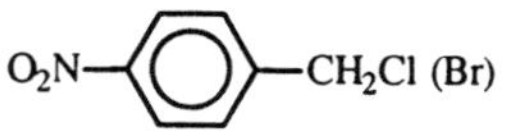 $O_2N-C_6H_4-CH_2Cl$ (Br)	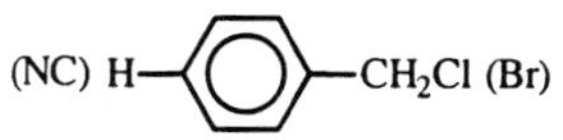 (NC) $H-C_6H_4-CH_2Cl$ (Br)	30
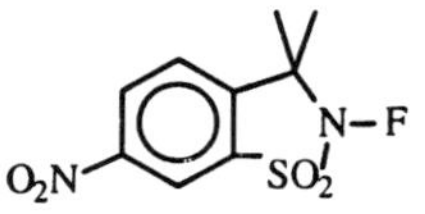 O_2N-substituted 3,3-dimethylbenzisothiazole N–F (SO_2)	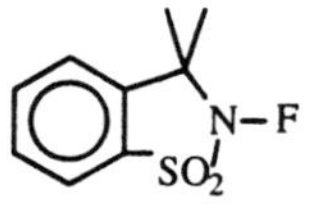 3,3-dimethylbenzisothiazole N–F (SO_2)	32
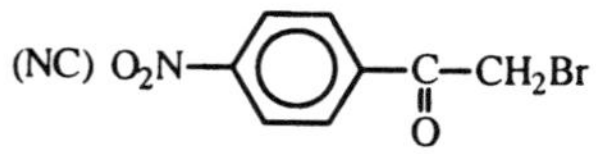 (NC) $O_2N-C_6H_4-C(=O)-CH_2Br$	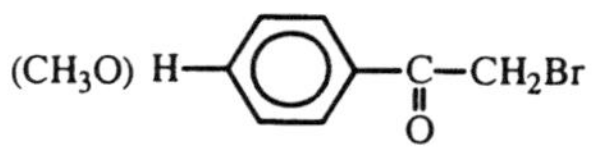 (CH_3O) $H-C_6H_4-C(=O)-CH_2Br$	159
Examples of the prevailing role of the bond dissociation energy (*D*)		
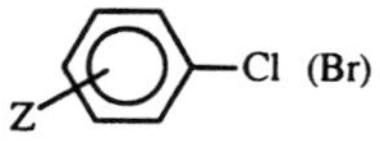 $Z-C_6H_4-Cl$ (Br)	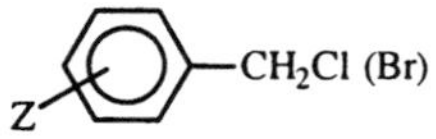 $Z-C_6H_4-CH_2Cl$ (Br) (except $Z = NO_2$)	1, 30
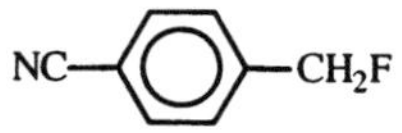 $NC-C_6H_4-CH_2F$	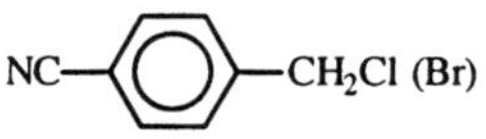 $NC-C_6H_4-CH_2Cl$ (Br)	31, 82
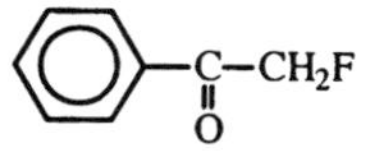 $C_6H_5-C(=O)-CH_2F$	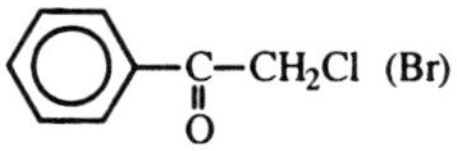 $C_6H_5-C(=O)-CH_2Cl$ (Br)	159
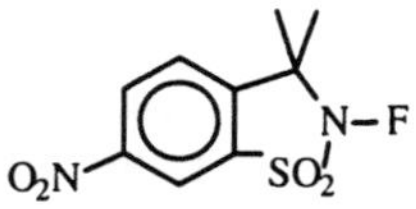 O_2N-substituted 3,3-dimethylbenzisothiazole N–F (SO_2)	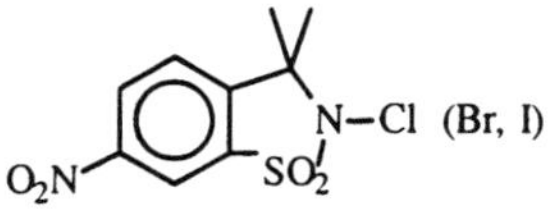 O_2N-substituted 3,3-dimethylbenzisothiazole N–Cl (Br, I) (SO_2)	32
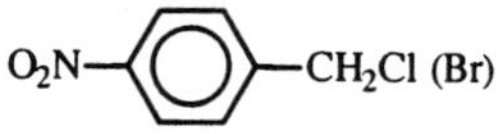 $O_2N-C_6H_4-CH_2Cl$ (Br)	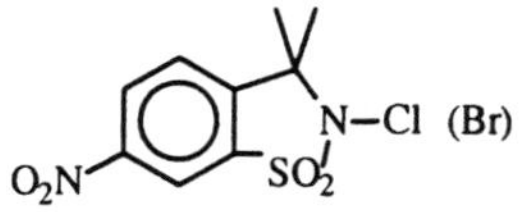 O_2N-substituted 3,3-dimethylbenzisothiazole N–Cl (Br) (SO_2)	30,32
Examples of the prevailing role of $E^0_{X^{\bullet}/X^-}$		
	$C_6H_5-C(=O)-CH_2X + e^- \longrightarrow C_6H_5-C(=O)-CH_2^{\bullet} + X^-$	
X = Br, Cl	X = OPh, OCH_3, OC_2H_5, SPh, SC_2H_5, $N(C_2H_5)_2$	159

When the reaction follows a concerted mechanism, an irreversible cyclic voltammogram is observed whatever the scan rate. Equations (17) are applicable in this case too. However, the reaction is endowed with a large intrinsic barrier resulting from the contribution of bond breaking (equation 8). Because a large driving force is required to overcome this large intrinsic barrier, the cyclic voltammetric peak appears at a potential much more negative than the standard potential $E^0_{RX/R^\bullet+X^-}$. It follows that small values of α_p much below 0.5 are indicative of a concerted mechanism. Some ambiguity may arise if such a situation is to be compared with a stepwise mechanism in which the initial electron transfer step is also endowed with a large intrinsic barrier albeit not involving any bond breaking. In such cases it is generally possible, by examining the behavior of structurally similar compounds, to obtain estimates of the standard potential of $E^0_{RX/RX^{\bullet-}}$ and of the corresponding intrinsic barrier, and then to see whether or not such values are consistent with the experimental results. This strategy has been applied in several of the experimental examples discussed below.

HOW MOLECULAR STRUCTURE CONTROLS THE MECHANISM

Both thermodynamic and kinetic factors are involved in the competition between concerted and stepwise mechanisms. The passage from the stepwise to the concerted situation is expected to arise when the ion radical cleavage becomes faster and faster. Under these conditions, the rate-determining step of the stepwise process tends to become the initial electron transfer. Then thermodynamics will favor one or the other mechanism according to equation (18). ΔG^0_{cleav} is also the standard free energy of cleavage of the ion radical.

$$\Delta G^0_{cleav} = -E^0_{RX/R^\bullet+X^-} + E^0_{RX/RX^{\bullet-}} \underset{\text{concerted}}{\overset{\text{stepwise}}{\gtrless}} D - E^0_{X^\bullet/X^-} + E^0_{RX/RX^{\bullet-}} - T\Delta S_{RX\to R^\bullet+X^\bullet} \qquad (18)$$

Thus, one passes from the stepwise to the concerted mechanism as the driving force for cleaving the ion radical becomes larger and larger. It may thus be predicted that a weak R–X bond, a negative value of $E^0_{RX/RX^{\bullet-}}$ and a positive value of $E^0_{X^\bullet/X^-}$ will favor the concerted mechanism and vice versa. All three factors may vary from one RX molecule to another. However, there are families of compounds where the passage from one mechanism to the other is mainly driven by one of them. Illustrating examples are given in Table 1.

The electrochemical reduction of sulfonium cations in acetonitrile according to Scheme 4[33] offers a striking example of the combined roles of the bond

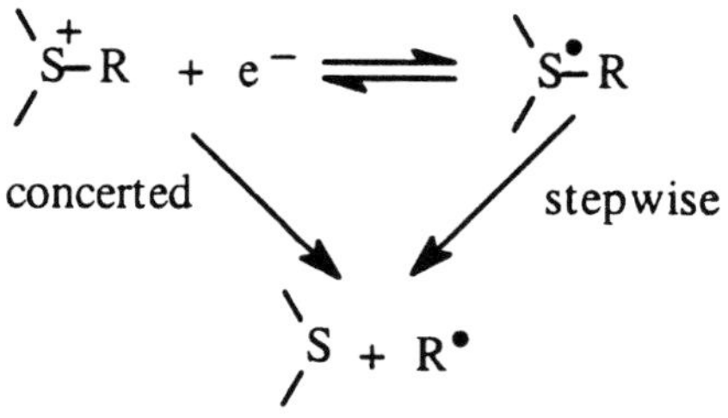

$$>\overset{+}{S}-R + e^- \rightleftharpoons >\overset{\bullet}{S}-R$$

concerted stepwise

$$>S + R^\bullet$$

Scheme 4

dissociation energy and of $E^0_{RX/RX^{\bullet -}}$, the latter parameter being a measure of the energy of the π^* orbital in which the incoming electron may be accommodated (see Scheme 5).

TRANSITION BETWEEN MECHANISMS UPON CHANGING THE DRIVING FORCE

Electrochemical examples

The same family of compounds shows another interesting feature, namely, the existence of borderline cases exhibiting an intermediate behavior between the concerted and stepwise mechanisms. More precisely, the width of the cyclic voltammetric peak and the variation of its location with scan rate change from a concerted to a stepwise behavior as the scan rate is raised (Fig. 4 and Scheme 6).

It may seem strange at first sight that a change in a parameter used to investigate the mechanism, such as the scan rate, could induce a change in the mechanism. In fact, an increase in scan rate shifts the cyclic voltammetric peak toward negative values, thus offering more driving force to the reaction, which consequently permits the possibility that the mechanism changes from concerted to stepwise as sketched in Fig. 3.

Another means of changing the driving force is to change the temperature. As seen from equation (10), decreasing the temperature decreases the activation free energy at the peak potential. More driving force is thus required to reach the peak potential, the decrease of temperature thus having a similar effect to raising the scan rate. This is the reason that, in the reduction of the sulfonium cation $[Ph(CH_3)SCH_2Ph]^+$ in acetonitrile, the passage between concerted and stepwise mechanisms observed at 20°C from the variation of the peak width with the scan rate disappears at 0°C where the stepwise mechanism is observed over the whole range of scan rates.[57]

Another example of passage from one mechanism to the other upon raising the scan rate has been reported, although the diagnosis was not as clear-cut as with the sulfonium reduction depicted above. It concerns the reduction of 1,2-

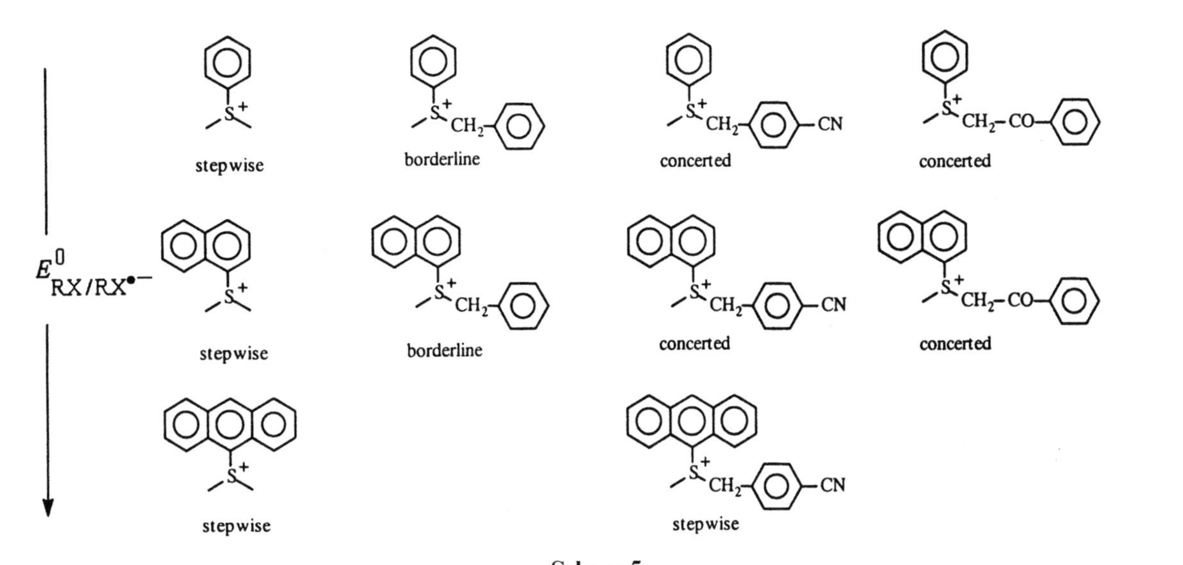
D
E0 RX/RX•−
stepwise
borderline
concerted
concerted
stepwise
borderline
concerted
concerted
stepwise
stepwise

Scheme 5

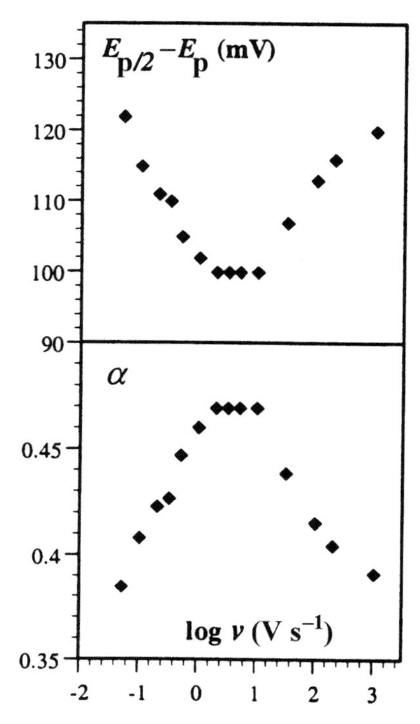

Fig. 4 Electrochemical reduction of a sulfonium cation (Scheme 6) showing the transition from the concerted to the stepwise mechanism as driving force increases upon raising the scan rate.[33] The apparent transfer coefficient, α, is derived from the peak width according to equation (17).

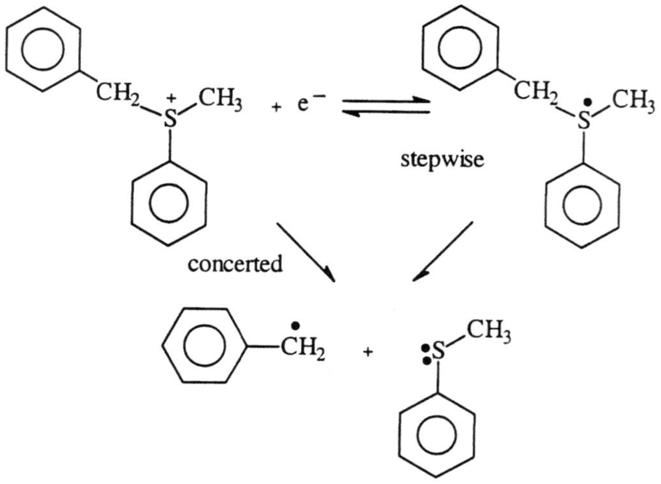

stepwise

concerted

Scheme 6

dibromo-3-(4-cyano)phenyl propane and 1,2-dibromo-3-(4-carboethoxy)-phenyl propane in acetonitrile.[58] Another, and much clearer, example of the mechanism transition has been reported recently.[59] It deals with the reduction of an organic peroxide in DMF as sketched in Scheme 7. The use of convolution voltammetry in addition to conventional cyclic voltammetry allowed a better accuracy of mechanism assignment. More examples of such transitions in the same class of compounds have very recently been analyzed with the same technique.[60]

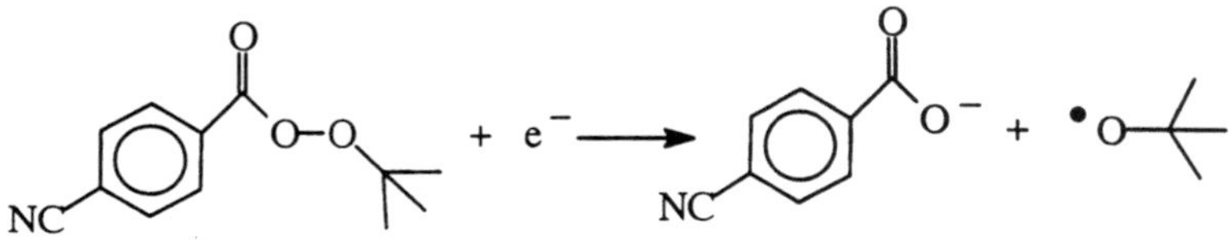

Scheme 7

Another striking example, recently discovered, is the reduction of iodobenzene in DMF.[61] The variation of the apparent transfer coefficient with the scan rate indicates that the mechanism passes from concerted to stepwise as the driving force increases (Fig. 5). As expected, the zone where the concerted mechanism prevails enlarges as one raises the temperature. In contrast, bromobenzene and 1-iodonaphthalene exhibit the characteristics of a stepwise mechanism over the whole range of scan rate.

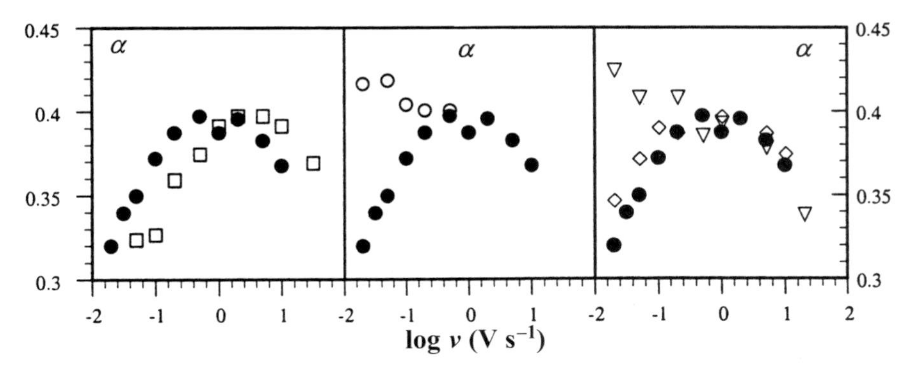

Fig. 5 Electrochemical reduction of aryl halides showing the variation of the apparent transfer coefficient with the scan rate. ●: iodobenzene, ○: bromobenzene, ▽: 1-iodonaphthalene, ◇: 4-methyliodobenzene, at 298 K, □: iodobenzene at 329 K.

Homogeneous examples

The variation in driving force that can be induced by a change in scan rate cannot be larger than a few hundred meV. The observation of the mechanism transition thus requires that the molecular properties of the investigated system make it a borderline case that is easy to push from one situation to the other with a relatively modest change of the driving force. Such mechanism transitions are easier to observe in electrochemical reactions than in homogeneous reactions, thanks to the fine tuning of the driving force offered by the variation of the scan rate and of the potential variation along the wave (which is exploited when using convolution voltammetry). Nevertheless, the first example of a passage between a concerted and a stepwise mechanism was suggested in a homogeneous case, namely the reduction of triphenylmethyl phenyl sulfide by a series of aromatic anion radicals.[62] Note, however, that the change of slope in the $\log k$ versus driving force plot observed with triphenylmethyl phenyl sulfide has recently been reported not to appear with other compounds of the same family.[63]

A clear-cut example of the transition between a stepwise and a concerted mechanism in homogenous reactions has recently been found in the area of $S_{RN}1$ chemistry.[64] $S_{RN}1$ substitutions (Scheme 8) form a well documented class of reactions.[10,11,65–70] Unlike in conventional nucleophilic substitutions, the nucleophile, Nu^-, does not react with the substrate, RX, but with the radical $R^\bullet$ resulting from the cleavage of the R—X bond (X is very often a halogen atom, but other leaving groups have been used). The key step in the $S_{RN}1$ reaction is the coupling of the radical $R^\bullet$ with the nucleophile leading to the anion radical of the substitution product, $RNu^{\bullet-}$. The chain process (Scheme 8) is completed by a reaction in which one electron is transferred from the product anion radical to the substrate, leading to the substitution product and closing the propagation loop.

In most cases, the reaction requires external stimulation in which a catalytic amount of electrons is injected into the solution. Solvated electrons in liquid ammonia,[66] sodium amalgam in the same solvent[71] light,[66,68] electrodes[11,67] and

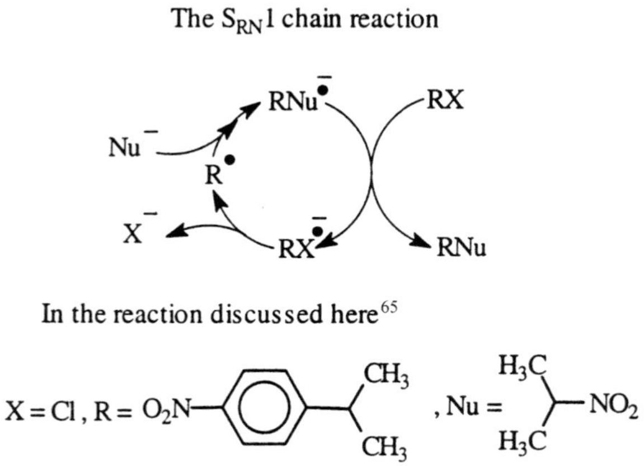

The $S_{RN}1$ chain reaction

In the reaction discussed here[65]

Scheme 8

ferrous salts[72] have been used for this purpose. However, there are several examples where the reaction works without external stimulation. They mostly concern Kornblum reactions,[65] in which the substrates are benzyl or cumyl derivatives activated by one or more electron-withdrawing substituents on the phenyl ring (generally nitro groups). In these so-called "thermal reactions", the only species that might provide stimulating electrons to the system is the nucleophile itself. However, the nucleophile seems such a poor electron donor that electron transfer from the nucleophile to the substrate is expected to be extremely slow, apparently too slow to serve as a viable initiation step of the chain process. Careful determination of the rate constant of each propagation step and of the various termination steps (dimerization of the radicals and reduction by the anion radicals) followed by the simulation of the overall kinetics showed that an outer-sphere electron transfer from the 2-nitropropanate ion to 4-nitrocumyl chloride (Scheme 8) is indeed far too slow (because of a largely unfavorable driving force of the order of 1.2 eV) to initiate the chain reaction so as to match the experimental kinetics. However, a viable initiation route is provided by an electron transfer from the nucleophile to the substrate concerted with the cleavage of the C—Cl bond as shown by application of the dissociative electron transfer theory. As summarized in Fig. 6, we thus observe the passage from a stepwise mechanism to a concerted mechanism upon decreasing the driving force by 0.55 eV. This reaction thus provides an unambiguous example of how the driving force controls the passage from one mechanism to the other as it does in electrochemical reactions. The two rate

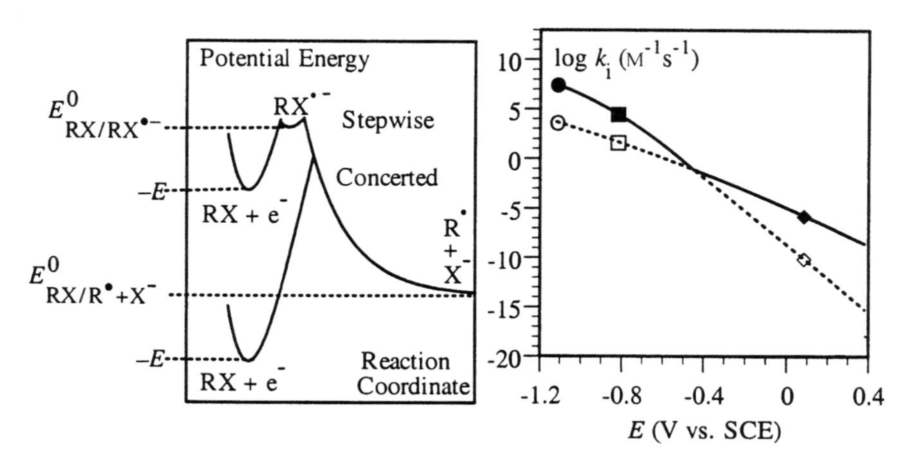

Fig. 6 Passage from the stepwise to the concerted mechanism upon decreasing the driving force. Left: potential energy profiles. Right: reaction of 4-nitrocumyl chloride with homogeneous donors; diamonds: 2-nitropropanate ion, squares: duroquinone anion radical, circles: $RNu^{\bullet-}$. E: electrode potential or standard potential of a homogeneous donor.

constants indeed differ by as much as 5 orders of magnitude owing to the very unfavorable driving forces. The fact that the reaction rate can still be measured at such low driving forces results from the kinetic amplification caused by the chain process (the half-reaction time is less than one hour, whereas the initiation step alone has a half-reaction time of more than 5 days).

We may conclude this section by noting that the observation of the transition between the two mechanisms confirms the existence of two different mechanisms in cases where the distinction between the two is not easy according to the criteria recalled earlier.

DOES CONCERTED MECHANISM MEAN THAT THE INTERMEDIATE "DOES NOT EXIST"?

The occurrence of a concerted mechanism may result from a transition state free energy advantage over the stepwise mechanism or from the fact that the intermediate "does not exist", i.e., its lifetime is shorter than one vibration.[73] Even if examples exist of the latter alternative,[74,75] there is no reason to view it as an absolute requirement for the mechanism to be concerted, as is sometimes claimed.[2] What is true is that it is a sufficient condition for the concerted mechanism to occur. However, it is not a necessary condition. In the examples discussed above, the intermediate "exists". It transpires along the reaction pathway at high driving forces. Nevertheless, a concerted mechanism is followed at low driving forces because it is then energetically more advantageous. Such competition between stepwise and concerted mechanisms is usually not easy to characterize. This is why the above examples have a general bearing in the theory of chemical reactivity.

Note, however, that the concerted/stepwise dichotomy discussed here concerns cases in which the intermediate is unstable toward cleavage that occurs by means of intramolecular dissociative electron transfer. As discussed in the foregoing sections, for primary radicals that cleave homolytically, criteria based on the life-time of the intermediate may be pertinent.

REACTION COORDINATES IN CONCERTED AND STEPWISE REACTIONS

In the potential energy diagrams used so far to visualize the transition between concerted and stepwise pathways (Figs 3 and 6), the potential energy is plotted against a "reaction coordinate" which has not been clearly defined. It is important to emphasize that the reaction coordinate is not the same for the three reactions represented. The cleaving bond length is a common ingredient for the three reactions, but solvation reorganization requires a specific coordinate for each of the reactions.[76]

In the electrochemical case, starting with a molecule R—X, the dissociative electron transfer reaction involves the solvent reorganization corresponding to the charging of the X portion of the molecule. For the formation of the ion radical, solvent reorganization corresponds to the charging of the R portion of the molecule. For the cleavage of the ion radical, solvent reorganization corresponds to the transfer of the charge between two locations in the molecule, namely the R portion and the X portion.

Describing the dissociative electron transfer step, $RX \pm e^- \Leftrightarrow R^\bullet + X^\mp$, involves determining the saddle point on the intersection of the two following free energy surfaces.[22,31]

$$G_{RX \pm e^-} = G^0_{RX \pm e^-} + \lambda_{0,1} X_1^2 + D Y_1^2 \tag{19}$$

$$G_{R^\bullet + X^\mp} = G^0_{R^\bullet + X^\mp} + \lambda_{0,1}(1 - X_1)^2 + D(1 - Y_1)^2 \tag{20}$$

where X_1 is a fictitious charge borne by the X portion of the molecule serving as solvation index for solvent reorganization around X and $\lambda_{0,1}$ is the corresponding solvent reorganization energy.

$$Y_1 = 1 - \exp[-\beta(y - y_{RX})] \tag{21}$$

with $\beta = v\,(2\pi^2\mu/D)^{1/2}$ (y = bond length, y_{RX} = equilibrium value of y in RX, v = frequency of the cleaving bond, μ = reduced mass) is a variable representing the stretching of the cleaving bond. The coordinates of the saddle point are

$$X_1^\ddagger = Y_1^\ddagger = \frac{1}{2}\left(1 + \frac{G^0_{R^\bullet + X^\mp} - G^0_{RX \pm e^-}}{\lambda_{0,1} + D}\right) \tag{22}$$

$$\Delta G_1^\ddagger = \frac{\lambda_{0,1} + D}{4}\left(1 + \frac{G^0_{R^\bullet + X^\mp} - G^0_{RX \pm e^-}}{\lambda_{0,1} + D}\right)^2 \tag{23}$$

and the steepest descent pathway is given by the following two equations:

$$0 \le X_1 \le X_1^\ddagger: \qquad \left(\frac{X_1}{X_1^\ddagger}\right)^{1/\lambda_{0,1}} = \left(\frac{Y_1}{Y_1^\ddagger}\right)^{1/D} \tag{24}$$

$$X_1^\ddagger \le X_1 \le 1: \qquad \left(\frac{1 - X_1}{1 - X_1^\ddagger}\right)^{1/\lambda_{0,1}} = \left(\frac{1 - Y_1}{1 - Y_1^\ddagger}\right)^{1/D} \tag{25}$$

An example is given in Fig. 7. The dotted line in Fig. 7a shows the projection of the steepest descent pathway on the X_1–Y_1 plane while the full line is the projection of the intersection of the two surfaces (19) and (20). A three-dimensional representation of the steepest descent pathway is given in Fig. 7b. We may combine X_1 and Y_1 so as to define a reaction coordinate using equations (26) and (27), adjusting normalization so as to obtain $Z_1 = 1$ for the product state, i.e., for $X_1 = Y_1 = 1$.

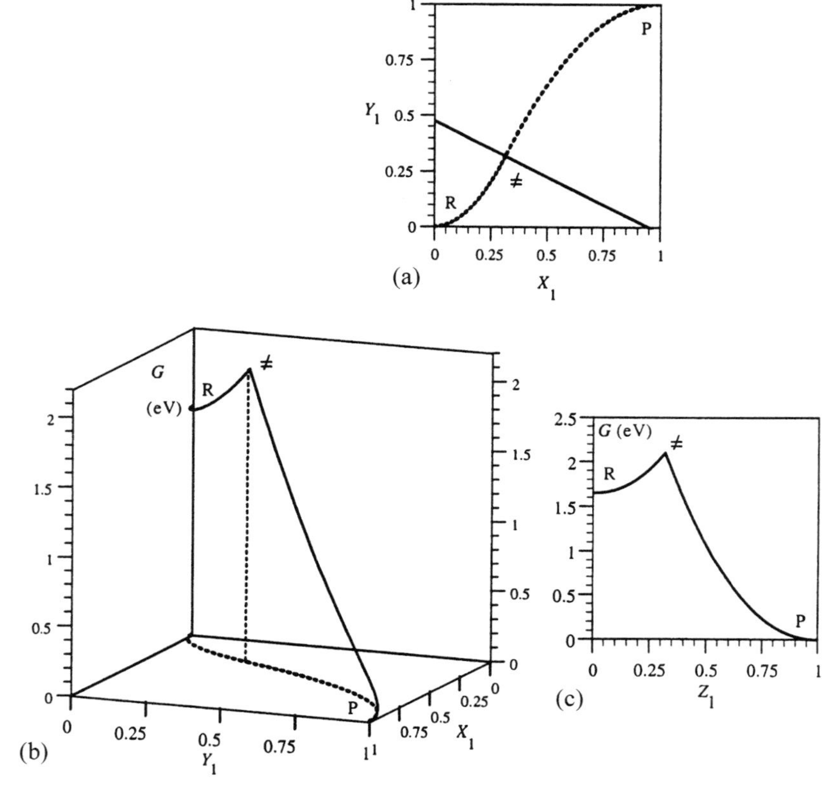

Fig. 7 $RX \pm e^- \Leftrightarrow R^\bullet + X^\mp$. In eV, $D = 3$, $\lambda_{0,1} = 1.5$, $G^0_{RX \pm e^-} = 1.65$, $G^0_{R^\bullet + X^\mp} = 0$. (a) Full line: projection of the intersection of the two surfaces (19) and (20); dotted line, projection of the steepest descent pathway. (b) Steepest descent reaction pathway. (c) Steepest descent reaction profile as a function of the reaction coordinate Z_1. $R = RX \pm e^-$, $P = R^\bullet + X^\mp$.

$0 \le X_1 \le X_1^\ddagger$:

$$Z_1 = X_1^\ddagger \int_0^{X_1/X_1^\ddagger} \sqrt{1 + \left(\frac{D}{\lambda_{0,1}} \eta^{(D/\lambda_{0,1})-1}\right)^2} \mathrm{d}\eta \Big/ \int_0^1 \sqrt{1 + \left(\frac{D}{\lambda_{0,1}} \eta^{(D/\lambda_{0,1})-1}\right)^2} \mathrm{d}\eta \quad (26)$$

$X_1^\ddagger \le X_1 \le 1$:

$$Z_1 = (1 - X_1^\ddagger) \int_0^{X_1/X_1^\ddagger} \sqrt{1 + \left(\frac{D}{\lambda_{0,1}} \eta^{(D/\lambda_{0,1})-1}\right)^2} \mathrm{d}\eta \Big/ \int_0^1 \sqrt{1 + \left(\frac{D}{\lambda_{0,1}} \eta^{(D/\lambda_{0,1})-1}\right)^2} \mathrm{d}\eta \quad (27)$$

The resulting reaction profile is shown in Fig. 7c.

The same type of analysis can be repeated for the two other reactions using appropriate free energy surfaces.[77]

For $RX \pm e^- \Leftrightarrow RX^{\bullet\mp}$,

$$G_{RX\pm e^-} = G^0_{RX\pm e^-} + D\{1 - \exp[-\beta(y - y_{RX})]\}^2 + \lambda_{0,2}X_2^2 \tag{28}$$

$$G_{RX^{\bullet\mp}} = G^0_{RX^{\bullet\mp}} + D'\{1 - \exp[-\beta(y - y_{RX^{\bullet\mp}})]\}^2 + \lambda_{0,2}(1 - X_2)^2 \tag{29}$$

after introduction of another coordinate, X_2, depicting the solvent reorganization around the R portion of the molecule and of the corresponding solvent reorganization energy $\lambda_{0,2}$. The effect of bond stretching is represented by a Morse curve (D' being the homolytic bond dissociation energy of $RX^{\bullet\mp}$), which has the same repulsive part as the RX Morse curve.[77] This condition implies that the two equilibrium values of the bond length are related by equation (30)

$$y_{RX^{\bullet\mp}} - y_{RX} = \frac{1}{\beta}\ln\left(\frac{\sqrt{D}}{\sqrt{D'}}\right) \tag{30}$$

Another stretching coordinate, Y_2, may thus be introduced:

$$Y_2 = \frac{\sqrt{D}}{\sqrt{D} - \sqrt{D'}}Y_1 \tag{31}$$

leading to

$$G_{RX\pm e^-} = G^0_{RX\pm e^-} + \left(\sqrt{D} - \sqrt{D'}\right)^2 Y_2^2 + \lambda_{0,2}X_2^2 \tag{32}$$

$$G_{RX^{\bullet\mp}} = G^0_{RX^{\bullet\mp}} + \left(\sqrt{D} - \sqrt{D'}\right)^2(1 - Y_2)^2 + \lambda_{0,2}(1 - X_2)^2 \tag{33}$$

and

$$X_2^\ddagger = Y_2^\ddagger = \frac{1}{2}\left[1 + \frac{G^0_{RX^{\bullet\mp}} - G^0_{RX\pm e^-}}{\lambda_{0,2} + (\sqrt{D} - \sqrt{D'})^2}\right] \tag{34}$$

$$\Delta G_2^\ddagger = \frac{\lambda_{0,2} + (\sqrt{D} - \sqrt{D'})^2}{4}\left[1 + \frac{G^0_{RX^{\bullet\mp}} - G^0_{RX\pm e^-}}{\lambda_{0,2} + (\sqrt{D} - \sqrt{D'})^2}\right]^2 \tag{35}$$

The reaction coordinate, Z_2, is obtained by application of equations that are identical to equations (27) and (28) in which X_2 , $\lambda_{0,2}$ and $(\sqrt{D} - \sqrt{D'})^2$ replace X_1 , $\lambda_{0,1}$ and D, respectively.

For $RX^{\bullet\mp} \rightleftharpoons R^{\bullet} + X^{\mp}$, a similar analysis leads to the following expressions:

$$G_{RX^{\bullet\mp}} = G^0_{RX^{\bullet\mp}} + D'\{1 - \exp[-\beta(y - y_{RX^{\bullet-}})]\}^2 + \lambda_{0,3}X_3^2 \tag{36}$$

$$G_{R^{\bullet}+X^{\mp}} = G^0_{R^{\bullet}+X^{\mp}} + D\{\exp[-\beta(y - y_{RX})]\}^2 + \lambda_{0,3}(1 - X_3)^2 \tag{37}$$

after introduction of another coordinate, X_3, depicting the reorganization of the solvent upon shifting the charge from the R to the X portion of the molecule and of the corresponding solvent reorganization energy $\lambda_{0,3}$. The effect of bond stretching is represented by a Morse curve for $RX^{\bullet\mp}$ and by a purely repulsive Morse curve for the fragments. Thus, introducing a new stretching coordinate, Y_3, one obtains the following governing equations:

$$Y_3 = 1 - \exp[-\beta(y - y_{RX^{\bullet\mp}})] = 1 - \frac{\sqrt{D}}{\sqrt{D'}}(1 - Y_1) \tag{38}$$

$$G_{RX^{\bullet\mp}} = G^0_{RX^{\bullet\mp}} + D'Y_3^2 + \lambda_{0,3}X_3^2 \tag{39}$$

$$G_{R^{\bullet}+X^{\mp}} = G^0_{R^{\bullet}+X^{\mp}} + D'(1 - Y_3)^2 + \lambda_{0,3}(1 - X_3)^2 \tag{40}$$

$$X_3^{\ddagger} = Y_3^{\ddagger} = \frac{1}{2}\left[1 + \frac{G^0_{R^{\bullet}+X^{\mp}} - G^0_{RX^{\bullet\mp}}}{\lambda_{0,3} + D'}\right] \tag{41}$$

$$\Delta G_3^{\ddagger} = \frac{\lambda_{0,3} + D'}{4}\left[1 + \frac{G^0_{R^{\bullet}+X^{\mp}} - G^0_{RX^{\bullet\mp}}}{\lambda_{0,3} + D'}\right]^2 \tag{42}$$

The reaction coordinate, Z_3, is obtained by application of equations that are identical to equations (27) and (28) in which X_3, $\lambda_{0,3}$ and D' replace X_1, $\lambda_{0,1}$ and D, respectively.

It follows that a proper description of the stepwise and concerted reaction pathways requires a three-dimensional representation as illustrated by Figs 8a and 8b.

The example chosen in Fig. 8 corresponds to the passage from a concerted to a stepwise mechanism as observed by means of cyclic voltammetry upon increasing the scan rate and/ or decreasing temperature. At the peak, the free energy of activation is given by equation (43)[30,32]

$$\Delta G^{\ddagger} = \frac{RT}{F}\left[\ln\left(Z^{el}\sqrt{\frac{RT}{\alpha F v D_i}}\right) - 0.78\right] \tag{43}$$

$Z^{el} = \sqrt{RT/2\pi M}$ (M, molar mass) is the electrochemical collision frequency, v is the scan rate and D_i the diffusion coefficient. Taking typical values, $Z^{el} = 4 \times 10^3$ cm s^{-1}, $D_i = 10^{-5}$ cm^2 s^{-1} leads to a bracketing of the free energy of activation at the peak between 0.385 eV (for $v = 0.1$ V s^{-1},

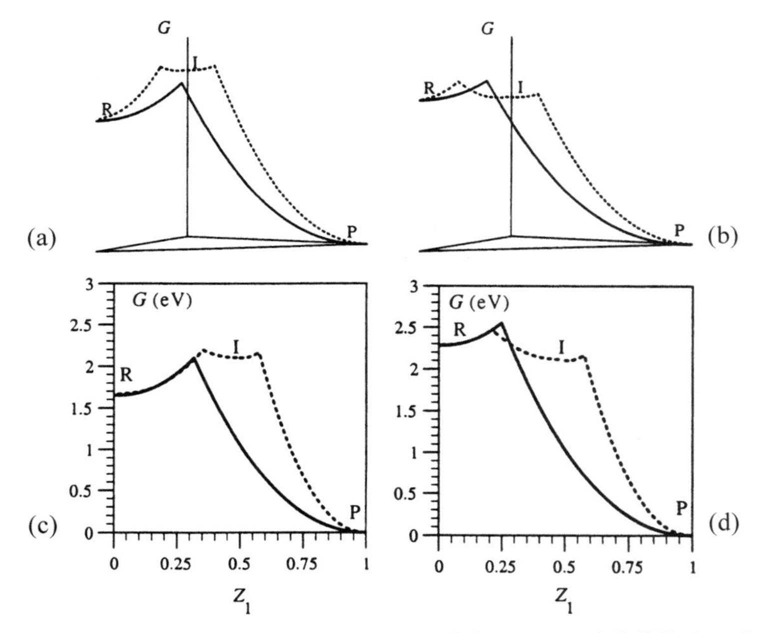

Fig. 8 3D (a, b) and 2D (c, d) representations of the concerted (full line) and stepwise (dotted lines) reaction pathways.
$RX \pm e^- \Leftrightarrow R^\bullet + X^\mp$: $D = 3$ eV, $\lambda_{0,1} = 1.5$ eV, $G^0_{RX \pm e^-} = 1.65$ (a, c), 2.28 eV (b, d).
$RX \pm e^- \Leftrightarrow RX^{\bullet\mp}$: $(\sqrt{D} - \sqrt{D'})^2 = 0.1$ eV, $\lambda_{0,2} = 1$ eV, $G^0_{RX^{\bullet\mp}} = 2.1$ eV.
$RX^{\bullet\mp} \Leftrightarrow R^\bullet + X^\mp$: $D' = 2$ eV, $\lambda_{0,3} = 1$ eV, $G^0_{R^\bullet + X^\mp} = 0$.
R = reactants ($RX \pm e^-$), I = intermediate ($RX^{\bullet\mp}$), P = products ($R^\bullet + X^\mp$).

$T = 301$ K) and 0.185 eV (for $v = 10^3$ V s^{-1}, $T = 253$ K), from which the values of the standard free energies of reaction used in Fig. 8 were derived.

Figures 8c and 8d represent the projection of the reaction pathways on the same plane, namely the front plane in which the dissociative electron transfer step is represented. This two-dimensional representation is easier to decipher than the 3D representation for determining the preferred pathway. They may however be misleading if it is not borne in mind that, in the 2D representation, the crossings between the three curves should not be considered as actual crossings of reaction pathways.

4 Cleavage of primary radicals (often ion radicals)

Ion radicals or, more generally, the primary radicals resulting from single electron transfer to or from a parent molecule are often frangible species that decompose more readily than their parents. Starting from an ion radical (Scheme 9), it is useful to distinguish two cases according to the location of the

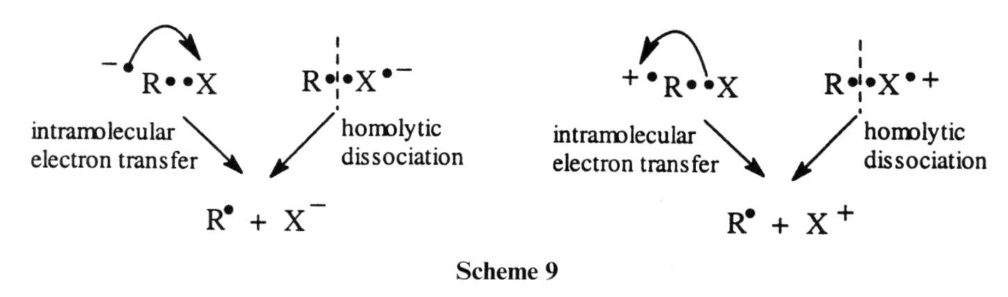

Scheme 9

unpaired electron in the ion radical. In the first case, the unpaired electron is located on the R portion of the molecule. The cleavage of the bond forming $R^\bullet$ and X^- or X^+ is heterolytic and involves an intramolecular electron transfer from the R to the X moieties in the case of an anion radical and an intramolecular electron transfer in the reverse direction in the case of a cation radical. In the second case, the unpaired electron is located on the X portion of the molecule. The cleavage of the bond forming $R^\bullet$ and X^- or X^+ involves a homolytic dissociation of the R—X bond. This distinction between two modes of cleavage also applies to uncharged primary radicals, resulting either from the reduction of cationic substrates or to the oxidation of anion substrates, producing a neutral secondary radical, $R^\bullet$, and a neutral leaving group, X.

Bond cleavage in ion radicals often involves the expulsion of an ionic moiety from the parent molecules. Rearrangements of ion radicals involving an intramolecular bond cleavage, as for example the opening of a small ring,[78–81] belong to the same class of reactions. Their dynamics should therefore lend themselves to the same type of analysis.

INTRAMOLECULAR DISSOCIATIVE ELECTRON TRANSFER

Modeling of the reaction dynamics

The dynamics of the reaction may be modeled by a simple extension of the dissociative electron transfer model in its intermolecular version. It follows[6,11,77] that the activation free energy for the cleavage of the primary radical is given by equations (7) and (8) in which the bond dissociation energy D of RX is replaced by the bond dissociation energy, $D_{(+)-\bullet \mathrm{RX}}$, of the ion radical, thus leading to the expression of the activation free energy of cleavage in equation (44):

$$\Delta G^{\ddagger}_{\mathrm{cleav}} = \frac{D_{(+)-\bullet \mathrm{RX}} + \lambda_0}{4}\left(1 + \frac{\Delta G^0_{\mathrm{cleav}}}{D_{(+)-\bullet \mathrm{RX}} + \lambda_0}\right)^2 \qquad (44)$$

where $\Delta G^0_{\mathrm{cleav}}$ is defined by equation (18) and the solvent reorganization energy is given by equation (45):

$$\lambda_0 = e_0^2\left(\frac{1}{2a_{(+)-\bullet RX}} + \frac{1}{2a_{\bullet R+X^{-(+)}}} - \frac{1}{d}\right)\left(\frac{1}{\varepsilon_{op}} - \frac{1}{\varepsilon_S}\right) \tag{45}$$

e_0 is the electron charge, $a_{(+)-\bullet RX}$ is the radius of the sphere approximating the region of the starting ion radical where the charge is initially located and $a_{\bullet R+X^{-(+)}}$ is the radius of the sphere approximating the location of the charge on the leaving group when the bond is broken; d is the distance between the centers of charge. The ε are the optical and static dielectric constants, respectively. $D_{(+)-\bullet RX}$ may be related to more readily accessible quantities through equation (46):

$$\begin{aligned} D_{(+)-\bullet RX} = D_{RX} + E^0_{RX/(+)-\bullet RX} - E^0_{R\bullet/(+)-\bullet(R\bullet)} \\ + T(\bar{S}_{RX} - \bar{S}_{(+)-\bullet RX} + \bar{S}_{(+)-\bullet(R\bullet)} - \bar{S}_{R\bullet}) \end{aligned} \tag{46}$$

(the E^0 and $\bar{S}$ are the standard potentials and the partial molar entropies of the subscript species). Because the model is based on a Morse curve representation of the homolytic dissociation of the R—X bond in the ion radical, the subscript $^{(+)-\bullet}(R^\bullet)$ means that one electron is added (or removed) from the radical $R^\bullet$ so that the electron (or the hole) resides on the portion of the molecule where it was introduced in the initial step.

If this model gives a correct representation of the dynamics of anion radical cleavage, it is also able to represent the dynamics of the reverse reaction, i.e., the coupling of radicals with nucleophiles. Cleavage of anion radicals and coupling of radicals with nucleophiles are the key steps of nucleophilic substitutions stimulated by single electron transfer thus occurring according to a $S_{RN}1$ mechanism (Scheme 8). The model may thus be used to rationalize the numerous experimental qualitative and quantitative observations made in the field of $S_{RN}1$ reactivity.[11,77]

Cleavage of anion radicals. Coupling of radicals with nucleophiles

Early examples of application of the above model were found in the cleavage of aryl halides' anion radicals. The observation that there is a roughly linear correlation, with a slope close to 0.5, between the cleavage activation free energy and the standard potential for the formation of the anion radical, $E^0_{RX/RX\bullet-}$ (see figure 1 in Ref. 77) may be explained as follows. The dissociation free energy of the aromatic carbon–halogen bond does not depend much on the particular nature of the aryl moiety. Thus, in each series, the cleavage standard free energy, ΔG^0_{cleav}, varies essentially as $E^0_{RX/RX\bullet-}$. In addition, the intrinsic barrier should not vary much in each series, because in the expression for $D_{RX\bullet-}$ (equation 46), D is about constant and the variations of $E^0_{RX/RX\bullet-}$ and $E^0_{R\bullet/(R\bullet)\bullet-}$ should compensate each other since both parallel the energy of the π^* orbital. The linear correlation with a 0.5 slope then derives from linearization of equation (7) (the point of zero driving force on the $E^0_{RX/RX^{\bullet-}}$ scale stands in the middle of the experimental data points). The approximate character of these various assumptions falls in line with the fact that the correlation indicates a general semi-quantitative trend rather than a precise depiction of the effect of fine changes in the molecular structure.

Many qualitative observations concerning the coupling between radicals and nucleophiles may likewise be rationalized in the framework of the model.[77] For example, why are phenyl radicals substituted by electron-withdrawing groups and polyaromatic radicals more reactive than plain phenyl radicals toward nucleophiles? The answer is simply that the driving force for coupling is more favorable because $E^0_{RX/RX^{\bullet-}}$ is more positive ($\Delta G^0_{coupl} = -\Delta G^0_{cleav}$), as results from equation (18). In more physical terms, the energy of the π^* orbital hosting the electron transferred during coupling is lower in the first cases than in the second. Why do benzyl radicals require strongly electron-withdrawing substituents to couple with nucleophiles whereas phenyl radicals do not? The forming bond is substantially weaker in the first case than in the second (by roughly 20 kcal mol^{-1}) and this unfavorable factor has to be compensated by an increase of the driving force offered by the more positive value of $E^0_{RX/RX^{\bullet-}}$ resulting from the presence of an electron-withdrawing substituent. It should be noted in this connection that increasing the strength of the forming bond has a favorable effect on the driving force but an unfavorable effect on the intrinsic barrier. In total, however, the favorable effect is predominant.

The coupling of cyanide ions with aryl radicals is an interesting example where quantitative kinetic data are available (Ref. 77 and references therein). The forming bond is strong, but this favorable factor is counteracted, in terms of driving force, by the fact that $E^0_{X^\bullet/X^-}$ (second term in equation 18) is very positive (in other words, CN^- is a hard nucleophile). In addition, the large value of D is unfavorable in terms of the intrinsic barrier. In total, the presence of electron-withdrawing substituents is necessary to allow the coupling reaction to compete successfully with side-reactions (electron transfer to the radical). The model allows a reasonably accurate quantitative description of the coupling kinetics (see figure 2 in Ref. 77).

The reductive defluorination of perfluorocarbons is an example where the above model helped to resolve a rather complicated mechanism and to rationalize the reactivity pattern.[82] The reductive elimination of the three fluorines occurs successively according to the following reaction sequence:

$$ArCF_3 \xrightarrow[-F^-]{2e^- + H^+} ArCHF_2 \xrightarrow[-F^-]{2e^- + H^+} ArCH_2F \xrightarrow[-F^-]{2e^- + H^+} ArCH_3$$

Carbenes do not transpire along the reaction pathway. Each defluorination reaction follows a stepwise mechanism:

$$Ar{-}\overset{|}{\underset{|}{C}}{-}F + e^- \rightleftharpoons {}^{-\bullet}Ar{-}\overset{|}{\underset{|}{C}}{-}F \longrightarrow Ar{-}\dot{C}{<} + F^-$$

As shown by direct cyclic voltammetry and redox catalysis, the cleavage rate constant increases considerably from CF_3 to CHF_2 and CH_2F, from 38 to 4×10^5 and 7×10^6 s^{-1} in DMF at 20°C for Ar = 4–CN–C_6H_4. The standard potential, $E^0_{RX/RX^{\bullet -}}$, decreases only slightly from CF_3 to CHF_2 and CH_2F. The main accelerating factor is the decrease of the bond dissociation energy in the series. The relative stability of the CF_3 anion radical opens an internal redox catalytic route in which it reduces the di- and monofluoro compounds, thus helping the global exchange of six electrons and elimination of the three fluoride ions to take place within a very narrow potential range. Similar observations and rationalizations can be made in liquid ammonia at −40°C. The cleavage activation free energies of the three anion radicals are smaller than in DMF, presumably because the more protic character of ammonia enhances the solvation of fluoride ions thus increasing the cleavage driving force.

The influence of the reaction medium on the kinetics of anion radical cleavage has been the subject of several other investigations. A first series of studies attempted to correlate the cleavage rate constants determined by means of cyclic voltammetry[83–85] or pulse radiolysis[86–90] with the Lewis acidity of the solvent as measured, for example, by the Gutmann acceptor number. The latter technique does not seem to produce reliable results as compared to the data derived from other sources, presumably because of problems in radiation absorption by the solvent. The cleavage rate constants of the anion radical of 9-chloroanthracene exhibit a good correlation with the Gutmann acceptor number,[83–85] as expected from the fact that the negative charge is more concentrated in the transition state (en route toward the formation of the halide ion) than in the starting molecule where the charge is delocalized over the substituted aromatic portion of the molecule. Whether this correlation essentially derives from a driving force effect resulting from a stabilization of the halide ion by acidic (Lewis) solvents or whether there is, in addition, an effect on the intrinsic barrier is not a settled question.

For the same reasons, addition of water to an acetonitrile solution of the anion radical of 9-chloroanthracene (and also of 9-fluoroanthracene), where the charge is spread out over a large molecular framework in the initial state, slightly accelerates the cleavage.[91] The situation is drastically different with anion radicals bearing a carbonyl or a nitro group, the charge then being concentrated on the oxygen atoms. The consequence is a strong decrease of the cleavage rate constant upon addition of water, which stabilizes the anion radical more than the leaving halide ion. The effect is so important that it entails a qualitative change in the evolution of the anion radical at high water concentration. The halide ion is no longer cleaved off; rather, protonation occurs at one oxygen atom, resulting eventually in the reduction of the carbonyl or of the nitro group. The effect of adding Li^+ or Mg^{2+} ions is similar, namely, little effect on anion radicals of the first category and a strong decrease in the second case because of ion-pairing with the carbonyl or nitro oxygen. There is again a change of mechanism upon increasing the cation concentration; that is, cleavage is replaced by dimerization into the corresponding pinacol in the case of a carbonyl group.

The results of a recent investigation of the dependence of the cleavage rate constant upon the solvent of two similar anion radicals, those of 3-nitrobenzyl chloride and 3-chloroacetophenone,[85] may likewise be interpreted as the outcome of a competition between the Lewis acid solvation of the developing halide ion and of the negatively charged oxygen atoms in the initial state.

Carbonyl compounds are also suited to the investigation of the role of solvent reorganization in the dynamics of intramolecular dissociative electron transfer as observed in a series of phenacyl derivatives bearing various leaving groups.[199]

Whether orbital symmetry restrictions are important in intramolecular dissociative electron transfers is another question of interest. Cleavage of aromatic anion radicals as well as coupling of aromatic radicals with nucleophiles are expected to be subject to such restrictions, since the reaction involves the transfer of one electron from a π^* to a σ^* orbital and vice versa in the second case. The fact that the coupling step is close to the diffusion limit in many $S_{RN}1$ aromatic nucleophilic substitution reactions indicates that this restriction is not too severe. A more quantitative analysis of the problem was carried out by comparing the cyclic voltammetric reductive behaviors in DMF of the two compounds shown in Scheme 10.[92] The anion radical of compound R_1X, which is subject to orbital symmetry restriction, is considerably more stable (lifetime ≈ 100 s) than the anion radical of compound R_2X (lifetime $\approx 10^{-9}$ s), where these restrictions do not exist. However, the largest part of this difference of reactivity is related to the huge decrease of the cleavage driving force when passing from $R_2X^{\bullet -}$ to $R_1X^{\bullet -}$ owing to the lack of conjugation of the unpaired electron with the aromatic ring in $R_2^{\bullet}$ as opposed to $R_1^{\bullet}$. Taking this effect into account, orbital symmetry restriction is responsible for only about three orders of magnitude in the difference of cleavage reactivity.

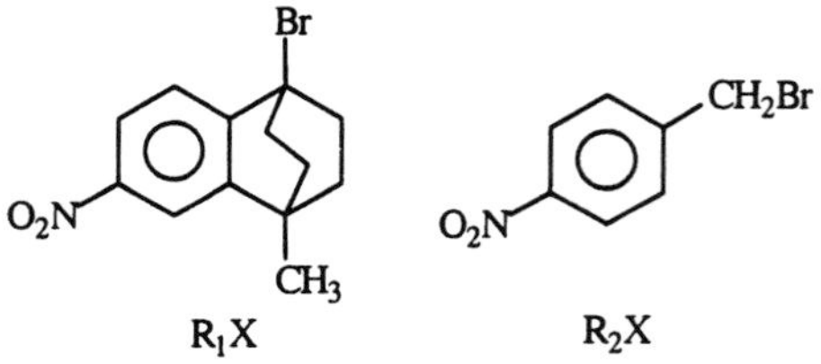

Br
O_2N
CH_3
R_1X

CH_2Br
O_2N
R_2X

Scheme 10

The problem has also been addressed by means of *ab initio* quantum-chemical calculations, showing with the example of the anion radical of 4-chlorotoluene how out-of-plane vibrations may remove the symmetry constraint.[93]

Electronic factors related to orbital overlap also appear to interfere significantly in the dynamics of concerted electron transfer/bond breaking reactions in donor-spacer-cleaving acceptor systems.[94]

Deprotonation at carbon in cation radicals

Deprotonation of hydrocarbons requires some form of activation of the carbon atom for the reaction to be amenable to thermodynamic and kinetic characterization. There are two main ways for decreasing the pK_a of carbon–hydrogen bonds. One involves the decrease of the electron density on the carbon atom by means of an electron-withdrawing group directly borne by the carbon, or located in a conjugated position to it, on an unsaturated substituent.[95–102] The other consists in oxidizing the substrate so as to produce the corresponding cation radical. The resulting decrease of the electron density induces a considerable decrease of the pK_a as compared to the parent molecule.[103–107] The deprotonation rate constants have been determined in several families of cation radicals.[108–131] Among them the NADH analogues series[132–136] offers two advantageous features. One is that the cation obtained after deprotonation of the cation radical and further electron abstraction (Scheme 11) is stable. The rates of deprotonation can thus be determined unambiguously in the framework of the mechanism depicted in Scheme 11. Combining the use of direct electrochemistry, redox catalysis and laser flash photolysis, they could be determined up to the diffusion limit thus allowing the construction of extended Brønsted plots, where the cation radicals are opposed to oxygen or nitrogen bases. Another advantage of this series of cation radicals is that key thermodynamic parameters, such as their pK_a values and homolytic bond dissociation energies can be derived from experimentally measurable quantities through appropriate thermodynamic cycles. In all cases, the slopes of the Brønsted plots indicate that the deprotonation reaction is activation controlled in contrast with oxygen or nitrogen acids, where the deprotonation or the protonation is under diffusion control. Since the pK_a of each cation radical is known, their intrinsic barriers, $\Delta G_0^{\ddagger}$, can be determined in each case, which is seldom the case for other cation radicals. As seen in Fig. 9, there is a correlation between the intrinsic barrier and the homolytic bond dissociation energy of the cation radical, close to a proportionality to $D_{AH^{\bullet+} \rightarrow A^+ + H^{\bullet}}/4$. Thus, rather than a proton transfer, the deprotonation of the cation radicals may be viewed as a concerted electron transfer/bond breaking reaction in which the C—H bond is broken while an electron is transferred from the hydrogen atom to the ring system. This unusual notion allows the use of equation (44) to express the activation/driving force

$$AH - e^- \rightleftharpoons AH^{\bullet+}$$

$$AH^{\bullet+} + B \rightleftharpoons A^{\bullet} + BH^+$$

$$A^{\bullet} - e^- \longrightarrow A^+$$

$$A^{\bullet} + AH^{\bullet+} \longrightarrow A^+ + AH$$

Scheme 11

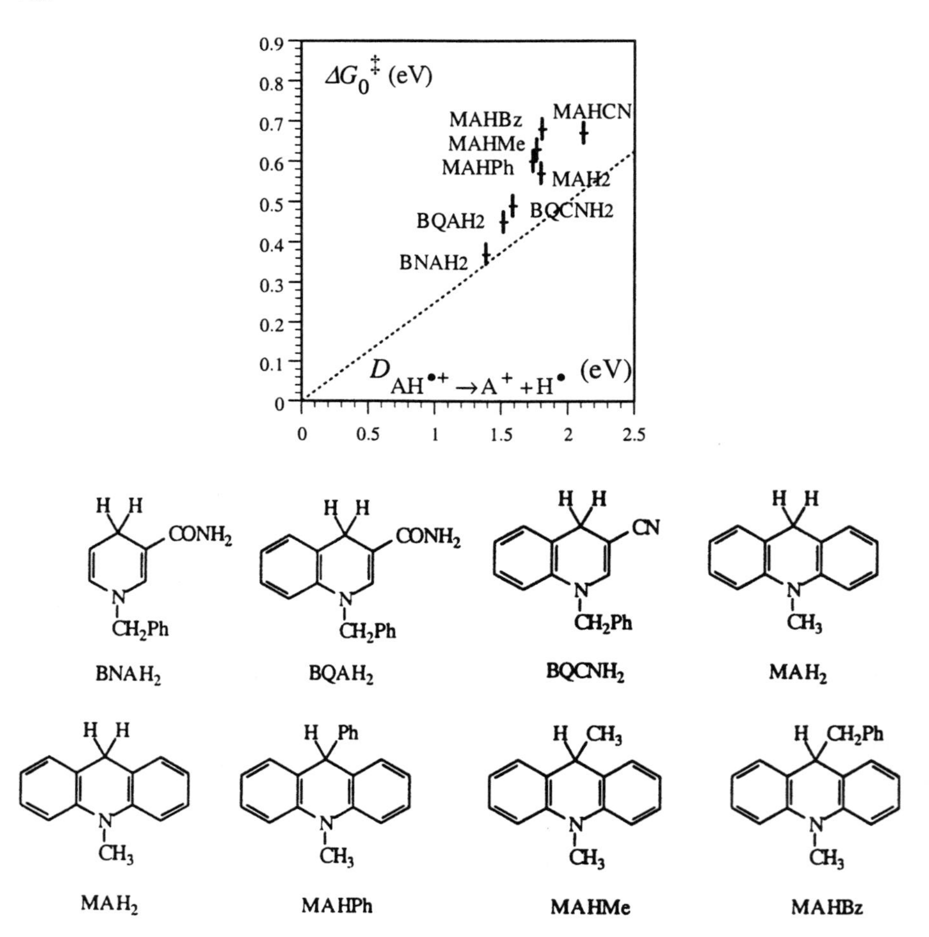

Fig. 9 Deprotonation of cations radicals of synthetic analogs of NADH by oxygen or nitrogen bases in acetonitrile. Correlation between the intrinsic barrier and the homolytic bond dissociation energy of the cation radical ($AH^{\bullet+} \rightarrow A^+ + H^\bullet$).

relationship. The deviations from the proportionality to $D_{AH^{\bullet+} \rightarrow A^+ + H^\bullet}/4$, particularly apparent when bulky substituents are present, may be explained by steric hindrance in the successor complex.[136]

HOMOLYTIC CLEAVAGE

Modeling of the reaction dynamics

The cleavage of ion radicals may be a homolytic process rather than an intramolecular dissociative electron transfer (Scheme 9). Equation (44) still applies, equation (46) being replaced by equation (47).

$$\begin{aligned} D_{\mathrm{RX}^{\bullet-(+)}} &= D_{\mathrm{RX}} + E^0_{\mathrm{RX/RX}^{\bullet-(+)}} - E^0_{\mathrm{X}^\bullet/(\mathrm{X}^\bullet)^{\bullet-(+)}} \\ &\quad + T(\bar{S}_{\mathrm{RX}} - \bar{S}_{\mathrm{RX}^{\bullet-(+)}} + \bar{S}_{(\mathrm{X}^\bullet)^{\bullet-(+)}} - \bar{S}_{\mathrm{R}^\bullet}) \end{aligned} \quad (47)$$

In a number of cases, the relaxation from $(\mathrm{X}^\bullet)^{\bullet-(+)}$ to $\mathrm{X}^{-(+)}$ does not require much energy , in which case equation (48) applies.

$$\begin{aligned} D_{\mathrm{RX}^{\bullet-(+)}} &= D_{\mathrm{RX}} + E^0_{\mathrm{RX/RX}^{\bullet-(+)}} - E^0_{\mathrm{X}^\bullet/\mathrm{X}^{-+}} \\ &\quad + T(\bar{S}_{\mathrm{RX}} - \bar{S}_{\mathrm{RX}^{\bullet-(+)}} + \bar{S}_{\mathrm{X}^\bullet/\mathrm{X}^{-(+)}} - \bar{S}_{\mathrm{R}^\bullet}) \end{aligned} \quad (48)$$

Under these conditions, as pictured in Fig. 10, the free energy of the separated fragments is $T\Delta S^0_{\mathrm{cleav}}$ ($\Delta S^0_{\mathrm{cleav}}$ = standard entropy of cleavage) distant from the homolytic bond dissociation energy. The standard entropy of cleavage is usually slightly negative because the positive fragmentation entropy slightly overcompensates for the negative entropy corresponding to the change in solvation (the leaving ion is more strongly solvated than the ion radical because of less delocalization of the charge in the former than in the latter). It follows that the activation barrier for the coupling of the radical with the leaving group is small as depicted in Fig. 10. The situation is then similar to what happens with homolytic cleavage of a standard molecule where the activation barrier for the reverse reaction, i.e., the coupling between the two radicals, is generally

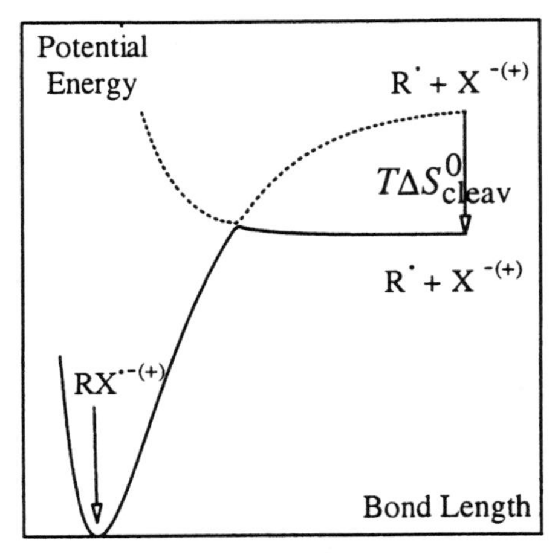

Fig. 10 Endothermic and endergonic homolytic dissociation of ion radicals.

regarded as negligible. With such small activation barriers for the reverse reaction, we may conceive systems for which the decay of the primary radical is kinetically controlled by a follow-up reaction such as diffusion of the fragments out of the solvent cage while the cleavage reaction acts as a pre-equilibrium. If this is indeed the case, we may expect that the rate constants for the forward, k_+, and the reverse, k_-, reaction (Scheme 12) should vary with the cleavage driving force as depicted by equations (49) and (50) and Fig. 11.[137]

$$\mathrm{A} \underset{k_-^{\mathrm{act}}}{\overset{k_+^{\mathrm{act}}}{\rightleftharpoons}} (\mathrm{BC}) \underset{k_{\mathrm{dif}}}{\overset{k'_{\mathrm{dif}}}{\rightleftharpoons}} \mathrm{B} + \mathrm{C}$$

Scheme 12

k_+^{act}, k_-^{act} are forward and reverse activation controlled rate constants; k'_{dif} is the rate constant for the diffusion of the fragments out of the solvent cage; k_{dif} is the bimolecular diffusion-limited rate constant.

$$k_+ = \frac{k'_{\mathrm{dif}} k_+^{\mathrm{act}}}{k'_{\mathrm{dif}} + k_-^{\mathrm{act}}} \qquad k_- = \frac{k_{\mathrm{dif}} k_-^{\mathrm{act}}}{k'_{\mathrm{dif}} + k_-^{\mathrm{act}}} \tag{49}$$

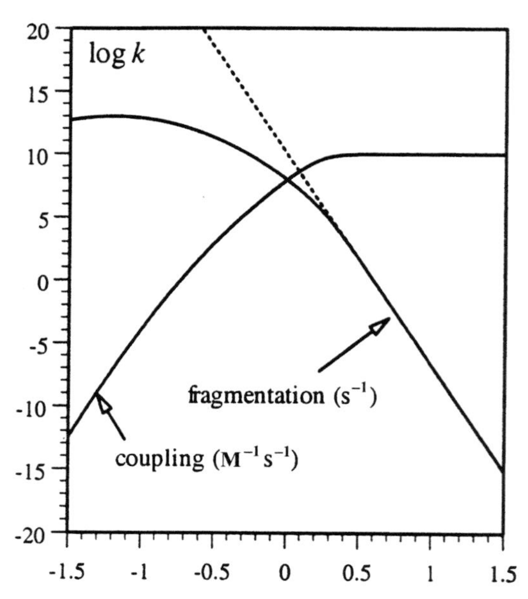

Fig. 11 Endergonic homolytic cleavage of primary radicals. Typical Brønsted plots combining the effect of diffusion and activation. $k_B T/h = 10^{13}\ \mathrm{s}^{-1}$, $k'_{\mathrm{dif}} = 10^{10}\ \mathrm{s}^{-1}$, $k_{\mathrm{dif}} = 10^{10}\ \mathrm{M}^{-1}\ \mathrm{s}^{-1}$, $D_{\mathrm{RX}\bullet-(+)} + \lambda_0 = 0.3$ eV, $T = 298$ K. Dotted line: $\log k'_{\mathrm{dif}} - (F/RT \ln 10)\Delta G^0_{\mathrm{cleav}}$.

with

$$k_{+}^{\mathrm{act}} = \frac{k_B T}{h} \exp\left(-\frac{F\Delta G_{\mathrm{cleav}}^{\ddagger}}{RT}\right) \tag{50}$$

$\Delta G_{\mathrm{cleav}}^{\ddagger}$ being given by equation (44) in which λ_0 is given by equation (45) and $D_{\mathrm{RX}^{\bullet-(+)}}$ by equation (48). With systems such as those represented in Fig. 10, we may thus expect that the log of the fragmentation rate constant varies with the driving force with a slope of 1/0.059 (eV) within a large portion of the accessible domain of driving forces. Experimental examples of such behavior are given in the next subsection.

It should, however, be emphasized that the relaxation from $(X^{\bullet})^{\bullet-(+)}$ to $X^{-(+)}$ does not always require a negligible energy. Counterexamples where the homolytic cleavage of an ion radical appears to be endowed with a sizable internal reorganization energy can be found in the electrochemical reduction of perbenzoates,[60] where the observation of a transition between a concerted and a stepwise mechanism fits with an exothermic cleavage of the anion radicals.

The kinetics of the electron transfer reaction leading to the homolytically dissociating primary radical is also a question of interest. It may be modeled using the Morse curve for the reactant and the Morse curve shown in Fig. 10 representing the homolytic dissociation of the primary radical. This point will be discussed in detail in Section 5.

Examples of homolytic cleavage of anion and cation radicals

From the results of exhaustive studies of the cleavage of bibenzyl anion and cation radicals[138–145] the log of the rate constants may be plotted against the driving force as shown in Fig. 12. It is striking that almost all the experimental points, from cations as well as for anion radicals series, stand on the same straight line of 1/0.059 (eV) slope as theoretically predicted for reactions that are governed by the diffusion of the fragments out of the solvent cage and where the cleavage of the ion radical interferes only by its thermodynamics.[137] Even more striking is the fact that the data points pertaining to cation radicals of an entirely different series of compounds, namely, *tert*-butyl derivatives of synthetic analogs of NADH, stand on the same straight line as the ion radicals in the bibenzyl series.[137]

HETEROLYTIC CLEAVAGE, HOMOLYTIC CLEAVAGE, DISSOCIATIVE ELECTRON TRANSFER

Bond cleavage may occur in ion radicals that contain more than one electron- (or hole-) hosting orbital, where the cleaving bond may or may not be separated by another bond (S_1, S_2) from the electron- (hole-) hosting portions of the molecule. As sketched in Scheme 13, there is a large variety of possible and

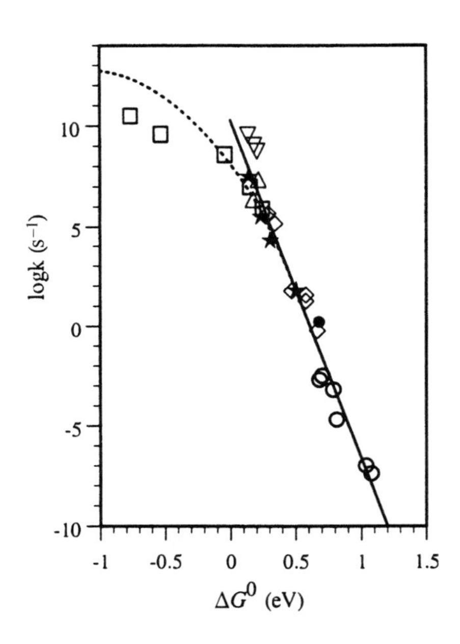

Bibenzyl Anion Radicals — From right to left: — Solvent:

○ : H–C_6H_4–C(R1)(R2)–C(R1)(R2)–C_6H_4–NO_2
R1: Me, Me, Me, Et, *n*-Bu, *n*-Pr
R2: Et, *n*-Pr, *i*-Bu, Et, *n*-Bu, *n*-Pr
(erythro)
Solvent: Me_2SO

□ : NC–C_6H_4–C–C(CN)(Y)–C_6H_4–X
X: MeO, H, CF_3, MeO, H
Y: Me, Me, Me , CN , CN
Solvent: MeCN

Bibenzyl Cation Radicals

◇ : H–C_6H_4–C(R1)(R2)–C(R1)(R2)–C_6H_4–$N(CH_3)_2$
R1: Me, Me, Me, Me, Et, *n*-Bu, *n*-Pr
R2: Me, Et, *n*-Pr, *i*-Bu, Et, *n*-Bu, *n*-Pr
(erythro)
Solvent: MeCN

▽ : MeO–C_6H_4–C(R1)(R2)–C(R1)(R2)–C_6H_4–OMe
R1: Me, Me, Et
R2: Me, Et , Et,
(meso)
Solvent: Et_2O

△: MeO–C_6H_4–C–C–C_6H_4–X
X: CN, H
Solvent: CH_2Cl

●: $(CH_3)_2N$–C_6H_4–C–C(OMe)(OMe)–C_6H_4–Me
Solvent: CH_2Cl

Cation Radicals of the *tert*-butyl Derivatives of NADH Synthetic Analogs

From right to left:

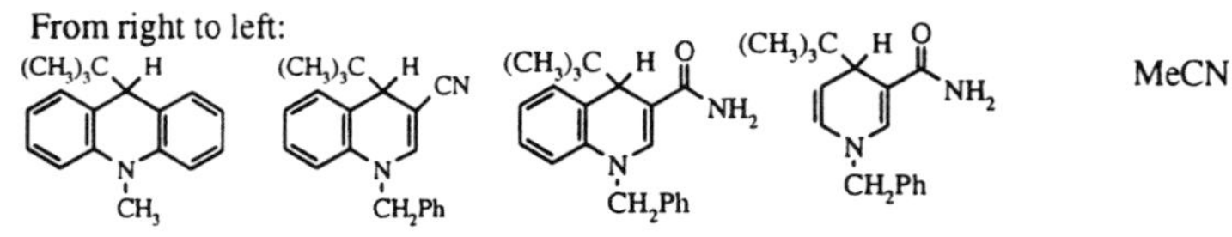

Solvent: MeCN

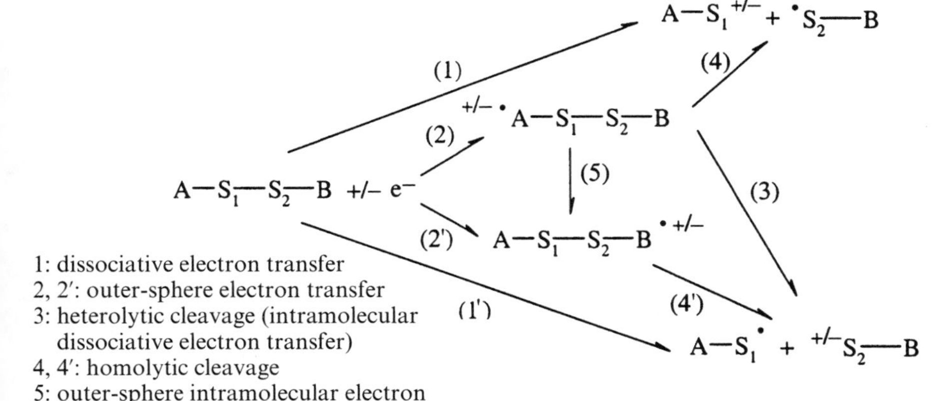

Scheme 13

potentially competing reaction pathways leading to a cleaved product. Several of these, like 1, 1′, 2 + 3 or 4, have been the object of active attention as discussed in the preceding sections. But some of the other possibilities depicted in Scheme 13 have begun to be investigated as, for example, in the electrochemical reduction of a series of nitrocumene derivatives:[146]

Z—C_6H_4—$C(CH_3)_2$—NO_2 (Z: H, CN, NO_2)

The mechanism changes from 2′ + 4′ to 2 + 3 when passing from Z = H and CN to Z = NO_2. One may also conceive that an intramolecular outer-sphere electron transfer (reaction (5) in Scheme 13) may be the prelude to a cleavage step. Such reactions (with no successive cleavage) have been described recently.[147] In such bifunctional systems, the electron is first transferred to (or from) the portion of the molecule that is the least accessible in the thermodynamic sense, but the "fastest" in terms of intrinsic barrier before being transferred to (or from) the second electrophore. This process is an intramolecular version of homogeneous redox catalysis,[38,55] as is also the stepwise mechanism in electron transfer/bond breaking reactions that we have discussed at length in the preceding sections.

The models discussed above may be adapted to analyze the dynamics of such

---

Fig. 12 Variation of the rate constant with the driving force in the homolytic cleavage of various types of anion and cation radicals. The open symbols refer to bibenzyl derivatives and the stars to cation radicals of the *tert*-butyl derivatives of synthetic analogs of NADH.

reaction sequences and the competition between the various pathways upon changing the molecular structure or the driving force offered to the reaction.

5 Interactions between fragments in the product cluster

There is indirect experimental evidence that such attractive interactions, of the charge–dipole type, may exist in the gas phase after injection of an electron in alkyl halides.[148,149] *Ab initio* calculations give contrasting results depending on the method used and approximations made.[28,150,151] It is usually assumed that these interactions vanish in polar solvents. One such case is the anionic state of CF_3Cl,[28] where the shallow minimum calculated in the gas phase disappears upon solvation, at least when a simple continuum solvation model is used. The questions that arise are whether the attractive interaction existing in the gas phase may persist, even weakened, in the caged product system in a polar solvent, how its magnitude depends on the structure of $R^{\bullet}$ and X^- and what its effect is on the dynamics of dissociative electron transfer.

INFLUENCE ON THE DYNAMICS OF DISSOCIATIVE ELECTRON TRANSFER

As shown by recent quantum-mechanical calculations concerning the reduction of carbon tetrachloride (see below), the interaction between the two fragments in the gas phase may be modeled by a Morse curve, with a shallow minimum, even if the nature of the interaction is more of a charge–dipole (and induced dipole) type rather than of a covalent bond. It also appears that the repulsive part of the Morse curve of the fragments is almost identical to the repulsive part of the Morse curve depicting the homolytic dissociation of the starting RX molecule. Assuming that this behavior is general, the reactant and product free energies may be expressed by equations (51) and (52), respectively.

$$G_{RX\pm e^-} = G^0_{RX\pm e^-} + D_R\{1 - \exp[-\beta(y - y_R)]\}^2 + \lambda_{0,1}X_1^2 \qquad (51)$$

$$G_{(R^\bullet,X^\mp)} = G^0_{(R^\bullet,X^\mp)} - \Delta G^0_{sp} + D_P\{1 - \exp[-\beta(y - y_P)]\}^2 + \lambda_{0,1}(1 - X_1)^2 \qquad (52)$$

D_R is the R—X BDE; y is the R—X distance; y_R and y_P are their values at equilibrium in the reactant and product system, respectively; $\beta = \nu(2\pi^2\mu/D)^{1/2}$, ν being the stretching frequency of the R—X bond and μ the corresponding reduced mass; X is a fictitious charge serving as solvation index in the depiction of the solvent reorganization; λ_0 is the Marcus solvent reorganization energy; ΔG^0 is the standard free energy of the reaction leading to the separated fragments and $\Delta G^0_{sp} = D_P - T\Delta S^0_{sp}$ is the difference between the standard free energies of the separated and the caged fragments. The assumption that the repulsive terms in the two Morse curves are approximately the same leads to equation (53), relating the difference in the equilibrium distances to the ratio of the dissociation energies.

$$y_{\mathrm{P}} = y_{\mathrm{R}} + \frac{1}{2\beta}\ln\left(\frac{D_{\mathrm{R}}}{D_{\mathrm{P}}}\right) \tag{53}$$

Equation (53) indicates that a shallow minimum ($D_{\mathrm{P}} \ll D_{\mathrm{R}}$) corresponds to a loose cluster ($y_{\mathrm{P}} \gg y_{\mathrm{R}}$). Equations (51) and (52) may be recast as follows:

$$G_{\mathrm{RX}\pm\mathrm{e}^-} = D_{\mathrm{R}} Y^2 + \lambda_0(Y)X^2 \tag{54}$$

$$G_{\mathrm{R}^\bullet,\mathrm{X}^\mp} = \Delta G^0 - \Delta G^0_{\mathrm{sp}} + D_{\mathrm{R}}\left[\left(1 - \sqrt{\frac{D_{\mathrm{P}}}{D_{\mathrm{R}}}}\right) - Y\right]^2 + \lambda_0(Y)(1-X)^2 \tag{55}$$

with $Y = 1 - \exp[-\beta(y - y_{\mathrm{R}})]$.

The activation free energy is then obtained by the usual minimization procedure, thus leading to equation (56).

$$\Delta G^{\ddagger} \approx \frac{(\sqrt{D_{\mathrm{R}}} - \sqrt{D_{\mathrm{P}}})^2 + \lambda_0}{4}\left[1 + \frac{\Delta G^0 - \Delta G^0_{\mathrm{sp}}}{(\sqrt{D_{\mathrm{R}}} - \sqrt{D_{\mathrm{P}}})^2 + \lambda_0}\right]^2 \tag{56}$$

Comparison between equation (56) and equations (7) and (8), which give the activation free energy in the case of a purely repulsive product profile, reveals that the effect of an attractive interaction between the fragments in the product cluster is not merely described by the introduction of a work term in the classical theory of dissociative electron transfer. Such a work term appears in the form $-\Delta G^0_{\mathrm{sp}}$, but there is also a modification of the intrinsic barrier. With the same λ_0, the change in the intrinsic barrier would simply be obtained by replacement of D_{R} by $(\sqrt{D_{\mathrm{R}}} - \sqrt{D_{\mathrm{P}}})^2$. It is noteworthy that small values of D_{P} produce rather strong effects of the intrinsic barrier (a similar conclusion was reached by Marcus[149] in the treatment of gas-phase electron attachment to aliphatic halides using a slightly different model). For example if D_{P} is 4% of D_{R}, a decrease of 20% of the intrinsic barrier results.

It is also interesting to note that the above model also applies to the kinetics of electron transfer leading to an ion radical that dissociates homolytically in an exothermic manner as represented in Fig. 10. D_{P} stands for $D_{\mathrm{RX}^{\bullet-(+)}}$ and ΔG^0_{sp} for $\Delta G^0_{\mathrm{cleav}}$. The smaller D_{P}, the larger the equilibrium distance, the larger the internal reorganization and the slower the electron transfer kinetics. Conversely, when D_{P} is close to D_{R}, the reactant BDE, the expression of the internal reorganization energy, λ_{i}, is, as shown below, the same as in the parabolic Marcus model.[25]

$$D_R = D + \Delta D, D_P = D - \Delta D \qquad \text{with} \qquad \Delta D \ll D$$

$$\lambda_i = \left(\sqrt{D_R} - \sqrt{D_P}\right)^2 \cong \frac{\Delta D^2}{D} \tag{57}$$

According to equation (53),

$$\Delta y = y_P - y_R = \frac{1}{\beta}\ln\left(\frac{\sqrt{D + \Delta D}}{\sqrt{D - \Delta D}}\right) \cong \frac{\Delta D}{\beta D} \tag{58}$$

It follows that

$$\lambda_i = D\beta^2\,\Delta y^2 = \frac{f\,\Delta y^2}{2} \tag{59}$$

where f is the force constant for stretching the R—X bond.

There is thus an apparent continuity between the kinetics of an electron transfer leading to a stable product and a dissociative electron transfer. The reason for this continuity is the use of a Morse curve to model the stretching of a bond in a stable product in the first case and the use of a Morse curve also to model a weak charge–dipole interaction in the second case. We will come back later to the distinction between stepwise and concerted mechanisms in the framework of this continuity of kinetic behavior.

DISSOCIATIVE ELECTRON TRANSFER TO CARBON TETRACHLORIDE

The problems discussed in the preceding section have been analyzed in detail in the case of carbon tetrachloride.[152] The analysis originated in the observation that the electrochemical reduction of CCl_4 in DMF is definitely faster than predicted by the classical dissociative electron transfer theory (a good fit between theory and experiment is obtained for a value of the bond dissociation energy of CCl_4, $D = 2.1$ eV while available independent data indicate a value of 2.8 eV). Is this discrepancy related to attractive interactions in the caged fragments that are neglected in the classical model? Figure 13 (curve a) shows the potential energy profile calculated in the gas phase for $CCl_3^\bullet, Cl^-$ (*ab initio* calculations at the MP2/6-31G* level). It exhibits a definite minimum, 0.4 eV deep in energy, and can be fitted with a Morse curve having the same repulsive term as the CCl_4 Morse curve (incidentally, a markedly different profile is found at short distances when a semi-empirical method such as PM3 is used as by Tikhomirov and German,[53] showing the lack of reliability of these semi-empirical techniques for this type of problem). Solvation, calculated by means of a Born-type continuum method with a uniform dielectric constant equal to the bulk value, suppresses the shallow minimum leading to a purely repulsive profile. The activation free energy estimated with this profile is still too large to fit the experimental data. It is

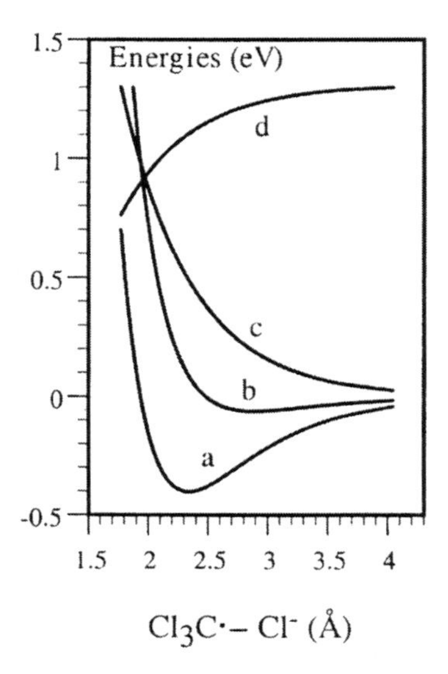

Fig. 13 Energy profiles for $Cl_3C^{\bullet}$, Cl^- in DMF. Curve a: potential energy in the gas phase. Curve b: potential energy in the solvent ($D'_P = 62$ meV). Curve c: variation of the solvation free energy. Curve d: solvent reorganization energy.

known that continuum dielectric methods tend to overestimate the solvation energy. One is thus led to envisage that a small interaction persists in solution and to attempt to relate the magnitude of this interaction to the effect it may have on the electron transfer kinetics according to the model developed in the preceding section, using D_P as an adjustable parameter. Actually, the model was refined by inclusion of the variation of the solvent reorganization energy, λ_0, with the R—X distance due to decrease of the solvation radius during the bond breaking process.

The difference between the energy profiles in the solvent and in the gas phase provides the solvation free energy, $G^{solv}_{R^{\bullet},X^-}$ (profile c in Fig. 13). The solvent reorganization energy, λ_0, may then be obtained from equation (60) as a function of the R—X distance (profile d in Fig. 13).

$$\lambda_0 = G^{solv}_{R^{\bullet},X^-} \frac{3.00}{3.38} \frac{\dfrac{1}{\varepsilon_{op}} - \dfrac{1}{\varepsilon_S}}{1 - \dfrac{1}{\varepsilon_S}} \tag{60}$$

(ε_{op} and ε_S are the optical and static dielectric constants of the solvent, respectively).

Application of this treatment (Fig. 13) led to the conclusion that a value of the interaction energy, D_P, as small as 0.062 eV suffices to fit the experimental data. The fact that this deviation from the classical dissociative electron transfer model appears without ambiguity in the particular case of CCl_4 is most likely related to the fact that the $CCl_3^{\bullet}$ radical bears a significant positive charge density on the carbon, because of the electron-withdrawing properties of the three Cl atoms, which allows the charge dipole interaction to be strong enough to survive in the presence of a polar solvent.

OTHER EXAMPLES

Another interesting case concerns a homogeneous electron transfer reaction, namely, the initiation step of the classical Kornblum–Russell reaction[153–155] of 2-nitropropanate ions with 4-nitrobenzyl chloride (Scheme 14).

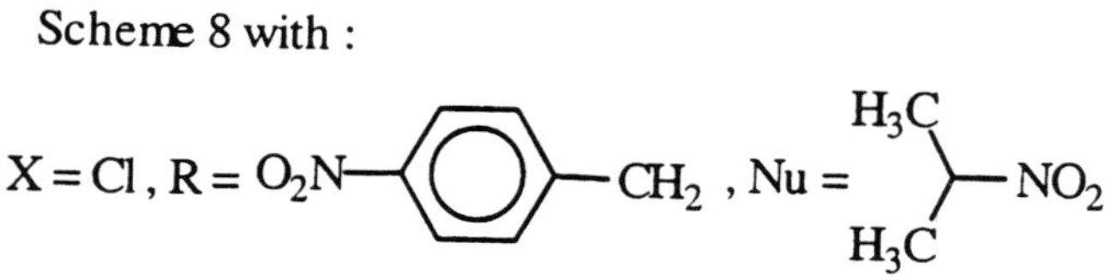

Scheme 8 with :

X = Cl, R = O_2N–C_6H_4–CH_2 , Nu = $(H_3C)_2C$–NO_2

Scheme 14

This is the first discovered $S_{RN}1$ reaction and the same question of how it is possible that such a poor electron donor as the 2-nitropropanate ion can initiate the reaction arises as in the case of the reaction of the same nucleophile with 4-nitrocumyl chloride that was discussed in the preceding section. A recent detailed study[156] of the reaction kinetics led to the following results. The reaction is faster than in the case of 4-nitrocumyl chloride. Thus, outer-sphere electron transfer from the nucleophile according to a stepwise mechanism is likewise much too slow to serve as a viable initiation step. However, unlike the 4-nitrocumyl chloride case, a classical dissociative electron transfer pathway predicts that the reaction is significantly slower than observed experimentally. Careful ruling out of other mechanisms involving the parallel S_N2 O-substitution pathway led to the conclusion that a small interaction (of the order of 0.1 eV) exists between the caged fragments, while it is negligible in the 4-nitrocumyl case. The rationale for the difference between the benzyl and the cumyl derivatives is that the attractive interaction in the cage is prevented in the latter case by steric hindrance and by electron donation from the two methyl groups borne by the benzylic carbon.

Other recent observations concerning substituted benzyl halides may also be rationalized within the same framework. Whereas the electrochemical reduction of 4-nitrobenzyl bromide in DMF is clearly a stepwise reaction, a concerted mechanism is observed with unsubstituted benzyl and 4-cyanobenzyl bromides.[30] The cyclic voltammetric peak potential of 4-cyanobenzyl bromide is significantly more positive than the cyclic voltammetric peak potential of benzyl bromide (by 250 mV at a scan rate of 0.1 V s^{-1}). It was inferred from these observations that the bond dissociation energy increases by 0.15 eV from the first to the second compound, in line with previous photoacoustic work[29] in which the substituent effect was regarded as concerning the starting molecule rather than the radical. However, further measurements using the same technique did not detect any substituent effect and the same conclusion was also reached in the gas phase by a low-pressure pyrolysis technique.[157] Recent quantum-chemical estimations[158] concluded that there is a small substituent effect of 0.07 eV, i.e., about half of the value derived from electrochemical experiments upon application of the classical dissociative electron transfer theory. These observations may be interpreted by a small attractive interaction in the caged product fragments that would be larger in the presence than in the absence of the cyano-substituent because of its electron-withdrawing character. An even larger similar effect is observed with phenacyl chloride and bromide, as expected from the electron-withdrawing effect of the carbonyl group. The apparent BDE values derived from cyclic voltammetry are 2.05 and 2.35 eV for the bromide and chloride, respectively,[159] whereas the values found by low-pressure pyrolysis are 2.75 and 3.13 eV, respectively.[160] The two sets of results may be reconciled after introduction of an interaction energy of 51 and 56 meV for the bromide and chloride, respectively.

The recently reported kinetics of the reduction of a series of organic disulfides[161,162] seem to enter this framework. Concerning diphenyl disulfide and a series of 4,4′-substituted derivatives, the values found for the electrochemical standard rate constant indicate a large intramolecular reorganization energy for the unsubstituted disulfide and for electron-donating substituents, close to the predictions of the classical dissociative electron transfer model. The standard rate constant increases rapidly upon introduction of more and more electron-withdrawing substituents, ending, in the case of the nitro-substituents, with values that are indicative of solvent reorganization with little intramolecular reorganization. These trends indicate that the interaction between the caged fragments is negligible in the unsubstituted derivative and in the case of electron-donating substituents. Because the positive charge on the functional sulfur is augmented, the interaction between the fragments increases with the electron-withdrawing character of the substituent. With the nitro-derivative, it has become a true bond. The unpaired electron and the negative charge are located on one nitrophenyl part of the molecule, and the dissociation of the anion radical is of the homolytic type.

IMPLICATIONS IN THE DICHOTOMY BETWEEN STEPWISE AND CONCERTED MECHANISMS

Figure 14 summarizes the effect of a small attractive interaction between the caged fragments in the product cluster on the dichotomy between stepwise and concerted mechanisms and on the passage from one mechanism and the other. The case shown in Fig. 14 corresponds to a situation in which the intermediate is more unstable than the final products, as is the case when it results from the injection of an electron or a hole in an orbital different from the orbital of the bond being broken, its cleavage then being an intramolecular dissociative electron transfer. Under these conditions, the distinction between stepwise and concerted mechanisms is not affected by the existence of a small attractive interaction between the caged fragments. The driving force offered to the reaction is expected to be a governing factor of the passage from the concerted to the stepwise mechanism, as it is when the interaction between fragments is negligible. In particular, the possibility of characterizing the transition from one mechanism to the other by changing the scan rate in cyclic voltammetry in its plain or convoluted versions remains in the presence of a significant

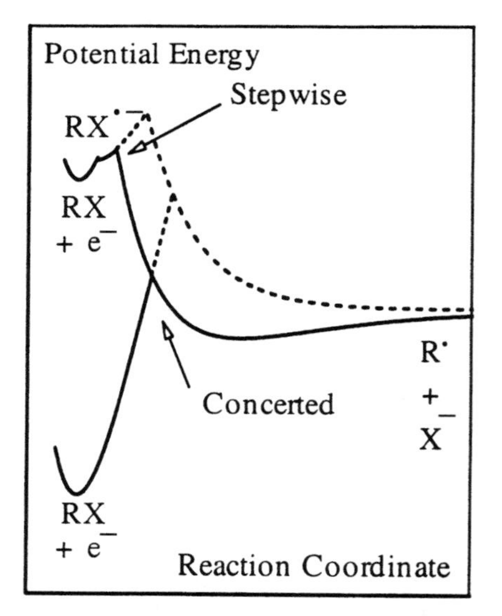

Fig. 14 Potential energy profile for stepwise and concerted mechanisms with (solid lines) and without (dotted lines) an attractive interaction between the caged fragments in the product cluster. The case of the reduction of a neutral substrate is represented. It can be transposed for reductions of a positively charged substrate or for oxidations of neutral or negatively charged substrates.

interaction. It is also worth noting that the dynamics of the concerted reaction is affected by the magnitude of the interaction (as depicted earlier in this section) but, in the case of a stepwise mechanism, the dynamics of the primary radical cleavage is also affected in a similar manner. The intramolecular electron transfer model depicted in Section 4 may indeed be adapted to this situation by replacing, in equation (44), ΔG^0_{cleav} by $\Delta G^0_{cleav} - \Delta G^0_{sp}$ and $D_{(+)-\bullet RX}$ by $(\sqrt{D_{(+)-\bullet RX}} - \sqrt{D_P})^2$.

The situation is different when the primary radical cleaves homolytically into separated fragments that are in all practical cases of higher energy, if not of higher free energy, than the primary radical itself. Under these conditions, the passage from a stepwise to a concerted mechanism is progressive, simply depending upon the nature and magnitude of the interaction between the fragments according to whether they are relevant to a true bond or are of the charge–dipole type. The observation of a transition between two mechanisms upon changing the driving force (as, for example, changing the scan rate in cyclic voltammetry) is not expected in such cases.

6 Photoinduced dissociative electron transfers

PHOTOINDUCED VERSUS THERMAL PROCESSES

In the preceding sections we have discussed thermal heterogeneous and homogeneous reductive and oxidative cleavages. These reactions may also be triggered photochemically by means of a sentitizer, as illustrated in Scheme 15 in the case of the reductive cleavage of a neutral substrate (transposition to oxidation and to charged substrate is straightforward).

$$D + h\nu \longrightarrow D^*$$
$$D^* + RX \longrightarrow D^{\bullet +} + RX^{\bullet -} \longrightarrow R^\bullet + X^- \text{ (stepwise)}$$
$$D^* + RX \longrightarrow D^{\bullet +} + R^\bullet + X^- \text{ (concerted)}$$

Scheme 15

An attractive way of fighting energy-wasting back electron transfer in photoinduced electron transfer reactions is to use a system in which either the acceptor or the donor in the resulting ion-pair, or both, undergo a fast cleavage reaction.[163–165] The occurrence of a concerted electron transfer/bond-breaking reaction rather than a stepwise reaction (Scheme 15) thus intuitively appears as an extreme and ideal situation in which the complete quenching fragmentation quantum yield, Φ_∞, should be unity.[163–165] From a diagnostic standpoint, the observation of a quantum yield smaller than unity would thus rule out the occurrence of a dissociative electron transfer mechanism.[166–168] So far no example of unity quantum yields involving an acceptor or a donor containing a frangible bond has been reported. Based on the aforementioned intuition, the intermediacy of a discrete ion radical was thus inferred in these systems.[165,166,170] In contrast, as discussed earlier, the occurrence of thermal, electrochemical or homogeneous dissociative electron transfers is well documented. The dichotomy between the concerted and stepwise mechanisms and the passage from one mechanism to the other have been observed unambiguously.

There are a few, but remarkable, examples where, for the same cleaving acceptor, the photochemical reaction was deemed, on the basis of a quantum yield smaller than unity, to follow a stepwise mechanism whereas the electrochemical reaction was reported to follow a concerted mechanism. One concerns the reaction of benzyl and 4-cyanobenzyl bromides with the excited state of the diphenylmethyl radical[166] or of pinacols;[167] the other concerns the reductive cleavage of the 4-cyanobenzylmethylphenyl sulfonium cation by the excited singlets of a series of aromatic compounds.[168] For all these compounds, the electrochemical reductive cleavage appears to follow a concerted mechanism.[33,58] The latter example is particularly interesting because great care was taken, by appropriate choice of the donors, to avoid the occurrence of electron transfer between the donor cation radical and the 4-cyanobenzyl radical, followed by regeneration of the starting sulfonium cation by coupling of the resulting 4-cyanobenzyl cation with methyl phenyl sulfide side-reactions, thus wasting the photochemical energy. Avoiding this type of side-reaction is crucial in studies aiming at relating the quantum yield and the dissociative character of the electron transfer/bond-breaking process.

How can these photochemical and electrochemical data be reconciled? With the benzylic molecules under discussion, electron transfer may involve the π^* or the σ^* orbital, giving rise to stepwise and concerted mechanisms, respectively. This is a typical case where the mechanism is a function of the driving force of the reaction, as evoked earlier. Since the photochemical reactions are strongly down-hill whereas the electrochemical reaction is slightly up-hill at low scan rate, the mechanism may change from stepwise in the first case to concerted in the second. However, regardless of the validity of this interpretation, it is important to address a more fundamental question, namely, *whether it is true, from first principles, that a purely dissociative photoinduced electron transfer is necessarily endowed with a unity quantum yield* and, more generally, to establish what are the expressions of the quantum yields for concerted and stepwise reactions.

QUANTUM YIELD FOR CONCERTED REACTIONS

The problem[54] is simplified if one may consider that bond stretching and cleavage is the only significant reorganization factor. Then on the diagram

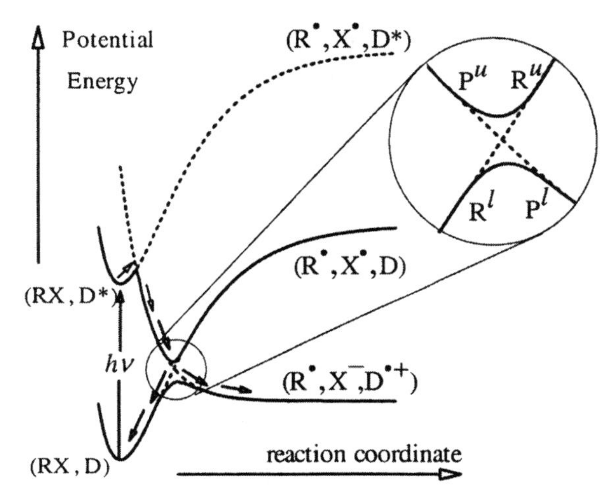

Fig. 15. Section of the zero-order (...) and first-order (—) potential energy surfaces along the reaction coordinate in cases where stretching of the cleaving bond is the dominant factor of nuclei reorganization.

depicting the potential energy curve of the three states involved, Fig. 15, one can see that after overcoming a small activation barrier, the system approaches the region where the ground state and fragment zero-order surfaces cross each other, i.e., the region where the transition state of the thermal electron transfer between the donor ground state and the cleaving acceptor is located. The system would proceed exclusively to the fragmented products only if the electronic matrix coupling element at the intersection, H, were nil, i.e., if the ground-state electron transfer were purely non-adiabatic. This is not usually the case and the system will thus partition between back electron transfer and fragmentation through the funnel offered by the upper first-order surface, leading to a value of the quantum yield smaller than unity.

A semi-classical treatment[171–175] of the model depicted in Fig. 15, based on the Morse curve theory of thermal dissociative electron transfer described earlier, allows the prediction of the quantum yield as a function of the electronic matrix coupling element, H.[54] The various states to be considered in the region where the zero-order potential energy curves cross each other are shown in the insert of Fig. 15. The treatment of the whole kinetics leads to the expression of the complete quenching fragmentation quantum yield, Φ_∞, given in equation (61)

$$\Phi_\infty = \frac{1}{(1+P)\left(1 + \dfrac{2P}{1+P}\dfrac{k_{-\mathrm{act}}}{k_{\mathrm{sp}} + k_{\mathrm{cc}}}\right)} \tag{61}$$

where k_{sp} is the rate constant at which the three fragments, $D^{\bullet +}$, $R^{\bullet}$, and X^{-}, diffuse away from the solvent cage and k_{cc} is the rate constant for the coupling of $R^{\bullet}$ and $D^{\bullet +}$. P is the probability that the system remains on the first-order potential energy surfaces formed by combination of the zero-order potential surfaces near their intersection.

The term $k_{-act}/(k_{sp} + k_{cc})$ reflects the possibility that back electron transfer could take place within the caged fragment cluster. In a number of cases, the activation barrier for this reaction is too high for it to compete with the diffusion of the fragments out of the cage. The quantum yield then takes its maximal value, being given simply by equation (62), which reflects the funnel partitioning between products and back electron transfer.

$$\Phi_{\infty} = \frac{1}{1 + P} \tag{62}$$

The probability P in equations (61) and (62) may be related to the electronic coupling matrix element through equation (63) by application of the Landau–Zener model:

$$P = 1 - \exp[-\pi^{3/2}H^2/h\nu(RTD)^{1/2}] \tag{63}$$

(D is the dissociation energy of the cleaving bond and ν is the frequency at which the system crosses the intersection region.)

In fact, besides bond cleavage, solvent reorganization may play an important role in the dynamics of dissociative electron transfer. It is thus necessary to generalize the preceding treatment, replacing the potential energy profiles (Fig. 15) by potential energy surfaces involving two coordinates that represent solvent reorganization and bond stretching, respectively. In this connection, expressions for the potential energy (or free energy) surfaces similar to those used above in the modeling of thermal dissociative electron transfer (equations 19 and 20) can be used here for the excited reactant state, R^*, the ground state reactant system, R, and the fragments system, P (the G^0 values are the free energies of the standard states at equilibrium):

$$G_{R^*} = G^0_{R^*} + \lambda_{0,1}X_1^2 + DY_1^2 \tag{64}$$

$$G_{R} = G^0_{R} + \lambda_{0,1}X_1^2 + DY_1^2 \tag{65}$$

$$G_{P} = G^0_{P} + \lambda_{0,1}(1 - X_1)^2 + D(1 - Y_1)^2 \tag{66}$$

Past the transition state of the photoinduced reaction, $\ddagger^*$, the system goes down the repulsive product surface, P, until P crosses the intersection with the reactant surface R. These two surfaces are sketched in Fig. 16a. Their intersection is a parabola (the central line in Fig. 16c) whose projection in the X_1–Y_1 plane (straight line in Fig. 16b) is defined by equation (67).

$$G^0_P - G^0_R = \lambda_0(2X_1 - 1) + D(2Y_1 - 1) \tag{67}$$

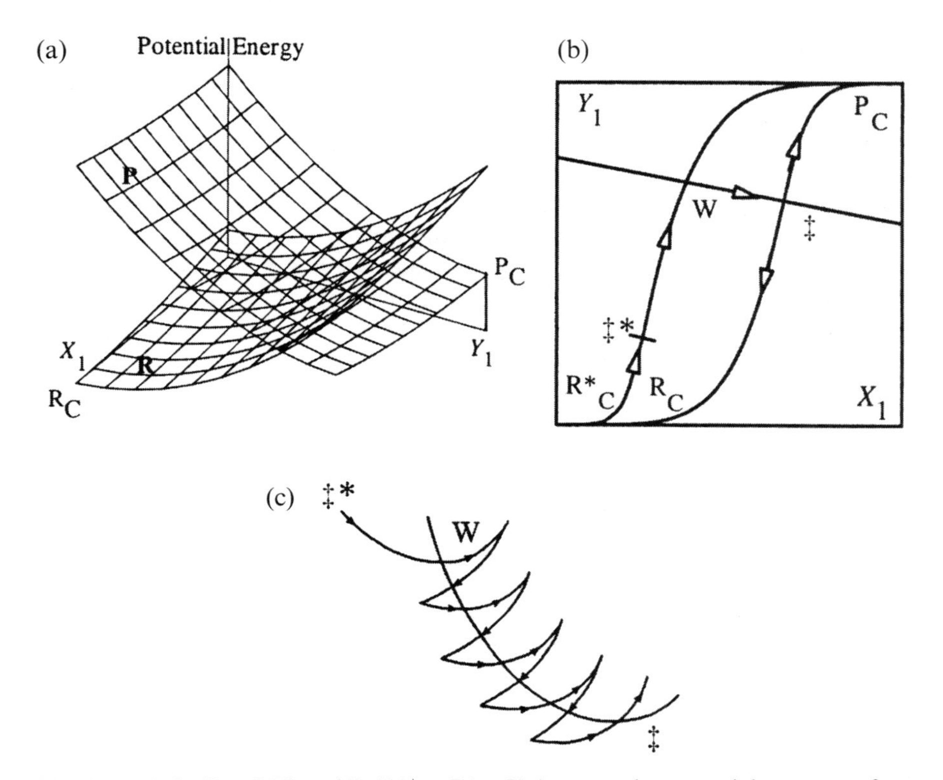

Fig. 16 (a): R (D + RX) and P ($D^{\bullet +} + R^{\bullet} + X^-$) zero-order potential energy surfaces. R_C and P_C are the caged systems. (b) Projection of the steepest descent paths on the X–Y plane; ‡*, transition state of the photoinduced reaction; ‡, transition state of the ground state reaction; W, point where the photoinduced reaction path crosses the intersection between the R and P zero-order surfaces; R^*_C, caged reactant system. (c) Oscillatory descent from W to ‡ on the upper first-order potential energy surface obtained from the R and P zero-order surfaces.

The minimum on the intersection parabola is the saddle point corresponding to the transition state of the dark reaction, denoted ‡ in Figs 16b and 16c. The first-order potential energy surfaces involve an upper surface associating the portions of the R and P zero-order potential energy surfaces situated above the intersection parabola and a lower surface associating the portions of the R and P zero-order potential energy surfaces situated below the intersection parabola.

The projection on the X_1–Y_1 plane of the steepest descent path followed by the system during the photoinduced reaction (calculated as in Section 3) is shown in Fig. 16b. Past the saddle point corresponding to the transition state of the photoinduced reaction, ‡*, the system follows the steepest descent path on the P surface en route to the caged product state, $X_1 = 1$, $Y_1 = 1$, until it reaches the intersection between the P and R surfaces in a point denoted W in Figs 16b and 16c. The fraction of the system that remains on the first-order potential energy surface then bounces down from point W to the minimum ‡, as represented schematically in Fig. 16c. Along this path, part of the system goes continuously to the lower first-order potential energy surface, thus partitioning between two paths, one leading to the caged product, P_C, and the other to the caged ground-state reactants, R_C. The projection on the X_1–Y_1 plane of the steepest descent paths from ‡ to P_C and R_C (calculated as in Section 3) is shown in Fig. 16b. Analysis of the dynamics along this reaction path[54] showed that equations (61)–(63) remain applicable.

QUANTUM YIELD FOR STEPWISE REACTIONS

The reaction pathways derive from the consideration of the following free energy surfaces according to the procedure depicted in Section 3. For simplicity in the use of symbols, the treatment is for a reductive process.

The photoinduced electron transfer step and the back electron transfer step involve three states, excited reactant (R^*), ground-state reactant (R) and intermediate (I):

$$G_{R^*} = G^0_{R^*} + (\sqrt{D} - \sqrt{D'})^2 Y_2^2 + \lambda_{0,2} X_2^2 \tag{68}$$

$$G_R = G^0_R + (\sqrt{D} - \sqrt{D'})^2 Y_2^2 + \lambda_{0,2} X_2^2 \tag{69}$$

$$G_I = G^0_I + (\sqrt{D} - \sqrt{D'})^2 (1 - Y_2)^2 + \lambda_{0,2}(1 - X_2)^2 \tag{70}$$

For the cleavage step, the states involved are I and the fragments, P:

$$G_I = G^0_I + D' Y_3^2 + \lambda_{0,3} X_3^2 \tag{71}$$

$$G_P = G^0_P + D'(1 - Y_3)^2 + \lambda_{0,3}(1 - X_3)^2 \tag{72}$$

Two situations may be distinguished according to whether the electron transfer step of the stepwise mechanism lies in the normal region or in the inverted region.

Back electron transfer in the inverted region

Past the transition state of the photoinduced reaction, the point representing the system reaches the ion-radical intermediate, I, before crossing the potential energy surface of the ground state as illustrated in Fig. 17. The intermediate I may then competitively undergo back electron transfer going over the inverted region barrier (rate constant k_{-act}) and cleavage (rate constant k_c) according to Scheme 16.

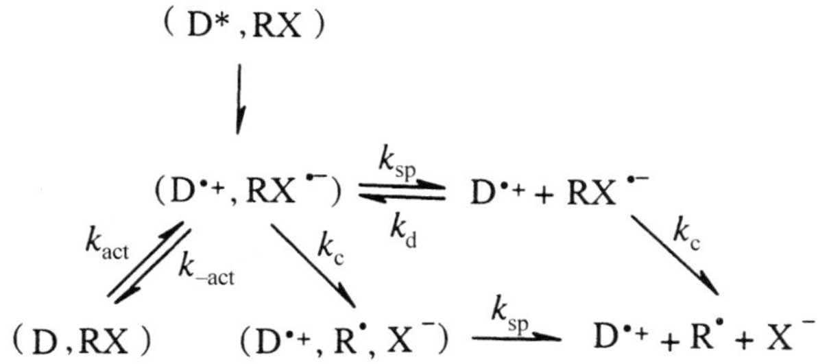

$$
\begin{array}{c}
(D^*, RX) \\
\downarrow \\
(D^{\bullet +}, RX^{\bullet -}) \underset{k_d}{\overset{k_{sp}}{\rightleftharpoons}} D^{\bullet +} + RX^{\bullet -} \\
k_{act} \nearrow\!\!\swarrow k_{-act} \qquad \searrow k_c \qquad\qquad\qquad \searrow k_c \\
(D, RX) \qquad (D^{\bullet +}, R^{\bullet}, X^-) \xrightarrow{k_{sp}} D^{\bullet +} + R^{\bullet} + X^-
\end{array}
$$

Scheme 16

The photoinduced and the back electron transfer pathways depend on the two coordinates, X_2 and Y_2, representing the solvent reorganization and bond stretching, respectively (Fig. 17a). Their projections on the X_2/Y_2 plane (Fig. 17a′) cross the straight lines representing the projections of the intersection between the excited reactant state surface and the intermediate surface (full line) and of the intersection between the ground-state reactant surface and the intermediate surface (dotted line). Although the photoinduced and back electron transfer pathways have different projections, the element of arc length, dZ_2, is the same function of $\lambda_{0,2}$ and $(\sqrt{D} - \sqrt{D'})^2$ in both cases. The same reaction coordinate, Z_2, as defined in the preceding section, may thus be used for the two pathways, bearing in mind that their traces on the X_2–Y_2 plane are not the same. For cleavage, the reaction coordinate is the same Z_3 as defined is Section 3. Using these two coordinates, the whole stepwise mechanism, involving photoinduced electron transfer, back electron transfer and cleavage of the intermediate, is illustrated by Figs 17 b, b′ and b″.

The quantum yield thus results simply from the treatment of the reaction kinetics in Scheme 16:

$$\Phi_\infty = \frac{k_{sp} + k_c}{k_{sp} + k_c + k_{-act}} \tag{73}$$

Back electron transfer in the normal region

In this case, past the transition state of the photoinduced reaction, ‡*, the system reaches the intersection between the I surface and the ground state (R) surface in a point W (Fig. 18a and its projection on the X_2–Y_2 plane, Fig. 18a′) before reaching the minimum corresponding to I. The transition state of the ground-state electron transfer reaction, ‡, is also located on this intersection. The system thus bounces down from W to ‡ while passing from the upper to the lower first-order surface, similarly to the previously described case of a dissociative electron transfer reaction, thus partitioning between back electron transfer and formation of the intermediate as a function of the magnitude of the electronic coupling matrix element between the I and R states, H. The ion radical in I then

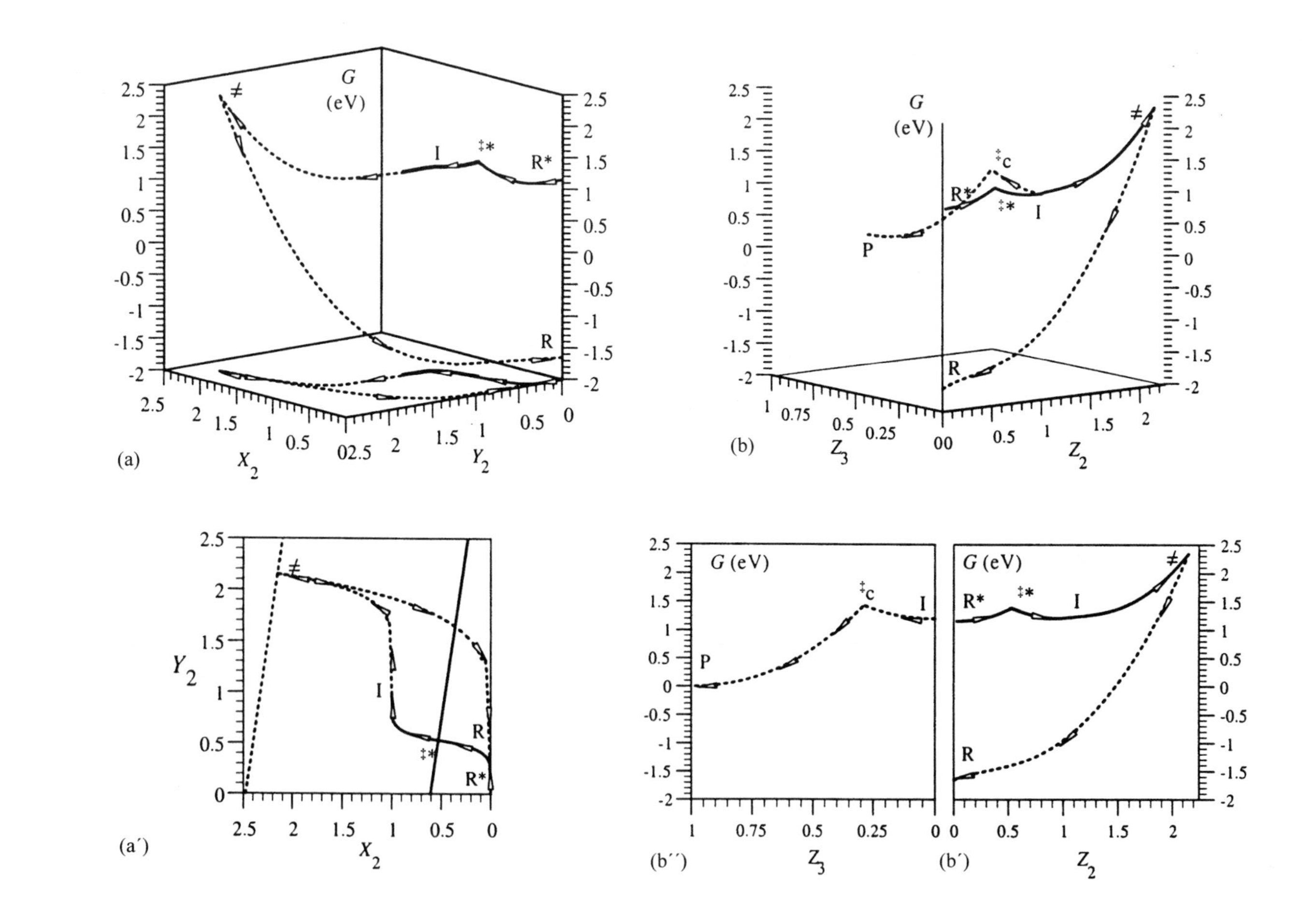

(a)
G (eV)
≠
‡*
I
R*
R
X2
Y2
(b)
G (eV)
‡c
R*
‡*
I
P
R
Z3
Z2
(a′)
≠
I
R
‡*
R*
Y2
X2
(b′′)
G (eV)
‡c
I
P
Z3
(b′)
G (eV)
R*
‡*
I
≠
R
Z2

competitively undergoes back electron transfer and cleavage. Although the photoinduced and back electron transfer pathways have different projections, the element of arc length, dZ_2, is the same function of $\lambda_{0,2}$ and $(\sqrt{D} - \sqrt{D'})^2$ in both cases. The same reaction coordinate, Z_2, as defined in the preceding section may be used for the two pathways, bearing in mind that their traces on the X_2/Y_2 plane are not the same. For cleavage, we introduce the reaction coordinate Z_3, as defined in the preceding section. Using these two coordinates, Figs 18b, b′ and b″ give a representation of the whole stepwise mechanism involving photoinduced electron transfer, back electron transfer and cleavage of the intermediate. We see in these representations as well as in Scheme 3, which summarizes all the reactions involved, that back electron transfer interferes twice: once at the intersection between the zero-order I and R surfaces and, a second time, from the intermediate through back crossing of the ground-state electron transfer barrier. P is the probability that the system remains on the first-order potential energy surfaces formed by combination of the zero-order potential surfaces near their intersection. Thus, pursuing the treatment of the kinetics depicted in Scheme 17 exactly in the same manner as for dissociative electron transfer leads to the expression for the quantum yield given in equation (74):

$$\Phi_\infty = \frac{1}{(1+P)\left(1 + \frac{2P}{1+P}\frac{k_{-\mathrm{act}}}{k_{\mathrm{sp}} + k_{\mathrm{c}}}\right)} \tag{74}$$

In the framework of the Landau–Zener model, P is related to H by means of equation (75). These equations are also valid when both the stretching and solvent reorganization coordinates are taken into account as in the case of dissociative electron transfer.

$$P = 1 - \exp\left[\frac{\pi^{3/2}H^2}{h\nu(RT)^{1/2}[(\sqrt{D} - \sqrt{D'})^2 + \lambda_{0,2}]^{1/2}}\right] \tag{75}$$

where ν is the effective vibration frequency of the reactants.

Fig. 17 Reaction pathway for induced reaction following a stepwise mechanism in which the electron transfer is in the inverted region. In eV, $G^0_{\mathrm{D^*+RX}} = 1.15$, $G^0_{\mathrm{D+RX}} = -1.65$, $G^0_{\mathrm{D^{\bullet+}+RX^{\bullet-}}} = 1.2$, $G^0_{\mathrm{D^{\bullet+}+R^\bullet+X^-}} = 0$, $(\sqrt{D} - \sqrt{D'})^2 = 0.1$, $D' = 1.8$, $\lambda_{0,2} = 0.75$, $\lambda_{0,3} = 1$. $\mathrm{R^* = D^* + RX}$; $\mathrm{R = D + RX}$; $\mathrm{I = D^{\bullet+} + RX^{\bullet-}}$; $\mathrm{P = D^{\bullet+} + R^\bullet + X^-}$. ‡*, ‡, ‡c represent transition states of the photoinduced electron transfer, back electron transfer and cleavage, respectively.

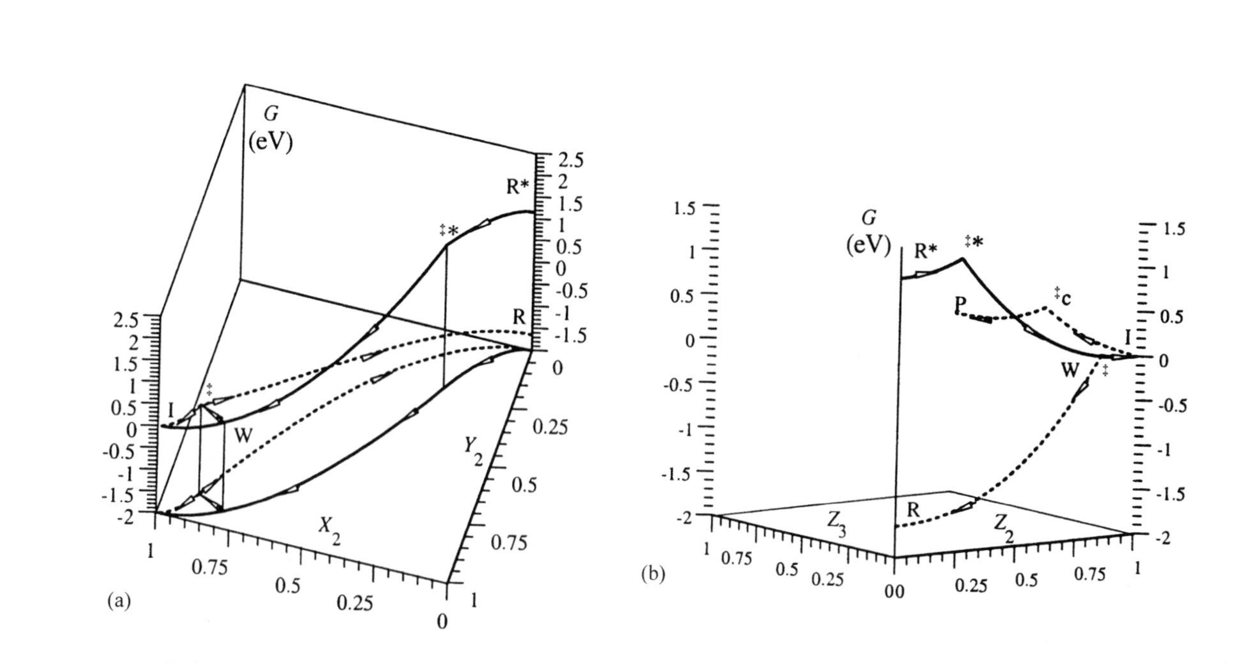

(a)
G
(eV)
R*
‡*
R
I
‡
W
X_2
Y_2
(b)
G
(eV)
R*
‡*
‡c
P
I
W
‡
R
Z_3
Z_2

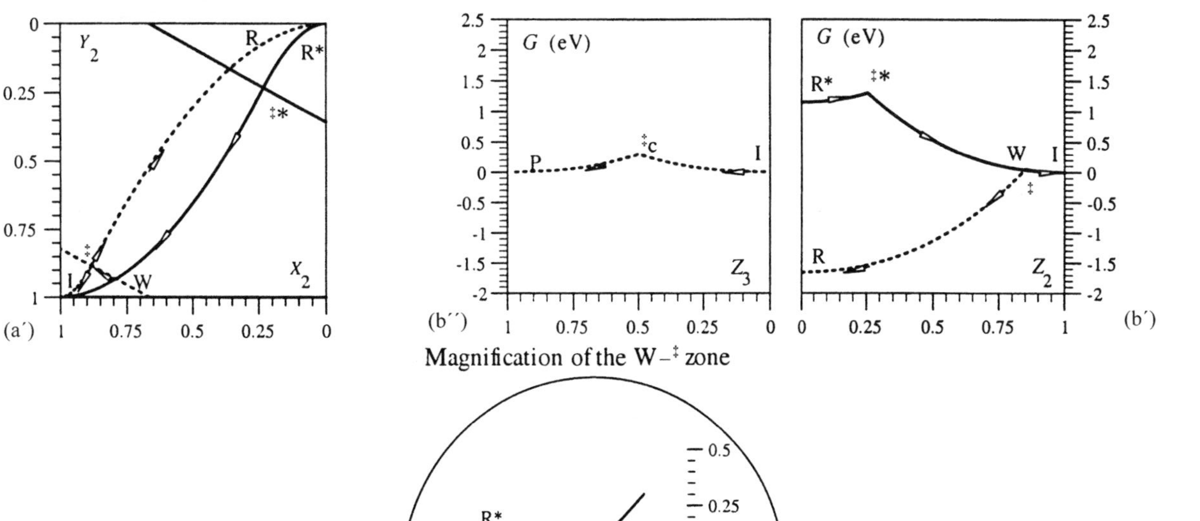

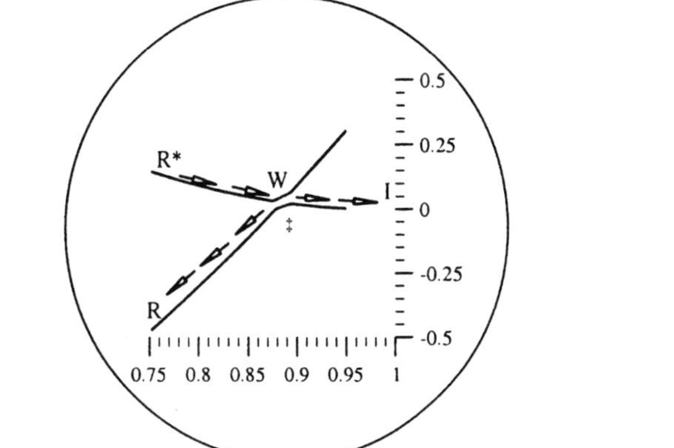

Fig. 18 Reaction pathway for a photoinduced reaction following a stepwise mechanism in which the electron transfer is in the normal region. In eV, $G^0_{D^*+RX} = 1.15$, $G^0_{D+RX} = -1.65$, $G^0_{D^{\bullet+}+RX^{\bullet-}} = 0$, $G^0_{D^{\bullet+}+R^{\bullet}+X^-} = 0$, $(\sqrt{D} - \sqrt{D'})^2 = 1.4$, $D' = 0.22$, $\lambda_{0,2} = 0.75$, $\lambda_{0,3} = 1$. $R^* = D^* + RX$; $R = D + RX$; $I = D^{\bullet+} + RX^{\bullet-}$; $P = D^{\bullet+} + R^{\bullet} + X^-$.

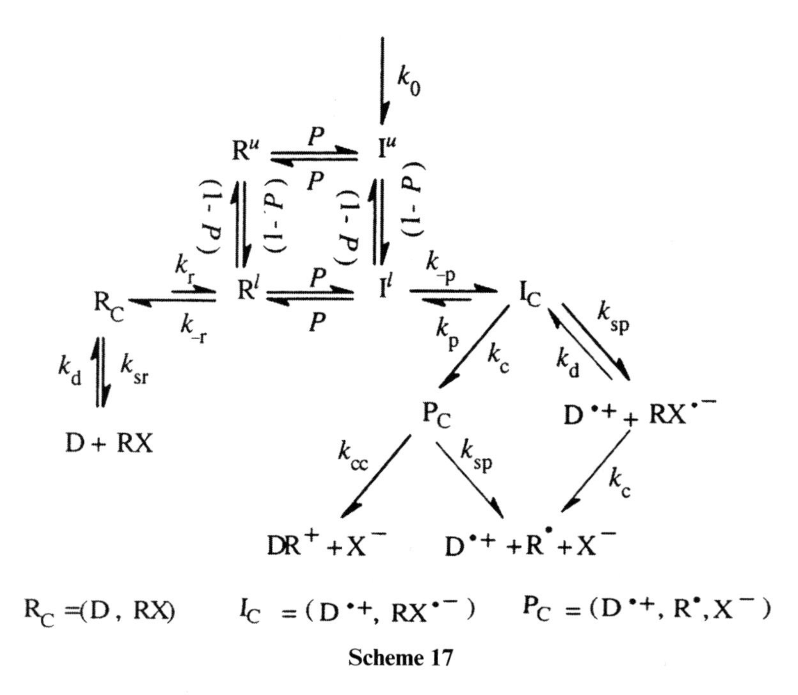

Scheme 17

In total, the situation is less favorable in terms of quantum yields than in the preceding case. Indeed, even if the cleavage rate constant is large enough to overcome back electron transfer from the ion pair ($k_c \gg k_{-act}$), the maximal value of the quantum yield is $1/(1 + P)$, which would reach unity only in the unlikely case where the ground-state electron transfer was entirely non-adiabatic ($H = 0$).

EXAMPLES

As noted before, an interesting example is the reductive cleavage of the 4-cyanobenzylmethylphenyl sulfonium cation by the excited singlets of a series of aromatic compounds.[168] Another example has recently been analyzed in detail. It deals with the reduction of carbon tetrachloride (yielding $Cl_3C^{\bullet}$ and Cl^-) by the excited singlets of perylene and of 2-ethyl-9,10-dimethoxyanthracene (EDA).[152] Although large (0.70 ± 0.05 and 0.77 ± 0.05, respectively), the complete quenching quantum yields are clearly below unity.

It was checked that neither Cl^- (standard potential of the $Cl^{\bullet}/Cl^-$ couple in DMF = 1.81 V versus SCE) nor $Cl_3C^{\bullet}$ (standard potential of the $Cl_3C^+/Cl_3C^{\bullet}$

couple = 1.33 V versus SCE) are able to reduce the EDA and perylene cation radicals (standard potentials 1.00 and 1.03 V versus SCE, respectively) to any significant extent. Electron transfer between Cl^- or $Cl_3C^\bullet$ to the cation radical of the sentitizer, followed by combination of $Cl_3C^\bullet$ with $Cl^\bullet$ or of Cl_3C^+ with Cl^- to regenerate CCl_4, thus cannot be taken as responsible for the fact that the quantum yields are less than unity.

In the application of equation (61), k_{-act} may be estimated using the model developed in Section 5, keeping the same value of the interaction energy between caged fragments as determined to fit the electrochemical data, leading to $k_{-act} = 8.5 \times 10^7\ s^{-1}$ and $2.1 \times 10^8\ s^{-1}$ for EDA and perylene, respectively. k_{sp} is estimated as $8 \times 10^8\ s^{-1}$, taking into account the value of the interaction energy between caged fragments. For both sentitizers, the main reason for observing a less-than-unity quantum yield is partitioning of the system at the intersection of the product and ground-state potential energy surfaces as opposed to back electron transfer from the fragment cluster. Application of equation (61) thus leads to $P = 0.27$ and 0.34 for EDA and perylene, respectively (the values would have been 0.3 and 0.43, respectively, if back electron transfer had been neglected as in equation (62)).

Thus, from equation (63), the magnitude of the electronic coupling matrix element may finally be estimated, leading to values of 21 and 24 meV for EDA and perylene, respectively. That these values are quite reasonable derives from the observation that they correspond to moderately non-adiabatic electron transfer at the ground state (with electronic factors of $2P/(1 + P) = 0.5$ and 0.6 with EDA and perylene, respectively).

7 Dichotomy and connections between S_N2 reactions and dissociative electron transfers

The S_N2 reaction is a good example of the dichotomy and connection between electron-pair transfer chemistry and single electron transfer (ET) chemistry and this is why many investigations and much debate have centered around the S_N2 reaction in this respect. From an experimental standpoint, a starting observation was that alkyl halides may alkylate aromatic anion radicals prepared as ion pairs from the reaction of the parent aromatic compound with an alkali metal.[176,177] Successive stereochemical studies[178] concerned the reaction of the anion radical of anthracene electrochemically generated in *N,N'*-dimethylformamide (DMF) in the presence of a quaternary ammonium cation, with several optically active 2-octyl halides. Most of the product was found to be racemic. However, a small but distinct amount of inverted product was also detected (of the order of 10% and slightly dependent on the halogen atom). This observation was interpreted as indicating the existence of competition between an S_N2

pathway, leading to the inverted product, and a single electron transfer pathway (ET), yielding the racemic product as depicted in Scheme 18.

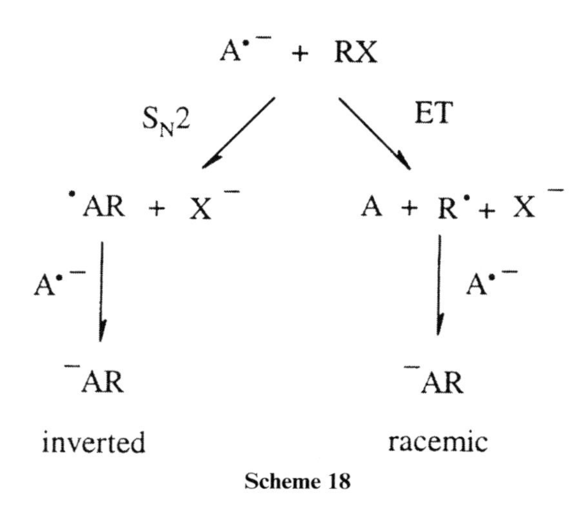

Scheme 18

Further experimental studies involved the determination of the rate constant of the reaction of several alkyl halides with a series of electrochemically generated anion radicals so as to construct activation driving force plots.[39,40,179] Such plots were later used to test the theory of dissociative electron transfer (Section 2),[22,49] assuming, in view of the stereochemical data,[178] that the S_N2 pathway may be neglected before the ET pathway in their competition for controlling the kinetics of the reaction.

Aromatic anion radicals thus appeared as prototypes of outer-sphere electron donors in dissociative electron transfer reactions as they were considered to be involved in simple outer-sphere reactions where no bond is broken concertedly with electron transfer. Based on this outer-sphere character, activation/driving force plots obtained with a series of aromatic anion radicals were used to judge the S_N2 or ET character of the reaction of an unknown nucleophile (viz. electron donor) with the same cleaving acceptor according to a kinetic advantage approach.[41,130,179–181] In this connection, classical S_N2 reactions such as halide self-exchange reactions exhibit rate constants that are much larger than predicted from the ET activation/driving force plot, thus confirming unambiguously their S_N2 character.[182]

However, a more accurate comparison between the experimental reaction kinetics and the predictions of the dissociative electron transfer theory revealed that the agreement is good when steric hindrance is maximal (tertiary carbon acceptors) and that the reaction is increasingly faster than predicted as steric hindrance decreases.[31] These results were interpreted as indicating an increase of the ET character of the reaction as steric hindrance increases. Similar conclusions were drawn from the temperature dependence of the kinetics, which revealed that the entropy of activation increases with steric hindrance, paralleling the increase of the ET character of the reaction.[41–43]

From these experimental data and/or from general considerations, two conceptions of the S_N2/ET dichotomy and of the passage between one mechanism and the other have emerged. One involves competition between two distinct pathways, implying the existence of two distinct transition states on the potential energy hypersurface representing the reacting system, each connected to the S_N2 and ET products, respectively.[1,31,41,178] The S_N2 pathway is accordingly favored in terms of energy by bonded interactions in the transition state that do not exist in the ET transition state. However, the latter has a looser structure and the ET reaction is less demanding in terms of directionality. The ET pathway thus possesses an entropy advantage over the S_N2 pathway. The effect of steric hindrance is to diminish the energy gain offered by bonded interactions in the S_N2 transition state and thus to favor the ET pathway.

The other conception is that the ET and S_N2 mechanisms are the ends of a continuous mechanistic spectrum, implying that a single transition state is to be found on the potential energy hypersurface of the reacting system.[40,181,183] In line with this conception is the contention that, with very few exceptions, homogeneous electron transfer reactions should be regarded as inner-sphere rather than outer-sphere processes with strong avoided crossing energies and tight geometries of their transition state.[184,185] As pointed out earlier,[31] entropy factors as well as energy factors are essential to the description of the reactions dynamics and product distribution. An extremely detailed analysis of the problem has been carried out in the framework of a Morse curve model of the reaction,[186] generalizing the Morse curve model of dissociative electron transfer reactions (Section 2).

Several *ab initio* quantum-chemical studies have been carried out on simple systems, much simpler than the experimental systems, aiming at determining which of the two conceptions evoked earlier best represents the course of these reactions. Contrasting results were obtained, leading to diverging conclusions (for recent reviews, see Speiser[8] and Zipse[9]). One example is the reaction of $H_2C{=}O^{\bullet-}$ with CH_3Cl, which leads to electron transfer products ($H_2C{=}O$ + $CH_3^{\bullet}$ + Cl^-) and to two types of substitution products deriving from O-substitution and C-substitution, respectively. One study finds an O- and a C-substitution transition state,[187] with the first transition state leading to the O-substitution product while the second is connected to electron transfer products by a reaction path obtained with ROHF calculations in non-mass-weighted coordinates. In a second study of the same system,[188] three transition states were found, namely the same O- and C-substitution transition states as before and, in addition, an outer-sphere electron transfer transition state. Following the steepest descent reaction pathways by the IRC method using,

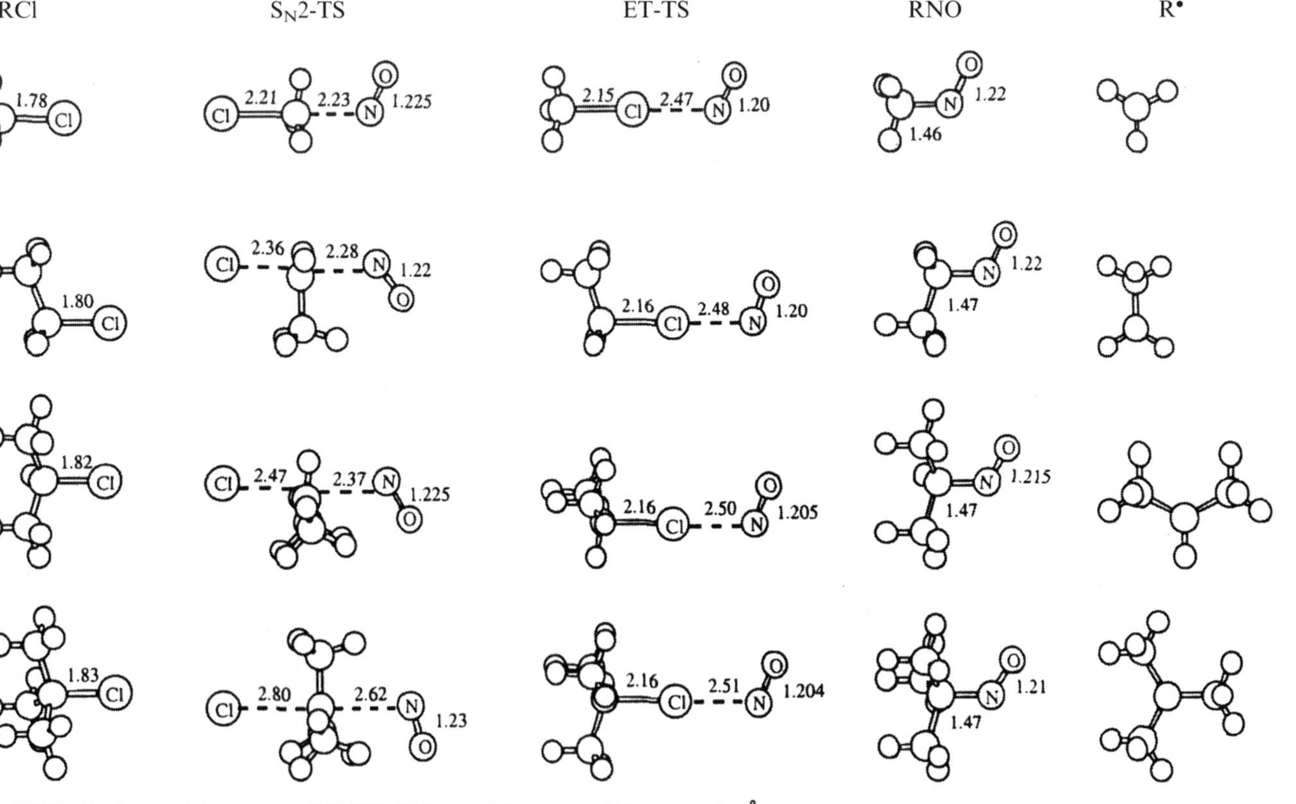

S_N2-TS=S_N2 transition state; ET-TS=ET transition state. Distances in Å.

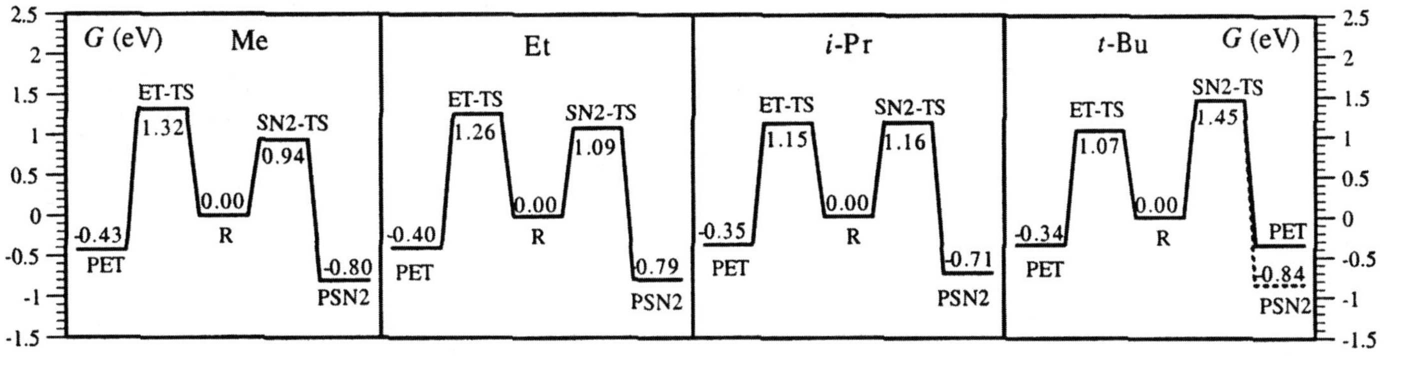

Fig. 19 UHF-MP2 *ab initio* calculation of the reaction of NO^- with alkyl chlorides. Energetic and geometric characteristics of the reactant, product and transition state systems as functions of steric hindrance.

more correctly, UHF calculations in mass-weighted coordinates, revealed that all three transition states are connected to the corresponding products. The same behavior was found with two other model systems, namely, $H_2C{=}O^{\bullet -}$ + CH_3F and $H_2C{=}CH_2^{\bullet -}$ + CH_3F.[188]

Further investigations indicated that, after passing through the C-substitution transition state, the system may bifurcate (for general analyses of mechanism bifurcation see Refs 189–194) and partition between the C-substitution products and the ET products.[195] However, the structure of the potential energy hypersurface in the vicinity of the C-substitution transition state and the very fact that the IRC–UHF pathway in mass-weighted coordinates leads to the C-substitution products, indicate that the percentage of ET products obtained by this mechanism should be much smaller than the percentage of C-substitution products.

In a recent study,[196] the effect of steric hindrance is treated in the framework of a mechanism in which a bifurcation follows the passage through an S_N2-like transition state. Bifurcation then appears to be increasingly favorable to the ET products as steric hindrance increases. According to this conception, there is a single transition state (the S_N2 transition state) for both reactions and product selection would be the result of a bifurcation following the transition state. The effect of steric hindrance would thus be to increase the energy of the S_N2 transition state and to make the bifurcation more and more favorable to the ET reaction. These views contrast with earlier results[188] in the sense that they do not take into account existence of an outer-sphere ET pathway going through a distinct ET transition state. Even if it is true that this ET transition state has a higher energy than the S_N2 transition state, the difference is less in terms of free energy and one might guess that steric hindrance will increasingly favor the ET transition state, possibly making the corresponding pathway predominant with tertiary alkyl derivatives.

This problem has been addressed recently[197] with model reactions in which the nucleophile (viz. electron donor) is NO^-. It has been found that outer-sphere transition states are easier to compute with NO^- than with other electron donors, thus allowing a complete characterization of the competition between the various pathways.

Figure 19 summarizes the geometric and free energy characteristics of the reactions investigated. The free energies, G, were calculated at the UHF-MP2 level and include the contribution of solvation.

The thermodynamics of the ET reaction, but also of the S_N2 reaction, do not vary much upon increasing steric hindrance. The reactions under discussion therefore essentially provide an illustration of the effect of steric hindrance on transition states. It is also worth noting that, although the S_N2 product is more stable than the ET product in each case, the difference in stability is not large ($\sim$0.4 eV), thus providing a good opportunity for the ET pathway to compete successfully with the S_N2 pathway as steric hindrance increases. In this sense, although not an anion radical, NO^- behaves more like an anion radical than like a classical nucleophile, leading to the formation of a strong bond.

In terms of free enthalpies, the S_N2 transition state is more stable than the ET transition state for the methyl derivative. As steric hindrance increases, the difference becomes smaller and smaller. The free enthalpies of the two transition states are practically equal for the isopropyl derivative, while the balance turns in favor of the ET transition state for the *t*-butyl derivative.

The geometries of the S_N2 and ET transition states are strikingly different. While the nitrogen atom of the electron donor stands on the side of the central carbon opposite to the chlorine atom in the former case, the chlorine stands between the reacting carbon and the nitrogen in the latter case. The S_N2-TS geometry reflects the interaction between the negative charge borne by the electron donor and the partial positive charge borne by the reacting carbon. In the ET-TS, the predominating factor is the overlap between the σ^* orbital of the C—Cl bond, which is mostly located on Cl, and the electrons on the NO^- nitrogen, which successfully counters electrostatic repulsion. Related to this difference in structure is the fact that the geometry of the S_N2 transition state is quite sensitive to steric hindrance, whereas the chlorine–nitrogen and carbon–chlorine distances in the ET transition state are about the same over the whole series. Another remarkable difference between the two transition states is the looser structure of the ET-TS compared to the S_N2-TS in the sense that the characteristic normal mode frequencies are smaller in the first case than in the second, which translates into the activation entropy being larger in the ET case than in the S_N2 case.

One may also examine the reaction pathways that connect each of the two transition states to products. The ET-TS is connected in all cases with the ET-PC (ET) cluster. Once the separated ET products have been formed from this cluster, the radicals $R^\bullet$ and $NO^\bullet$ may easily combine to yield the substitution product, RNO. Note that in the case of a chiral reacting carbon the racemic substitution product will be obtained along this reaction pathway, whereas inversion will take place along the direct S_N2 pathway. In the Me case, the S_N2 IRC pathway connects the transition state to the S_N2 product (Fig. 20). A ridge separates the S_N2 and ET product valleys. It starts from a point B′ situated above the S_N2 transition state (i.e., the saddle point) on the col separating the reactant and S_N2 product valleys, and goes through the transition state of the homolytic cleavage (HD-TS) of the S_N2 product into the ET product. There is thus a possibility for the system to partition statistically between the ET product valley and the S_N2 product valley. However, the driving force that attracts the system toward the S_N2 product is so much larger than the attraction toward the ET product that the IRC takes the S_N2 valley en route to the S_N2 product. As found earlier in the reaction of formaldehyde anion radical with methyl chloride,[188] electron transfer lies somewhat ahead of the formation of the bond but the formation of the new bond is already substantial at this point, making the two steps overlap each other so much that they must be viewed as a single event.

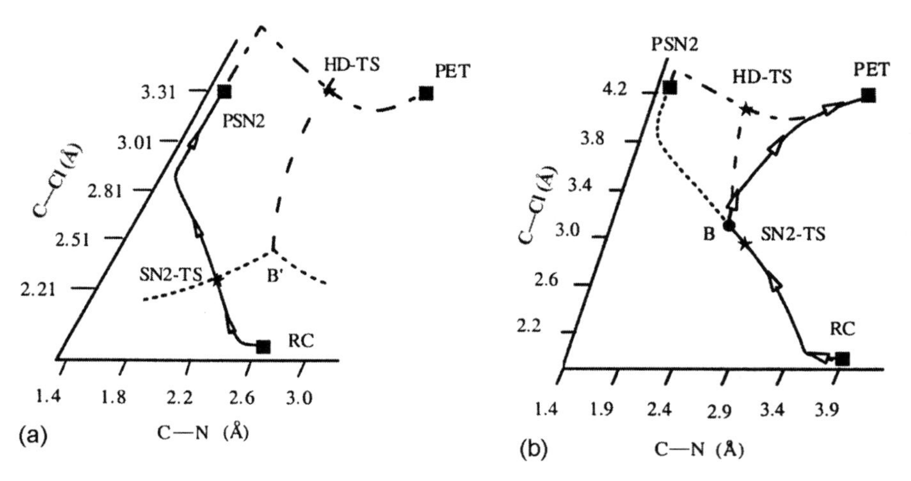

Fig. 20 Reaction pathways in the reduction of methyl (a) and *t*-butyl chloride (b) by NO^-. ■, reactant and products; ★, transition states. In (a) and (b), the full line is the mass-weighted IRC path from the reactant to the product states; the dashed line is a ridge separating the S_N2 and ET valleys; and the dotted-dashed line is the mass-weighted IRC path from the S_N2 product state to the ET product state (homolytic dissociation). The dotted line in (a) represents the col separating the reactant and the S_N2 product valleys. The dotted line in (b) represents the steepest descent path from the bifurcation point, B, to the S_N2 product. In (a), B′ is the point of the col separating the reactant and the S_N2 product valleys where the ridge separating the S_N2 and ET valleys starts.

A totally opposite situation is met with *t*-BuCl, where steric hindrance is maximal in the series. The IRC pathway connects the S_N2 transition state to the ET products rather than to the S_N2 products (Fig. 20). Past the S_N2 transition state, the N—O distance decreases abruptly to the value it has in $NO^•$ in a way that is very similar to what happens with the same compound along the direct ET pathway. Past the S_N2 transition state, the surface does not divide into an S_N2 valley and an ET valley until point B is reached. The ridge separating the two valleys (dashed line) starts from this point and goes through the transition state of the homolytic dissociation reaction (HD-TS). As expected, the IRC pathway stays on the ridge for a short while[194] and then bifurcates toward the ET valley. In spite of the fact that the S_N2 product cluster is lower in energy than the ET product cluster, the descent is steeper in the second case than in the first.

Electron transfer products are thus formed both by an indirect pathway after passing through the S_N2 transition state and by the direct ET pathway passing through the ET transition state, which now tends to be lower in energy than the S_N2 transition state. The electron transfer reaction thus takes place from the rear side, from the front side, and also from other sides since the *t*-Bu case is characterized by large vibrational entropies both for the ET and the S_N2 transition states. In other words, because of steric hindrance, all directions of attack lead to single electron transfer with similar activation energies, with similar reacting distances and negligible bonded interaction in the transition state. This corresponds precisely to the concept of outer-sphere electron transfer for which the Marcus–Hush model[23,24,26] was conceived when no bond is broken and for which the Morse curve dissociative electron transfer model[22] has been proposed when a bond is broken concertedly with electron transfer, as is the case here.

In the ethyl and isopropyl cases, the steepest descent pathway still connects the S_N2-TS to the S_N2 products but the formation of ET products along the bifurcation in the indirect ET pathway is expected to increase. These trends are likely to be at the origin of the stereochemistry of the reaction of the anion radical of anthracene with optically active 2-octyl halides recalled at the beginning of this section.

8 Conclusions

The coupling of single electron transfer, bond breaking and bond formation has received a great deal of attention over the past decade both from experimental and theoretical points of view, resulting in substantial progress in the understanding of dynamics and mechanisms.

Simplicity and generality are the main advantages of the dissociative electron transfer model discussed above and of its extensions to related reactions, even if their counterpart is some approximation in the predictions. They allow the rationalization of a large number of experimental observations and may serve as guidelines for designing new reactions. As an alternative approach, one may envisage treating the dynamics of these reactions by means of quantum-chemical calculations, thus making each particular system a case study. Actually, general semi-empirical approaches are made necessary by the fact that available quantum-chemical techniques are often unable to cope reliably with sufficiently large molecular frameworks and to describe solvent effects on reaction kinetics. Probably the best way is to combine the two types of approach, testing the validity of general trends on small model systems where reasonably reliable quantum-chemical results can be obtained as illustrated with a few examples in the preceding discussion. Semi-empirical modeling and cyclic voltammetric analysis of typical experimental examples are an illustration of how electrochemical approaches may be used to address important questions pertaining to chemical reactivity in general. Recent extension of the initial model to take into account interactions between cage fragments is an illustration of this mixed approach. The comprehension of the dynamics of ion radical cleavage (and of the reverse bond-forming reaction) in its heterolytic (intramolecular electron transfer) and homolytic versions has progressed greatly from a theoretical point of view as the number of successfully analyzed experimental reactions has increased. In this respect, deprotonation of cation radicals was a rather unexpected example of application of the dissociative electron transfer theory.

It is interesting to note the increasing number of systems in which the passage between stepwise and concerted pathways is observed as the result of an increase of the driving force offered to the reaction. These systems are quite different from one another and their observation not only confirms the original theoretical predictions but also provides unambiguous evidence of the existence of the two types of mechanisms when application of the classical diagnostic criteria fails to give a clear-cut answer.

It is also worth emphasizing that recent theoretical work on photoinduced stepwise and concerted electron transfer/bond-breaking reactions opens the route to a more systematic combination than before of the electrochemical and photochemical approaches to the same problems.

The long-debated nature of the dichotomy and relationships between single electron transfer (ET) and S_N2 reaction has made important recent progress. There is competition between the two reaction pathways, implying the existence of two distinct transition states on the same potential energy hypersurface. In the absence of steric hindrance, the S_N2 pathway is energetically favored over the ET pathway. Bifurcation toward the ET products, if any, remains negligible. In contrast, in the presence of strong steric hindrance, as for example when a tertiary reacting carbon is involved, not only is the ET pathway more favorable energetically but also the S_N2 pathway bifurcates, past the S_N2 transition state, toward the ET products. All directions of attack thus lead to single electron transfer with similar activation energies, with similar reacting distances and negligible bonded interaction in the transition state, i.e., the prototype of an outer-sphere dissociative electron transfer reaction. For less sterically congested systems, the competition is expected to involve, as well as the formation of the S_N2 products, electron transfer products that come in part from the ET transition state and in part from a bifurcation following the S_N2 transition state. An important question that remains open, among others, is the identification of the molecular factors, other than steric hindrance, that control the competition between the two pathways.

Acknowledgements

It is a pleasure to acknowledge the essential contribution of Dr. C. P. Andrieux to most of the work reported above as well as that of Dr. D. Lexa in the field of porphyrins, Professor Moiroux and Dr. A. Anne to cation radical reactivity, Dr. M. Robert to photoinduced dissociative electron transfer and to the stepwise/concerted competition and Drs. P. Hapiot and Médebielle to recent work on thermal $S_{RN}1$ reactions. Many students from our group have also contributed effectively to the work, namely, C. Costentin, G. Delgado, V. Grass, A. Le Gorande, C. Tardy and D. L. Wang. Fruitful and pleasant collaborations have been developed over the years with Dr. Z. Welwart (CNRS, Thiais, France), Dr. A. Thiébault (ESPCI, Paris, France), Professor H. Schäfer (Wilhelms Universität, Münster, Germany), Professor W. Adcock (Flinders University, Adelaide, Australia), Professor A. Tallec (Université de Rennes) and their coworkers as well as with Dr. E. Differding (Ciba Geigy, Basel, Switzerland), Dr. P. Neta (NIST, Washington DC, USA) and Dr. F. D. Saeva (Eastman Kodak Company, Rochester, NY, USA). In the field of quantum-chemical calculations, the sustained collaboration with Professor J. Bertran (Universitat Autonoma de Barcelona, Spain) and his coworkers, I. Gallardo and M. Moreno was particularly helpful and pleasant. I am particularly grateful to my young colleagues Cyrille Costentin and Marc Robert for their careful re-reading of the manuscript.

References

1. Savéant, J-M. (1990). *Advances in Physical Organic Chemistry* (ed. Bethell, D.), vol. 26, pp. 1–130. Academic Press, London
2. Eberson, L. (1999). *Acta Chem. Scand.* **53**, 751
3. Hush, N.S. (1999). *J. Electroanal. Chem.* **470**, 170
4. Maletin, Y.A. and Cannon, R.D. (1998). *Theoret. Exp. Chem.* **34**, 57
5. Savéant, J-M. (1993). *Acc. Chem. Res.* **26**, 455
6. Savéant, J-M. (1994). In *Advances in Electron Transfer Chemistry* (ed. Mariano, P.S.), vol. 4. pp. 53–116. JAI Press, New York
7. Lund, H., Daasbjerg, K., Pedersen, S.U. and Lund, T. (1995). *Acc. Chem. Res.* **28**, 313
8. Speiser, B. (1996). *Angew. Chem. Int. Ed. Engl.* **35**, 2471
9. Zipse, H. (1997) *Angew. Chem. Int. Ed. Engl.* **36**, 1697
10. Rossi, R.A., Pierini, A.B. and Peñéñory, A.B. (1995). In *The Chemistry of Halides, Pseudo-Halides and Azides* (ed. Patai, S. and Rappoport, Z.), vol. 24, pp. 1395–1485. Wiley, New York
11. Savéant, J-M. (1994) *Tetrahedron*, **50**, 10117
12. Hush, N.S. (1957). *Z. Elektrochem.* **61**, 734
13. Eberson, L. (1982). *Acta Chem. Scand. B* **36**, 533
14. Eberson, L. (1982). In *Advances in Physical Organic Chemistry* (eds. Gold, V. and Bethell, D.), vol. 18, pp. 78–185. Academic Press, London
15. Tolman, R.C. (1927). *Statistical Mechanics*, p. 247. Chemical Catalog Co., New York
16. Partington, J.R. (1967). *An Advanced Treatise on Physical Chemistry*, Vol. 1, p. 292. Longmans, London
17. Moore, J.W. and Pearson, R.N. (1981). *Kinetics and Mechanism*, pp. 130, 131. Wiley, New York
18. Glasstone, S., Laidler, K.J. and Eyring, H. (1941). *The Theory of Rate Processes*. McGraw-Hill, New York
19. Andrieux, C.P., Merz, A., Tomahogh, R. and Savéant J-M. (1984). *J. Am. Chem. Soc.* **106**, 1957
20. Andrieux, C.P., Merz, A. and Savéant, J-M. (1985). *J. Am. Chem. Soc.* **107**, 6097
21. Merz, A. and Tomahogh, R. (1979) *Angew. Chem. Int. Ed. Engl.* **18**, 938
22. Savéant, J-M. (1987). *J. Am. Chem. Soc.* **109**, 6788

23. Marcus, R.A. (1956). *J. Chem. Phys.* **24**, 966
24. Marcus, R.A. (1956). *J. Chem. Phys.* **24**, 979
25. Marcus, R.A. (1965). *J. Chem. Phys.* **43**, 679
26. Hush, N.S. (1958). *J. Chem. Phys.* **28**, 962
27. Hush, N.S. (1961). *Trans. Faraday Soc.* **57**, 557
28. Bertran, J., Gallardo, I., Moreno, M. and Savéant, J-M. (1992). *J. Am. Chem. Soc.* **114**, 9576
29. Clark, K.B. and Wayner D.D.M. (1991). *J. Am. Chem. Soc.* **113**, 9363
30. Andrieux, C.P., Le Gorande, A. and Savéant J-M. (1992). *J. Am. Chem. Soc.* **114**, 6892
31. Savéant, J-M. (1992). *J. Am. Chem. Soc.* **114**, 10595
32. Andrieux, C.P., Differding, E., Robert, M. and Savéant, J-M. (1993). *J. Am. Chem. Soc.* **115**, 6592
33. Andrieux, C.P., Robert, M., Saeva, F.D. and Savéant, J-M. (1994). *J. Am. Chem. Soc.* **116**, 7864
34. Lexa, D., Savéant, J-M., Schäfer, H., Su, K.B., Vering, B. and Wang, D.L. (1990). *J. Am. Chem. Soc.* **112**, 6162
35. Adcock, W., Clark, C., Houmam, A., Krstic, A.R., Pinson, J., Savéant, J-M., Taylor, D.K. and Taylor, J.F. (1994). *J. Am. Chem. Soc.* **116**, 4653
36. Workentin, M.S., Maran, F. and Wayner D.D.M. (1995). *J. Am. Chem. Soc.* **117**, 2120
37. Workentin, M.S. and Donkers, R.L. (1998). *J. Am. Chem. Soc.* **120**, 2664
38. Andrieux, C.P. and Savéant, J-M. (1986). In *Investigations of Rates and Mechanisms of Reactions* (ed. Bernasconi, F.), vol. 6, 4/E, Part 2, C, pp. 305–390. Wiley, New York
39. Andrieux, C.P., Gallardo, I., Savéant, J-M. and Su, K.B (1986). *J. Am. Chem. Soc.* **108**, 638
40. Lund, T. and Lund, H. (1986). *Acta Chem. Scand. Ser. B* **40**, 470
41. Lexa, D., Savéant, J-M., Su, K.B. and Wang., D.L. (1988). *J. Am. Chem. Soc.* **110**, 7617
42. Daasbjerg, K., Pedersen, S.U. and Lund H. (1991). *Acta Chem. Scand.* **45**, 470
43. Andrieux, C.P., Delgado, G., Savéant, J-M. and Su, K.B. (1993). *J. Electroanal. Chem.* **448**, 141
44. Antonello, S., Musumeci, M., Wayner, D.D.M. and Maran, F. (1997). *J. Am. Chem. Soc.* **119**, 9541
45. Imbeaux, J.C. and Savéant J-M. (1973). *J. Electroanal. Chem.* **44**, 169
46. Savéant, J-M. and Tessier, D. (1982). *Disc. Faraday Soc.* **74**, 57
47. Grimshaw, J., Langan, J.R. and Salmon, G.A. (1988). *J. Chem. Soc., Chem. Commun.* 1115
48. Grimshaw, J., Langan, J.R. and Salmon, G.A. (1994). *J. Chem. Soc., Faraday Trans.* **90**, 75
49. Andrieux, C.P., Savéant, J-M. and Tardy, C. (1998). *J. Am. Chem. Soc.* **120**, 4167
50. German, E.D. and Kuznetsov, A.M. (1994). *J. Phys. Chem.* **98**, 7120
51. German, E.D., Kuznetsov, A.M. and Tikhomirov, V.A. (1995). *J. Phys. Chem.* **99**, 9095
52. German, E.D., Kuznetsov, A.M. and Tikhomirov, V.A. (1997). *J. Electroanal. Chem.* **420**, 235
53. Tikhomirov, V.A. and German, E.D. (1998). *J. Electroanal. Chem.* **450**, 13
54. Robert, M. and Savéant, J-M. (1999). *J. Am. Chem. Soc.* **122**, 514
55. Andrieux, C.P., Hapiot, P. and Savéant, J-M. (1990). *Chem. Rev.* **90**, 723
56. Nadjo, L. and Savéant, J-M. (1973). *J. Electroanal. Chem.* **48**, 113
57. Andrieux, C.P., Savéant, J-M. and Tardy, C. (1997). *J. Am. Chem. Soc.* **119**, 11546

58. Andrieux, C.P., Le Gorande, A. and Savéant J-M. (1994). *J. Electroanal. Chem.* **371**, 191
59. Antonello, S. and Maran, F. (1997). *J. Am. Chem. Soc.* **119**, 12595
60. Antonello, S. and Maran, F. (1999). *J. Am. Chem. Soc.* **121**, 9668
61. Pause, L., Robert, M. and Savéant, J-M. (1999). *J. Am. Chem. Soc.* **121**, 7158
62. Severin, M.G., Farnia, G., Vianello, E. and Arévalo M.C. (1988). *J. Electroanal. Chem.* **251**, 369
63. Jakobsen, S., Jensen, H., Pedersen, S.U. and Daasbjerg, K. (1999). *J. Phys. Chem. B* **103**, 4141
64. Costentin, C., Hapiot, P., Medebielle, M. and Savéant, J-M. (1999). *J. Am. Chem. Soc.* **121**, 4451
65. Kornblum, N. (1975). *Angew. Chem. Int. Ed. Engl.* **14**, 734
66. Bunnett, J.F. (1978). *Acc. Chem. Res.* **11**, 413
67. Savéant, J-M. (1980). *Acc. Chem. Res.* **13**, 323
68. Rossi, R.A. and Rossi, R.H. (1983). *Aromatic Substitution by the $S_{RN}1$ Mechanism.* ACS Monograph 1978. The American Chemical Society, Washington, DC
69. Bowman, W.R. (1988). *Chem. Soc. Rev.* **17**, 283
70. Rossi, R.A., Pierini, A.B. and Palacios, S.M. (1990). In *Advances in Free Radical Chemistry*, Vol. 1, pp. 193–252. JAI Press, New York
71. Austin, E., Ferrayoli, C.G., Alonso, R.A. and Rossi, R.A. (1993). *Tetrahedron* **49**, 4495
72. Galli, C. and Gentili, P. (1993). *J. Chem. Soc., Perkin Trans., 2*, 1135
73. Jencks, W.P. (1981). *Chem. Soc. Rev.* **10**, 345
74. Richard, J.P. and Jencks, W.P. (1984). *J. Am. Chem. Soc.* **106**, 1383
75. Eldin, S.F. and Jencks, W.P. (1995). *J. Am. Chem. Soc.* **117**, 9415
76. Costentin, C., Robert, M. and Savéant, J-M. (2000). *J. Phys. Chem.*, in press
77. Savéant, J-M. (1994). *J. Phys. Chem.* **98**, 3716
78. Tanko, J.M. and Drumright, R.E. (1990). *J. Am. Chem. Soc.* **112**, 65362
79. Tanko, J.M. and Drumright, R.E. (1992). *J. Am. Chem. Soc.* **114**, 1844
80. Tanko, J.M., Drumright, R.E., Suleman, N.K. and Brammer, L.E. Jr. (1994). *J. Am. Chem. Soc.* **116**, 1795
81. Tanko, J.M. and Paige Phillips, J. (1999). *J. Am. Chem. Soc.* **121**, 6078
82. Andrieux, C.P., Combellas, C., Kanoufi, F., Savéant, J-M. and Thiébault, A. (1997). *J. Am. Chem. Soc.* **119**, 9527
83. Wipf, D.O. and Wightmann, R.M. (1990) *Anal. Chem.* **62**, 98
84. Jaworski, J.S., Leszczynski, P. and Tykarski, J.J. (1995). *J. Chem. Res. (S)*, 510
85. Jensen, H. and Daasbjerg, K. (1998). *Acta Chem. Scand.* **52**, 1151
86. Kimura, N. and Takamu, S. (1986). *Bull. Chem. Soc. Jpn* **59**, 3653
87. Kimura, N. and Takamu, S. (1987). *Rad. Phys. Chem.* **29**, 179
88. Kimura, N. and Takamu, S. (1991). *Bull. Chem. Soc. Jpn* **64**, 2433
89. Kimura, N. and Takamu, S. (1992). *Bull. Chem. Soc. Jpn* **65**, 1668
90. Kimura, N. and Takamu, S. (1995). *J. Am. Chem. Soc.* **117**, 8023
91. Andrieux, C.P., Robert, M. and Savéant, J-M. (1995). *J. Am. Chem. Soc.* **117**, 9340
92. Adcock, W., Andrieux, C.P., Clark, C.I., Neudeck, A., Savéant, J-M. and Tardy, C. (1995). *J. Am. Chem. Soc.* **117**, 8285
93. Fontanesi, C. (1997). *J. Mol. Struct.* **392**, 87
94. Antonello, S. and Maran, F. (1998). *J. Am. Chem. Soc.* **1120**, 5713
95. Bell, R.P. and Goodall, D.M. (1966). *Proc. R. Soc. London, Ser. A*, **294**, 273
96. Bordwell, F.G. and Boyle, W.J.J. (1972). *J. Am. Chem. Soc.* **94**, 3907
97. Albery, W.J., Campbell-Crawford, A.N. and Curran, J.S. (1972). *J. Chem. Soc., Perkin Trans.* 2206
98. Bernasconi, C.F. (1985). *Tetrahedron* **41**, 3219
99. Bernasconi, C.F. (1987). *Acc. Chem. Res.* **20**, 301

100. Bernasconi, C.F. (1992). *Adv. Phys. Org. Chem.* **27**, 119
101. Terrier, F., Boubaker, T., Xiao, L. and Farrell, P.G. (1992). *J. Org. Chem.* **57**, 3924
102. Moutiers, G., El Fahid, B., Goumont, R., Chatrousse, A.P. and Terrier, F. (1996). *J. Org. Chem.* **61**, 1978
103. Nicholas, A. and Arnold, D.R. (1982). *Can. J. Chem.* **60**, 2165
104. Nicholas, A., Boyd, R.J. and Arnold, D.R. (1982). *Can. J. Chem.* **60**, 3011
105. Bordwell, F.G. and Cheng, J.P. (1989). *J. Am. Chem. Soc.* **111**, 1792
106. Zhang, X. and Bordwell, F.G. (1992). *J. Org. Chem.* **57**, 4163
107. Zhang, X., Bordwell, F.G., Bares, J.E. and Cheng, J.P. (1993). *J. Org. Chem.* **58**, 3051
108. Schlesener, C.J., Amatore, C. and Kochi, J.K. (1984). *J. Am. Chem. Soc.* **106**, 7472
109. Schlesener, C.J., Amatore, C. and Kochi, J.K. (1986). *J. Phys. Chem.* **90**, 3747
110. Masnovi, J.M., Sankararaman, S. and Kochi, J.K. (1989). *J. Am. Chem. Soc.* **111**, 2263
111. Sankararaman, S., Perrier, S. and Kochi, J.K. (1989). *J. Am. Chem. Soc.* **111**, 6448
112. Bacciochi, E., Del Giacco, T. and Elisei, F. (1993). *J. Am. Chem. Soc.* **115**, 12290
113. Bacciochi, E., Bietti, M., Putignani, L. and Steenken, S. (1996). *J. Am. Chem. Soc.* **118**, 5952
114. Bacciochi, E., Bietti, M., Lanzalunga, L. and Steenken, S. (1998). *J. Am. Chem. Soc.* **120**, 11516
115. Bacciochi, E., Bietti, M., Del Giacco, T. and Steenken, S. (1998). *J. Am. Chem. Soc.* **120**, 11800
116. Bacciochi, E., Bietti, M., Manduchi, L. and Steenken, S. (1999). *J. Am. Chem. Soc.* **121**, 6624
117. Bacciochi, E., Bietti, M. and Steenken, S. (1999). *Chem. Eur. J.* **5**, 1785
118. Dinnocenzo, J.P. and Banach, T.E. (1989). *J. Am. Chem. Soc.* **111**, 8646
119. Dinnocenzo, J.P., Karki, S.B. and Jones, J.P. (1993). *J. Am. Chem. Soc.* **115**, 7111
120. Xu, W. and Mariano, P.S. (1991). *J. Am. Chem. Soc.* **113**, 1431
121. Xu, W., Zhang, X. and Mariano, P.S. (1991). *J. Am. Chem. Soc.* **113**, 8863
122. Reitstöen, B. and Parker, V.D. (1990). *J. Am. Chem. Soc.* **112**, 4698
123. Parker, V.D., Chao, Y. and Reitstöen, B. (1991). *J. Am. Chem. Soc.* **113**, 2336
124. Parker, V.D. and Tilset, M. (1991). *J. Am. Chem. Soc.* **113**, 8778
125. Fukuzumi, S., Kondo, Y. and Tanaka, T. (1984). *J. Chem. Soc., Perkin Trans. 2*, 673
126. Fukuzumi, S., Tokuda, Y., Kitano, T., Okamoto, T.and Otera, J. (1993). *J. Am. Chem. Soc.* **115**, 8960
127. Sinha, A. and Bruice, T.C. (1984) *J. Am. Chem. Soc.* **106**, 7291
128. Tolbert, L.M. and Khanna, R.K. (1987). *J. Am. Chem. Soc.* **109**, 3477
129. Tolbert, L.M., Khanna, R.K., Popp, A.E., Gelbaum, L. and Bottomley, L.A. (1990). *J. Am. Chem. Soc.* **112**, 2373
130. Tolbert, L.M., Bedlek, J., Terapane, M. and Kowalik, J. (1997). *J. Am. Chem. Soc.* **119**, 2291
131. Tolbert, L.M., Li, Z., Sirimanne, S.R. and Van Derveer, D.G. (1997). *J. Org. Chem.* **62**, 3927
132. Hapiot, P., Moiroux, J. and Savéant, J.-M. (1990). *J. Am. Chem. Soc.* **112**, 1337
133. Anne, A., Hapiot, P., Moiroux, J., Neta, P. and Savéant, J.-M. (1991). *J. Phys. Chem.* **95**, 2370
134. Anne, A., Hapiot, P., Moiroux, J., Neta, P. and Savéant, J.-M. (1992). *J. Am. Chem. Soc.* **114**, 4694
135. Anne, A., Fraoua, S., Hapiot, P., Moiroux, J. and Savéant, J.-M. (1995). *J. Am. Chem. Soc.* **117**, 7412
136. Anne, A., Fraoua, S., Grass, V., Moiroux, J. and Savéant, J.-M. (1998). *J. Am. Chem. Soc.* **120**, 2951

137. Anne, A., Fraoua, S., Moiroux, J. and Savéant, J.-M. (1996). *J. Am. Chem. Soc.* **118**, 3938
138. Maslak, P. and Asel, S.L. (1988). *J. Am. Chem. Soc.* **110**, 8260
139. Maslak, P. and Chapmann, W.H. (1989). *J. Chem. Soc., Chem. Commun.* **110**, 1810
140. Maslak, P. and Chapmann, W.H. (1990). *Tetrahedron* **46**, 2715
141. Maslak, P. and Chapmann, W.H. (1990). *J. Org. Chem.* **55**, 6334
142. Maslak, P. and Narvaez, J.N. (1990). *Angew. Chem. Int. Ed. Engl.* **29**, 283
143. Maslak, P., Vallombroso, T.M., Chapmann, W.H. and Narvaez, J.N. (1994). *Angew. Chem. Int. Ed. Engl.* **33**, 73
144. Maslak, P., Narvaez, J.N., Vallombroso, T.M. and Watson, B.A. (1995). *J. Am. Chem. Soc.* **117**, 12380
145. Maslak, P., Chapmann, W.H. and Vallombroso, T.M. (1995). *J. Am. Chem. Soc.* **117**, 12373
146. Zheng, Z-R., Evans, D.H., Soazara Chan-Shing, E. and Lessard, J. (1999). *J. Am. Chem. Soc.* **121**, 9429
147. Zheng, Z-R. and Evans, D.H. (1999). *J. Am. Chem. Soc.* **121**, 2941
148. Wentworth, W.E., George, R. and Keith, H. (1969). *J. Chem. Phys.* **51**, 1791
149. Marcus, R.A. (1998). *Acta Chem. Scand.* **52**, 858
150. Benassi, R., Bernardi, F., Bottoni, A., Robb, M.A. and Taddei, F. (1989). *Chem. Phys. Lett.* **161**, 79
151. Tada, T. and Yoshimura, R. (1992). *J. Am. Chem. Soc.* **114**, 1593
152. Pause, L., Robert, M. and Savéant, J-M. (2000). Submitted
153. Kerber, C.R., Urry, G.W. and Kornblum, N. (1965). *J. Am. Chem. Soc.* **87**, 4520
154. Kornblum, N., Michel, R.E. and Kerber, C.R. (1966). *J. Am. Chem. Soc.* **88**, 5664
155. Russell, G.A. and Danen, W.C. (1966). *J. Am. Chem. Soc.* **88**, 5666
156. Costentin, C., Hapiot, P., Medebielle, M. and Savéant, J-M. (2000). *J. Am. Chem. Soc.*, in press
157. Laarhoven, L.J.J., Born, J.G.P., Arends, I.W. and Mulder, P. (1997). *J. Chem. Soc., Perkin Trans. 2*, 2307
158. Pratt , D.A., Wright, J.S. and Ingold, K.U. (1999). *J. Am. Chem. Soc.* **121**, 4877
159. Andrieux, C.P., Savéant, J-M., Tallec, A., Tardivel, R. and Tardy, C. (1997). *J. Am. Chem. Soc.* **119**, 2420
160. Dorrestijn, E., Hemmink, S., Hultsmaan, G., Monnier, L., Van Scheppingen, W. and Mulder, P. (1999). *Eur. J. Org. Chem.* 607
161. Christensen, T.B. and Daasbjerg, K. (1997). *Acta Chem. Scand.* **51**, 317
162. Daasbjerg, K., Jensen, H., Benassi, R., Taddei, F., Antonello, S., Gennaro, A. and Maran, F. (1999). *J. Am. Chem. Soc.* **121**, 1750
163. Saeva, F.D. (1990). *Topics in Current Chemistry* **156**, 61
164. Saeva, F.D. (1994). In *Advances in Electron Transfer Chemistry* (ed. Mariano, P.S.), vol. 4, pp. 1–25. JAI Press, New York
165. Gaillard, E.R. and Whitten, D.G. (1996). *Acc. Chem. Res.* **29**, 292
166. Arnold, B.R., Scaiano, J.C. and McGimpsey, W.G. (1992). *J. Am. Chem. Soc.* **114**, 9978
167. Chen, L., Farahat, M.S., Gaillard, E.R., Gan, H., Farid, S. and Whitten, D.G. (1995). *J. Am. Chem. Soc.* **117**, 6398
168. Wang, X., Saeva, F.D. and Kampmeier, J.A. (1999). *J. Am. Chem. Soc.* **121**, 4364
169. Koper, M.T.M. and Voth, G.A. (1998). *Chem. Phys. Lett.* **282**, 1998
170. Chen, L., Farahat, M.S., Gaillard, E.R., Farid, S. and Whitten, D.G. (1996). *J. Photochem. Photobiol. A: Chem.* **95**, 21
171. Landau, L. (1932). *Phys. Z. Sowjet* **2**, 46
172. Zener, C. (1932). *Proc. R. Soc. London Ser. A* **137**, 696
173. Hush, N.S. (1968). *Electrochim. Acta* **13**, 1005
174. Newton, M. D. and Sutin, N. (1984) *Annu. Rev. Phys. Chem.* **35**, 437

175. Brunschwig, B.S., Logan, J., Newton, M.D. and Sutin, N. (1980). *J. Am. Chem. Soc.* **102**, 5798
176. Garst, J.F. (1971). *Acc. Chem. Res.* **4**, 400
177. Bank, S. and Juckett, D.A. (1976). *J. Am. Chem. Soc.* **98**, 7742
178. Herbert, E., Mazaleyrat, J-P., Welvart, Z., Nadjo, L. and Savéant, J-M. (1985). *Nouv. J. Chem.* **9**, 75
179. Lund, T. and Lund, H (1987). *Acta Chem. Scand. Ser. B* **41**, 93
180. Lexa, D., Mispelter, J. and Savéant, J-M. (1981). *J. Am. Chem. Soc.* **103**, 6806
181. Lund, T. and Lund, H (1988). *Acta Chem. Scand. Ser. B* **42**, 269
182. Eberson, L. (1987). *Electron Transfer Reactions in Organic Chemistry*. Springer-Verlag, Heidelberg
183. Pross, A. (1985). In *Advances in Physical Organic Chemistry* (ed. Bethell, D.), vol. 21, pp. 99–196. Academic Press, London
184. Eberson, L. and Shaik, S.S. (1990). *J. Am. Chem. Soc.* **112**, 4484
185. Cho, J.K. and Shaik, S.S. (1991) *J. Am. Chem. Soc.* **113**, 9890
186. Marcus, R.A. (1997). *J. Phys. Chem. A* **101**, 4072
187. Sastry, G.N. and Shaik, S.S. (1995). *J. Am. Chem. Soc.* **117**, 3290
188. Bertran, J., Gallardo, I., Moreno, M. and Savéant, J-M. (1996). *J. Am. Chem. Soc.* **118**, 5737
189. Garrett, B.C., Truhlar, D.G., Wagner, A.F. and Dunning, T.H. (1983). *J. Chem. Phys.* **78**, 4400
190. Valtazanos, P. and Ruedenberg, K. (1986). *Theor. Chim. Acta* **69**, 281
191. Baker, J. and Gill, P.M.W. (1988). *J. Comput. Chem.* **9**, 465
192. Bosch, E., Moreno, M., Lluch J.M. and Bertran, J. (1989). *Chem. Phys. Lett.* **1604**, 543
193. Natanson, G.A., Garrett, B.C., Truong, T.N., Joseph, T. and Truhlar, D.G. (1991). *J. Chem. Phys.* **94**, 7875
194. Schlegel, H.B. (1994). *J. Chem. Soc. Faraday Trans.* **90**, 1569
195. Shaik, S.S., Danovich, D., Sastry, G.N., Ayala, P.Y. and Schlegel, H.B. (1997). *J. Am. Chem. Soc.* **119**, 9237
196. Sastry, G.N. and Shaik, S.S. (1998). *J. Am. Chem. Soc.* **120**, 2131
197. Costentin, C. and Savéant, J-M. (2000) *J. Am. Chem. Soc.* **122**, 2329
198. Adcock, W., Andrieux, C.P. , Clark, C.I. , Neudeck, A., Savéant J-M. and Tardy, C. (1995). *J. Am. Chem. Soc.* **117**, 8285
199. Andrieux, C.P., Savéant, J-M., Tallec, A., Tardivel, R. and Tardy, C. (1996). *J. Am. Chem. Soc.* **118**, 9788
200. Calhoun, A., Koper, M.T.M. and Voth, G.A. (1999). *J. Phys. Chem. B* **103**, 3442
201. Donkers, R.L., Maran, F., Wayner, D.D.M. and Workentin, M.S. (1999). *J. Am. Chem. Soc.* **121**, 7239

Donor/Acceptor Organizations and the Electron-Transfer Paradigm for Organic Reactivity

RAJENDRA RATHORE and JAY K. KOCHI

Department of Chemistry, University of Houston, Houston, Texas, USA

ADVANCES IN PHYSICAL ORGANIC CHEMISTRY
VOLUME 35 ISBN 0-12-033535-2

1 Introduction

Organic chemistry – unlike inorganic chemistry – is beset with a large number of distinctive reactions and a wide variety of reaction types. So much so that a student who faces organic chemistry for the first time is commonly reduced to the tedious task of rote memorization. The difference is also reflected in the extensive use of "name" reactions in organic chemistry for the efficient categorization of different structural changes, whereas no such mnemonic aide is necessary for inorganic reactions. Furthermore, there is not even the equivalent of a Periodic Table of the elements to serve as a helpful guide for classification. The unfortunate fact is that *given the (physical) properties of an organic reactant, there is no quantitative means to predict a priori how it will react.* As a result, organic chemistry has depended heavily on linear free energy relationships (LFER) to merely *correlate* the various reactivity trends of structurally related compounds.

2 Diverse classifications of organic reactions

Early attempts to fathom organic reactions were based on their classification into ionic (heterolytic) or free-radical (homolytic) types.[1] These were later subclassified in terms of either electrophilic or nucleophilic reactivity of both ionic and paramagnetic intermediates – but none of these classifications carries with it any quantitative mechanistic information. Alternatively, organic reactions have been described in terms of acids and bases in the restricted Brønsted sense, or more generally in terms of Lewis acids and bases to generate cations and anions. However, organic cations are subject to one-electron reduction (and anions to oxidation) to produce radicals, i.e.,

$$R^+ \underset{}{\overset{e^-}{\rightleftarrows}} R\bullet \underset{}{\overset{e^-}{\rightleftarrows}} R:^-$$

Such a (formal) interchange of the reactive intermediates obscures the traditional heterolytic/homolytic classification as a rigid distinction.[2] Other descriptors suffer from an analogous ambiguity. Indeed, the parallelism between nucleophile and reductant was originally pointed out by Edwards and Pearson a number of years ago.[3] Furthermore, nucleophiles are often most effective as negatively charged anions, and they are also sometimes referred to as Brønsted and Lewis bases, or in terms of their softness on the hard and soft acid–base (HSAB) scale.[4] Since each of these classifications relates in some way to an intrinsic property that is vaguely considered in degrees of "electron richness",[5] we prefer the more inclusive description of *electron donors*, as originally defined by Mulliken.[6]

The contrasting descriptions can also be applied to cations as electrophiles and as oxidants – in reference to their behavior as *electron acceptors*.

3 Donor/acceptor organizations

The interchangeability of the various classifications under the general heading of electron donors (D) and electron acceptors (A) is emphasized in the direct comparisons in Chart 1.[7]

Electron donor (D)	*Electron acceptor* (A)
Anion	Cation
Reductant	Oxidant
Nucleophile	Electrophile
Base (Brønsted, Lewis, HSAB)	Acid
(Electron-rich)	(Electron-poor)

Chart 1

To illustrate the overall magnitude of the mechanistic problem, let us consider the varied reactivity of a prototypical carbonyl compound such as acetone, which is subject to many diverse reactions such as addition, substitution, cycloaddition, oxidation, reduction, etc., as illustrated in Chart 2.

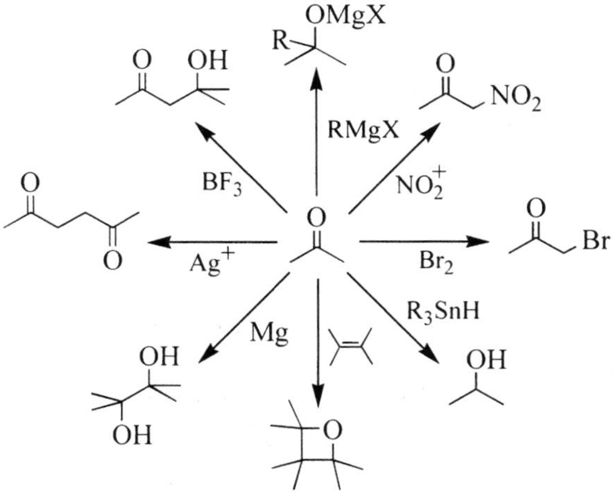

Chart 2

The products described in Chart 2 clearly derive from two sites of acetone reactivity that can be identified with the carbonyl and α-carbon centers, as they are revealed by the reversible transfer of a proton from the α-carbon to oxygen. As such, enolization of the carbonyl compound represents a most fundamental change – an umpolung in which the keto acceptor (A) is interconverted to the enol donor (D) (equation 1).

$$\underset{\text{(Acceptor)}}{H_3C-C(=O)-CH_3} \rightleftharpoons \underset{\text{(Donor)}}{H_3C-C(OH)=CH_2} \tag{1}$$

The electron-donor property of the enol form is indicated by the magnitude of its ionization potential (IP in the gas phase) or its oxidation potential (E^0_{ox} in solution).[8] Conversely, the relatively electron-poor keto form is a viable acceptor in measure with the magnitude of its electron affinity (EA in the gas phase) or its reduction potential (E^0_{red} in solution).

Electron donors (D) and electron acceptors (A) constitute reactant pairs that are traditionally considered with more specific connotations in mind – such as nucleophile/electrophile in bond formation, reductant/oxidant in electron transfer, base/acid in adduct production, and so on. In each case, the chemical transformation is preceded by a rapid (diffusion-controlled) association to form the 1:1 intermolecular complex[9] (equation 2).

$$D + A \underset{}{\overset{K_{EDA}}{\rightleftharpoons}} [D, A] \tag{2}$$

This association has its counterpart that was also variously described as an encounter complex, a nonbonded electron donor–acceptor (EDA) complex, a precursor complex, and a contact charge-transfer complex.[10] For electrically charged species such as anion/cation pairs (which are relevant to ion-pair annihilation), the pre-equilibrium association results in contact ion pairs (CIP)[7] (equation 3)

$$D^- + A^+ \underset{}{\overset{K_{CIP}}{\rightleftharpoons}} [D^-, A^+] \tag{3}$$

Likewise, cationic acceptors afford mixed (positively) charged complexes with electron-rich donors,[11] i.e., $[D, A^+]$; and anionic donors associate with electron-poor acceptors to form mixed (negatively) charged complexes,[12] i.e., $[D^-, A]$. In each case, the intermolecular (ionic) complexation or association represents the highly oriented organization of the donor/acceptor pair (independently of whether they bear positive, negative or no charge) that is often sufficient to afford crystalline complexes amenable to direct X-ray structure elucidation.[13]

4 The spectral probe for donor/acceptor organization

To develop the utility of donor/acceptor organization as a unifying theme in reaction mechanisms, we focus on the spectroscopic identification of the charge-transfer absorption bands of the EDA complex [D, A] and the contact ion pair $[D^-, A^+]$ in equations (2) and (3), respectively. According to Mulliken,[6] the charge-transfer transition ($h\nu_{CT}$) derives from an intracomplex transfer of a single electron from the donor to the acceptor. As such, the photoexcitation of the charge-transfer absorption band of the electrically neutral EDA complex generates the ion-radical pair[14,15] (equation 4),

$$[D, A]_{EDA} \underset{}{\overset{h\nu_{CT}}{\rightleftharpoons}} D^{+\bullet}, A^{-\bullet} \tag{4}$$

and the contact ion pair generates the geminate radical pair[15] (equation 5),

$$[D^{-}, A^{+}]_{CIP} \underset{}{\overset{h\nu_{CT}}{\rightleftharpoons}} D^{\bullet}, A^{\bullet} \tag{5}$$

Since the intensity of the charge-transfer absorption is directly related to the concentration of the EDA complex or contact ion pair in equations (4) and (5), respectively, it can be used as an analytical tool to quantify complex formation in equations (2) and (3). According to the commonly utilized Benesi–Hildebrand treatment,[16] the formation constants are quantitatively evaluated from the graphical plot of the CT absorbance change (A_{CT}) as the donor is progressively added to a solution of the acceptor (or vice versa) (equation 6)

$$\frac{[A]}{A_{CT}} = \frac{1}{K_{EDA}\varepsilon_{CT}}\frac{1}{[D]} + \frac{1}{\varepsilon_{CT}} \tag{6}$$

where ε_{CT} and K_{EDA} (or K_{CIP}) are the molar extinction coefficient and the formation constant of the EDA complex (or CIP), respectively.

A relatively strong organization of an electron donor by an acceptor is typically indicated by experimental values of K_{EDA} or $K_{CIP} > 10\ \text{M}^{-1}$. For intermediate values of the formation constant, i.e., $1 < K_{EDA} < 10\ \text{M}^{-1}$, the donor/acceptor organization is considered to be weak.[17] Finally, at the limit of very weak donor/acceptor organizations with $K_{EDA} \ll 1$, the lifetime of the EDA complex can be on the order of a molecular collision; these are referred to as *contact charge-transfer* complexes.[18]

Irrespective of whether one is dealing with a strong, a weak, or a very weak donor/acceptor organization, the charge-transfer probe ($h\nu_{CT}$) always signals a common electron-transfer process. As such, the intracomplex transfer of an electron from D to A in equation (4) corresponds to the (electron) population of the acceptor LUMO at the expense of the donor HOMO (see Chart 3). The resulting cation radical $D^{+\bullet}$ and anion radical $A^{-\bullet}$ are reactive intermediates, as are the analogous (uncharged) radicals $D^{\bullet}$ and $A^{\bullet}$ that are generated by charge-transfer activation of the contact ion pair $[D^{-}, A^{+}]$ in equation (5).

The further reactions of these transient intermediates modulate the overall (second-order) reactivity of the donor/acceptor reactants in the following way.

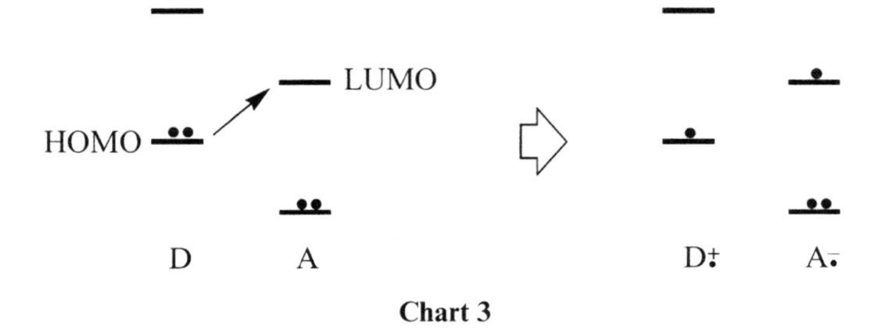

Chart 3

5 The electron-transfer paradigm

The electron donor/acceptor organization in equations (2) and (3) together with the charge-transfer activation in equations (4) and (5) act in tandem as a coupled set of pre-equilibria[19] (equation 7).

$$\mathrm{D} + \mathrm{A} \underset{}{\overset{\mathrm{EDA}}{\rightleftharpoons}} [\mathrm{D}, \mathrm{A}] \underset{}{\overset{\mathrm{CT}}{\rightleftharpoons}} \mathrm{D}^{+\bullet}, \mathrm{A}^{-\bullet} \qquad (7)$$

The extent to which the ion-radical pair suffers a subsequent (irreversible) transformation (with rate constant k_f characteristic of highly reactive cation radicals and anion radicals) that is faster than the reverse or back electron transfer (k_{BET}) then represents the basis for the electron-transfer paradigm that drives the coupled EDA/CT equilibria forward onto products (P)[20] (equation 8).

$$\mathrm{D} + \mathrm{A} \rightleftharpoons [\mathrm{D}, \mathrm{A}] \underset{k_{BET}}{\rightleftharpoons} \mathrm{D}^{+\bullet}, \mathrm{A}^{-\bullet} \xrightarrow{k_f} \mathrm{P} \qquad (8)$$

Scheme 1

If the EDA and CT pre-equilibria are fast relative to such a (follow-up) process, the overall second-order rate constant is $k_2 = K_{EDA}K_{CT}k_f$. In this kinetic situation, the ion-radical pair might not be experimentally observed in a thermally activated adiabatic process. However, photochemical (laser) activation via the deliberate irradiation of the charge-transfer absorption ($h\nu_{CT}$) will lead to the spontaneous generation of the ion-radical pair (equations 4, 5) that is experimentally observable if the time-resolution of the laser pulse exceeds that of the follow-up processes (k_f and k_{BET}). Indeed, *charge-transfer activation provides the basis for the experimental demonstration of the viability of the electron-transfer paradigm in Scheme 1.*[21]

The applicability of the electron-transfer paradigm to the thermal activation of electron donors and acceptors is more difficult to establish, since the ion-radical pair $\mathrm{D}^{+\bullet}$, $\mathrm{A}^{-\bullet}$ is unlikely to attain concentrations high enough for direct observation. The exception will be those metastable donor/acceptor pairs in which the HOMO–LUMO gap is sufficient to promote the very rapid (adiabatic) electron-transfer activation (k_{ET}) of the ground-state EDA complex in the dark. In the more common situation involving electron donors with unexceptional acceptors (or vice versa), the experimental proof of the electron-transfer paradigm is less forthcoming, and the mechanistic distinction between the electron-transfer paradigm in equation (8) and the direct bimolecular (k_2) reaction of D and A can be ambiguous. Since such a distinction between stepwise and concerted pathways is difficult to resolve with certainty, let us see how the electron-transfer paradigm can provide a single unifying mechanism for organic reactions in general.[22]

6 Case in point: the keto–enol umpolung

We return now to the representative reactions in Chart 2 to illustrate the utility of the electron-transfer paradigm as a unifying theme in the varied behavior of carbonyl compounds. We first recognize that the keto–enol tautomerism in equation (1) effectively converts an electron-poor carbonyl acceptor into an electron-rich (hydroxy-substituted) olefinic donor,[23] and such an umpolung allows the conventional *nucleophilic* additions at the carbonyl center (e.g., Grignard addition and hydride reduction) to be readily replaced by *electrophilic* substitution at the α-carbon (halogenation, etc.) of ketones, aldehydes, esters, etc.[24] Let us initially examine the diverse chemistry of electron-rich enol tautomers of various carbonyl compounds with a variety of electron-poor reagents. It is noteworthy that the fecund chemistry of enolates and its derivatives derives in large part from the facile preparation of the silyl ether and related derivatives which effectively "freeze" the enol form.[25]

ENOL SILYL ETHERS (ESE) AS ELECTRON DONORS

The facile α-substitution reactions of ESE donors with different electron acceptors is indicated by such diverse transformations as

- *halogenation* with bromine, *N*-bromosuccinimide, iodine monochloride, iodine, silver acetate/iodine, lead(IV) acetate/metal halides, and thallium(I) acetate/iodine;[26]
- *nitration* with nitric acid, acyl nitrate, alkyl nitrates, and tetranitromethane;[27]
- *nitrosation* with nitrosyl chloride and nitrosonium salts;[28]
- *oxygenation* with peracids (Rubottom reaction), hydrogen peroxide, sulfonyloxaziridine, and nitrogen oxides;[29]

- *oxidative addition* with quinones such as chloranil and DDQ, nitrobenzenes, electron-poor olefins such as tetracyanoethylene, carbenium ions, and diazonium salts, etc.[30]

The versatile reactivity of enol silyl ethers has also dominated the chemistry of C—C bond formation such as

- *oxidative coupling* to 1,4-diketones with silver oxide, lead(IV) acetate, vanadium oxytrichloride, ferric chloride, and ceric ammonium nitrate;[31]
- *Mukaiyama reaction* (Lewis acid-catalyzed Michael reaction) with electron-poor olefins, ketals and acetals, and enones;[32]
- *aldol condensation* with ketones and aldehydes catalyzed by electron-poor iron porphyrins, etc.[33]

The wide diversity of the foregoing reactions with electron-poor acceptors (which include cationic and neutral electrophiles as well as strong and weak one-electron oxidants) points to enol silyl ethers as electron donors in general. Indeed, we will show how the electron-transfer paradigm can be applied to the various reactions of enol silyl ethers listed above in which the donor/acceptor pair leads to a variety of reactive intermediates including cation radicals, anion radicals, radicals, etc. that govern the product distribution. Moreover, the modulation of ion-pair (cation radical and anion radical) dynamics by solvent and added salt allows control of the competing pathways to achieve the desired selectivity (see below).

The notion of enol silyl ethers (ESE) as electron donors was first provided by Gassman and Bottorff,[34] who showed that selective (carbonyl) deprotection can be readily achieved in the presence of an alkyl silyl ether group via an electron-transfer activation (e.g., equation 9).

OSiMe$_3$ / Me$_3$SiO (cyclohexenyl) $\xrightarrow[\text{MeCN–MeOH}]{h\nu,\ \text{1-cyanonaphthalene}}$ O / Me$_3$SiO (cyclohexanone) (9)

The success of this transformation depends upon the oxidation potential of the ESE group ($E_{ox} \sim 1.5$ V), which is lower than that of the alkyl silyl ether group ($E_{ox} \sim 2.5$ V). Recently, Schmittel *et al.*[35] showed (by product studies) that the enol derivatives of sterically hindered ketones (e.g., 2,2-dimesityl-1-phenylethanone) can indeed be readily oxidized to the corresponding cation radicals, radicals and α-carbonyl cations either chemically with standard one-electron oxidants (such as tris(p-bromophenyl)aminium hexachloroantimonate or ceric ammonium nitrate) or electrochemically (equation 10).

$$\mathrm{Mes_2C{=}C(OR)Ph} \xrightarrow[\mathrm{MeCN}]{-e^-} \mathrm{Mes_2C{=}C(OR)Ph}^{+\bullet} \xrightarrow{-R^+} \mathrm{Mes_2\dot{C}{-}C(O)Ph} \xrightarrow{-e^-} \mathrm{Mes_2C^+{-}C(O)Ph} \longrightarrow \text{2-Ph-3-Mes-benzofuran} \tag{10}$$

These authors also noted that the electron-donor ability of various derivatives of 2,2-dimesityl-1-phenylethenol decreases in the order: enolate > enol > enol silyl ether > enol phosphate > enol acetate. As such, a simple derivatization allows the ready modulation of the electron-donor properties of enols.

Electron donor/acceptor organization of enol silyl ethers

The visual indication of enol silyl ethers as effective electron donors derives from the vivid colorations that are observed when ESEs are exposed to a variety of well-known electron acceptors such as tetranitromethane (TNM), chloranil (CA), dichlorodicyanobenzoquinone (DDQ), tetracyanobenzene (TCNB), tetracyanoquinodimethane (TCNQ), tetracyanoethylene (TCNE), etc.[36,37] For example, intense orange colors develop immediately when chloranil is added to a solution of cyclohexanone ESE in dichloromethane. Figure 1a shows the progressive growth of the new (UV-vis) absorption band at $\lambda_{max} = 460$ nm with incremental addition of enol silyl ether to a solution of chloranil in dichloromethane. Spectrophotometric analysis of the absorbance data (A_{CT}) according to the Benesi–Hildebrand procedure (equation 6) yields a linear plot (see inset in Fig. 1a) from which the values of the formation constant ($K_{EDA} = 1.1\ \mathrm{M}^{-1}$) and extinction coefficient ($\varepsilon_{CT} = 416\ \mathrm{M}^{-1}\ \mathrm{cm}^{-1}$) of the charge-transfer transitions are readily extracted. The quantitative effects of such colorations are illustrated in Figure 1b by the systematic shift in the new absorption bands of chloranil with the enol silyl ethers of cyclohexanone, methylcyclohexanone and β-tetralone. (Note that neither chloranil nor ESE absorbs beyond $\lambda = 400$ nm.) These well-resolved featureless absorption bands in Fig. 1 are characteristic of weak intermolecular donor/acceptor [D, A] complexes in which the color arises from the charge-transfer transition ($h\nu_{CT}$) (equation 11).

The generality of such intermolecular [D, A] complexes is shown in Fig. 1c by the exposure of a colorless solution of α-tetralone enol silyl ether to different

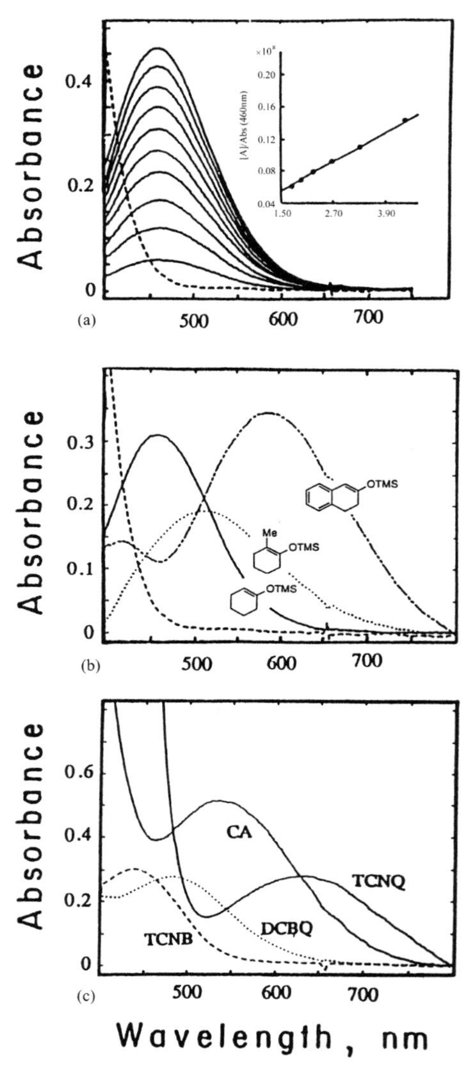

Fig. 1 Charge-transfer absorption spectra of enol silyl ethers complexes with π-acceptors. (a) Spectral changes accompanying the incremental additions of cyclohexanone enol silyl ether [2] to chloranil in dichloromethane. Inset: Benesi–Hildebrand plot. (b) Charge-transfer absorption spectra of chloranil complexes showing the red shift in the absorption maxima with decreasing IP of the enol silyl ethers. (c) Comparative charge-transfer spectra of EDA complexes of α-tetralone enol silyl ether [6] showing the red shift in the absorption maxima with increasing EAs of the acceptors: tetracyanobenzene (TCNB), 2,6-dichlorobenzoquinone (DCBQ), chloranil (CA), and tetracyanoquinodimethane (TCNQ). Reproduced with permission from Ref. 37.

$$\text{OSiMe}_3\text{-cyclohexene} + \text{Cl}_4\text{C}_6\text{O}_2 \overset{K_{\text{EDA}}}{\rightleftharpoons} \left[\text{OSiMe}_3\text{-cyclohexene},\ \text{Cl}_4\text{C}_6\text{O}_2\right]_{\text{EDA}} \xrightarrow{h\nu_{\text{CT}}} \text{OSiMe}_3\text{-cyclohexene}^{+\bullet},\ \text{Cl}_4\text{C}_6\text{O}_2^{-\bullet} \tag{11}$$

electron acceptors (of increasing acceptor strength such as tetracyanobenzene (TCNB), dichlorobenzoquinone (DCBQ), chloranil (CA), and tetracyanoquinodimethane (TCNQ) leading to orange (λ_{max} = 432 nm), red (λ_{max} = 486 nm), purple (λ_{max} = 538 nm) and green (λ_{max} = 628 nm) colorations, respectively. Such a spectral (red) shift in the absorption band of α-tetralone enol silyl ether with electron acceptors of increasing electron affinity is in accordance with Mulliken theory.[6]

A more precise determination of the energetics of electron detachment is also available from the gas-phase measurement of ionization potentials (IP) obtained from the He(I) photoelectron spectra (PES) of the enol silyl ethers.[36] Importantly, a linear correlation of the ionization potentials of ESE and the charge-transfer transition energies ($h\nu_{CT}$) in the chloranil [D, A] complexes illustrated in Fig. 2 further accords with Mulliken's prediction that $h\nu_{CT}$ = IP − EA − ω, where the electron affinity (EA) of the chloranil acceptor and the electrostatic interactions (ω) of the charge-transfer ion pair in equation (4) are constant for a related series of [D, A] complexes. In other words, Mulliken theory predicts that absorption maxima λ_{CT} of the [D, A] complexes with a given electron acceptor undergo bathochromic shift with increasing electron donicity or decreasing ionization potentials (and oxidation potentials) of the enol silyl ethers. The absorption maxima (λ_{CT}) of the [D, A] complexes of various substituted enol silyl ethers with representative electron acceptors together with the ionization potentials (IP) and anodic peak potentials (E_p) obtained from cyclic voltammetry are compiled in Table 1.

Having shown that the enol silyl ethers are effective electron donors for the [D, A] complex formation with various electron acceptors, let us now examine the electron-transfer activation (thermal and photochemical) of the donor/acceptor complexes of tetranitromethane and quinones with enol silyl ethers for nitration and oxidative addition, respectively, via ion radicals as critical reactive intermediates.

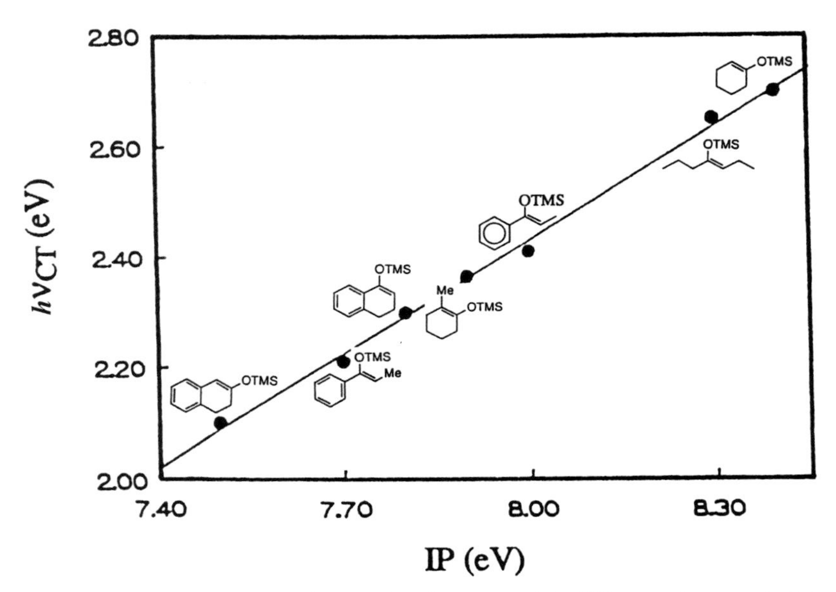

Fig. 2 Mulliken correlation of the ionization potentials (IP) of various enol silyl ethers with the charge-transfer transition energies ($h\nu_{CT}$) of their EDA complexes with chloranil. Reproduced with permission from Ref. 36.

Nitration of enol silyl ethers with tetranitromethane

When α-tetralone enol silyl ether in dichloromethane is mixed with an equivalent of tetranitromethane (TNM) at room temperature, the solution immediately takes on a bright red coloration owing to formation of the characteristic [D, A] complex[37] (equation 12).

$$\text{(α-tetralone enol }OSiMe_3\text{)} + C(NO_2)_4 \underset{}{\overset{K_{EDA}}{\rightleftharpoons}} [\text{(α-tetralone enol }OSiMe_3\text{)}, C(NO_2)_4]_{EDA} \quad (12)$$

Upon standing in the dark, the solution progressively bleaches to a pale yellow solution (~ 10 min), and a simple aqueous workup procedure affords crystalline 2-nitro-α-tetralone in quantitative yield. The analysis of the aqueous layer indicates the formation of one equivalent of trinitromethide according to the stoichiometry in equation (13).

$$[\text{(α-tetralone enol }OSiMe_3\text{)}, C(NO_2)_4] \xrightarrow[25^\circ C]{\text{dark}} \text{2-}NO_2\text{-α-tetralone} + Me_3SiC(NO_2)_3 \quad (13)$$

Table 1 EDA complex formation of enol silyl ethers with various electron acceptors in dichloromethane.

No.	Enol silyl ether	IP[a] (eV)	E_p[a] (V versus SCE)	λ_{CT} (nm)[b] TCNB	CA	TCNQ
1	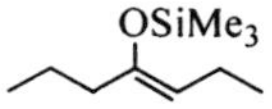 $OSiMe_3$	8.3	1.50	*c*	469	480
2	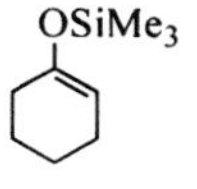 $OSiMe_3$	8.4	1.41	*c*	460	482
3	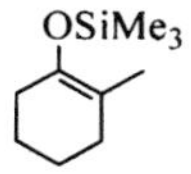 $OSiMe_3$	7.9	1.14	415	524	585
4	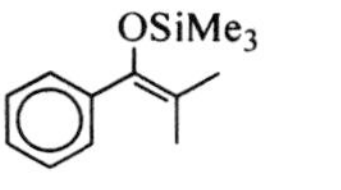 $OSiMe_3$	8.0	1.43	410	515	570
5	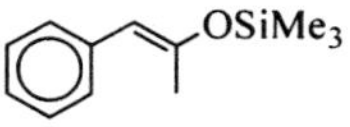 $OSiMe_3$	7.7	1.26	452	566	665
6	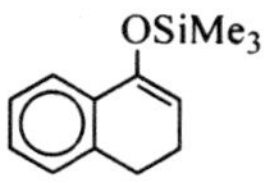 $OSiMe_3$	7.8	1.30	432	538	628
7	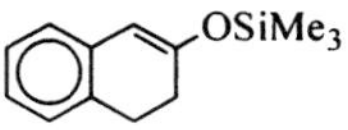 $OSiMe_3$	7.5	1.10	468	590	705
8	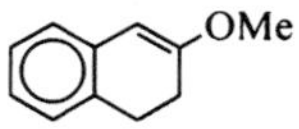 OMe	7.6	1.10	*d*	606	*d*
9	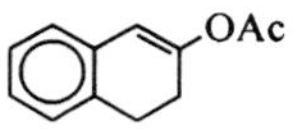 OAc	8.0	1.48	*d*	492	*d*

[a]From ref. 36. [b]Maxima of the charge-transfer band: TCNB, $E^0_{red} = 0.65$ V; CA, $E^0_{red} = 0.02$ V; TCNQ, $E^0_{red} = 0.19$ V versus SCE. [c]No new charge-transfer absorption band observed. [d]Not measured.

Table 2 Thermal and photochemical nitration of enol silyl ethers with tetranitromethane in dichloromethane.

No.	Enol silyl ether	IP (eV)	α-Nitroketone	Thermal nitration[a]		Photochemical nitration[b]	
				Time (h)	Yield (%)[c]	Time (h)	Yield (%)[c]
1	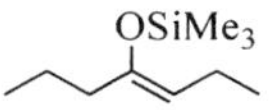	8.3	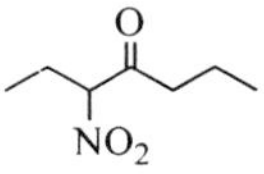	72	20	1.5	64
2	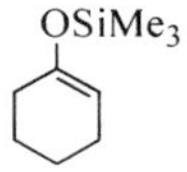	8.4	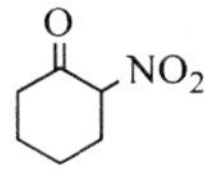	16	70	1.0	73
3	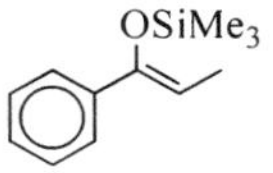	8.0	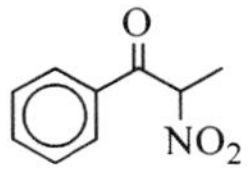	6	90	0.5	96
4	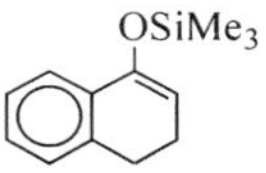	7.8	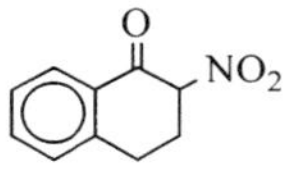	0.17	95	0.5	98
5	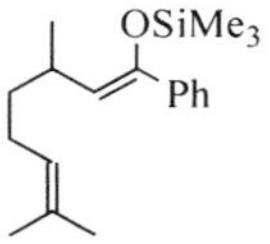	*d*	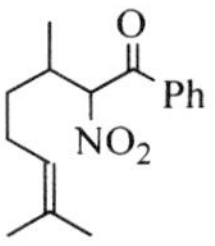	8	78	0.8	72

[a]Reaction carried out at 25°C. [b]Reaction at −78°C. [c]Isolated yield. [d]Not determined.

The facile nitration of a wide variety of ketones with TNM in Table 2 is illustrative of the synthetic utility of enol silyl ethers in facilitating α-substitution of carbonyl derivatives. It is necessary to emphasize here that the development of a strong charge-transfer (orange to red) coloration immediately upon the mixing of various ESEs with TNM invariably precedes the actual production of α-nitroketones in the thermal nitration (in the dark). The increasing conversion based on the time/yields listed in Table 2 qualitatively follows a trend in which electron-rich ESE from 6-methoxy-α-tetralone reacts faster than the relatively electron-poor ESE from cyclohexanone.

Photochemical activation of [ESE,TNM] complexes. Colored solutions of the ESE derived from the relatively unreactive cycloalkanones and acyclic ketones in Table 2 and TNM persist for prolonged periods, and they can thus be separately subjected to filtered actinic radiation with $\lambda_{exc} > 420$ nm. (Note that under these conditions, neither TNM nor ESE alone is excited since they absorb only below 400 nm.) The photonitrations of ESE with TNM are carried out at low temperatures (−40°C) at which the thermal reactions are too slow to compete (as established by appropriate control experiments). The photonitration in each case affords excellent yields of the corresponding nitroketones in relatively short periods of time (see Table 2, columns 7 and 8). For example, the irradiation of a dark red solution of α-tetralone enol silyl ether and TNM at low temperature (equation 14) results in a continuous bleaching of the red color; and aqueous workup affords a quantitative yield of corresponding 2-nitro-α-tetralone (identical to that from the thermal reaction in equation (13)).

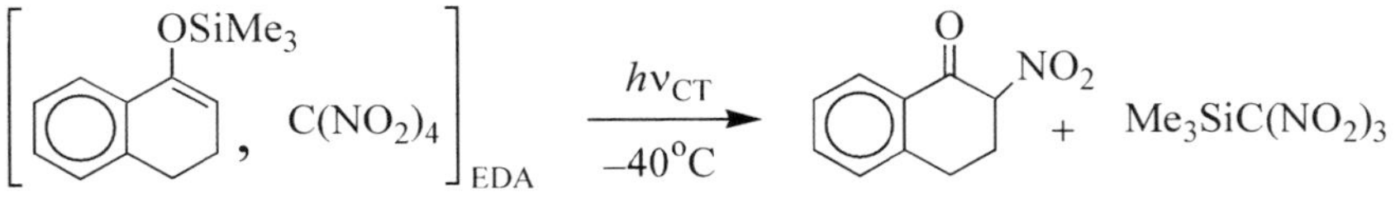

$$[\text{α-tetralone ESE } (OSiMe_3),\ C(NO_2)_4]_{EDA} \xrightarrow[-40^{\circ}C]{h\nu_{CT}} \text{2-nitro-α-tetralone } (NO_2) + Me_3SiC(NO_2)_3 \quad (14)$$

According to the Mulliken theory, the direct photoactivation of intermolecular [D, A] complex by selective irradiation of the charge-transfer absorption band ($h\nu_{CT}$) leads to a photoinduced electron transfer. As applied to [D, A] complexes of ESE and TNM, such a photoactivation is tantamount to the spontaneous formation of ion-radical pair in equation (16) (Scheme 2).[14] The rapid fragmentation of resulting tetranitromethane anion radical affords the ion-radical triad shown in equation (17).[37] As such, this mechanistic formulation is closely related to previous studies of the photoinduced electron transfer of TNM with arene (ArH) donors, in which the corresponding ion radical triad ($ArH^{+\bullet}$, $NO_2^{\bullet}$, $C(NO_2)_3^-$) is directly produced as the result of charge-transfer activation.[38] Moreover, the mechanistic formulation in Scheme 2 is further confirmed by the direct observation of the β-tetralone ESE cation radical (λ_{max} = 520 nm) by time-resolved spectroscopy. An excellent material balance obtains in photonitration, and the short lifetime of <3 ps for the anion radical $C(NO_2)_4^{-\bullet}$[39] suggests a fast homolytic coupling of the cation radical $ESE^{+\bullet}$ with $NO_2^{\bullet}$ within the solvent cage. This is followed by the loss of cationic trimethylsilyl moiety that leads to the α-nitroketones in Scheme 3.

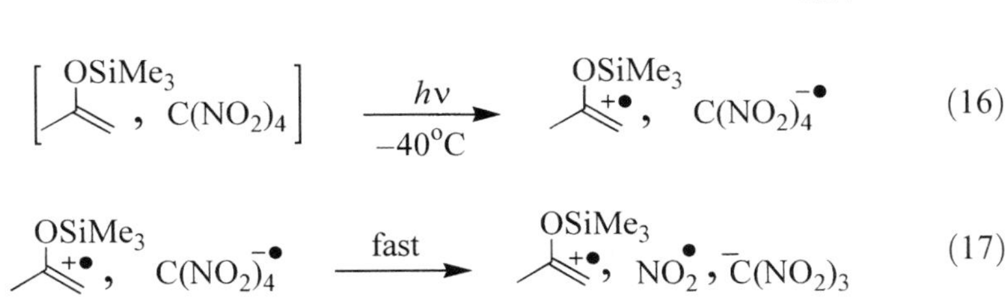

$$\text{ESE } (OSiMe_3) + C(NO_2)_4 \overset{K_{EDA}}{\rightleftharpoons} [\text{ESE } (OSiMe_3),\ C(NO_2)_4]_{EDA} \quad (15)$$

$$[\text{ESE } (OSiMe_3),\ C(NO_2)_4] \xrightarrow[-40^{\circ}C]{h\nu} \text{ESE}^{+\bullet},\ C(NO_2)_4^{-\bullet} \quad (16)$$

$$\text{ESE}^{+\bullet},\ C(NO_2)_4^{-\bullet} \xrightarrow{\text{fast}} \text{ESE}^{+\bullet},\ NO_2^{\bullet},\ {}^{-}C(NO_2)_3 \quad (17)$$

Scheme 2

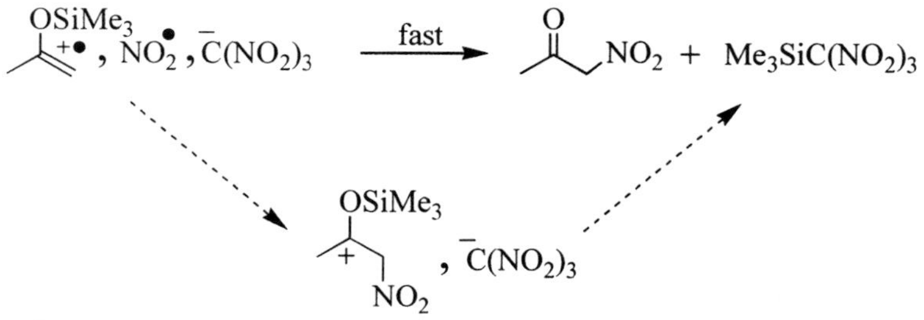

$$\text{ESE}^{+\bullet},\ NO_2^{\bullet},\ ^{-}C(NO_2)_3 \xrightarrow{\text{fast}} CH_3COCH_2NO_2 + Me_3SiC(NO_2)_3$$

Scheme 3

Comments on the thermal nitration of enol silyl ethers with TNM. The strikingly similar color changes that accompany the photochemical and thermal nitration of various enol silyl ethers in Table 2 indicates that the pre-equilibrium [D, A] complex in equation (15) is common to both processes. Moreover, the formation of the same α-nitroketones from the thermal and photochemical nitrations suggests that intermediates leading to thermal nitration are similar to those derived from photochemical nitration. Accordingly, the differences in the qualitative rates of thermal nitrations are best reconciled on the basis of the donor strengths of various ESEs toward TNM as a weak oxidant in the rate-limiting dissociative thermal electron transfer (k_{ET}), as described in Scheme 4.[40]

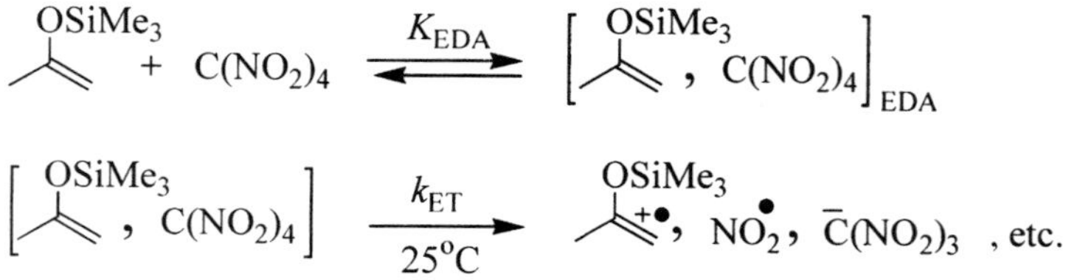

$$\text{ESE} + C(NO_2)_4 \overset{K_{EDA}}{\rightleftharpoons} [\text{ESE}, C(NO_2)_4]_{EDA}$$

$$[\text{ESE}, C(NO_2)_4] \xrightarrow[25^{\circ}C]{k_{ET}} \text{ESE}^{+\bullet},\ NO_2^{\bullet},\ ^{-}C(NO_2)_3 \text{, etc.}$$

Scheme 4

Indeed, the observed trend in the reactivities of various ESEs is directly related to their donor strengths (see Table 2).

The alternative electrophilic mechanism for the nitration of ESE with TNM requires a close approach of a hindered ESE to a NO_2 group on the quaternary carbon center of TNM. However, this transition state is sterically very demanding, and it will not readily account for the observed reactivity. Furthermore, the observed lack of regioselectivity in the nitration of the isomeric enol silyl ethers of 2-methylcyclohexanone that leads to the same 2-methyl-2-nitrocyclohexanone (in thermal as well as photochemical nitration) is not readily reconciled by a concerted (electrophilic) mechanism (equation 18).

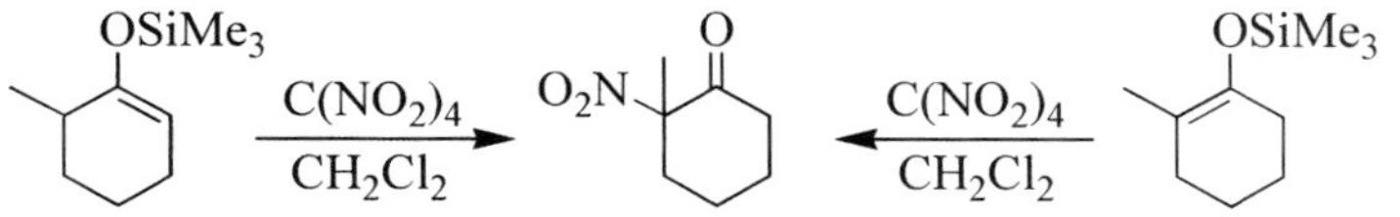

$$\text{(2-methyl-1-cyclohexenyl } OSiMe_3) \xrightarrow[CH_2Cl_2]{C(NO_2)_4} \text{2-nitro-2-methylcyclohexanone } (O_2N) \xleftarrow[CH_2Cl_2]{C(NO_2)_4} \text{(6-methyl-1-cyclohexenyl } OSiMe_3) \quad (18)$$

On the other hand, the cation radical of the kinetic ESE$^{+\bullet}$ is readily converted to the thermodynamic cation radical by a 1,3-prototropic shift in the course of the photoinduced electron transfer with chloranil.[41] As such, the efficient isomerization of the kinetic ESE cation radical as an intermediate in equation (19) accounts for the observed lack of regioselectivity in equation (18).[37]

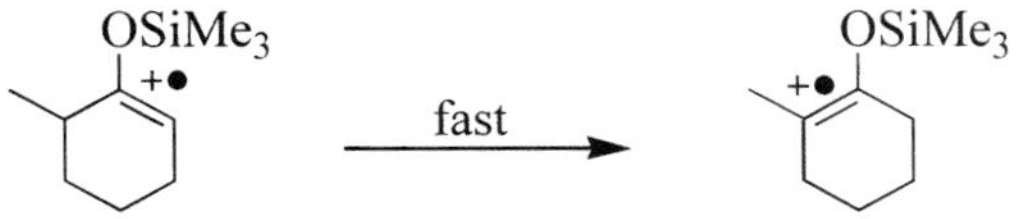

$$\text{(6-methyl-1-cyclohexenyl } OSiMe_3)^{+\bullet} \xrightarrow{\text{fast}} \text{(2-methyl-1-cyclohexenyl } OSiMe_3)^{+\bullet} \quad (19)$$

Dehydrosilylation versus oxidative addition of enol silyl ethers to quinones

Various enol silyl ethers and quinones lead to the vividly colored [D, A] complexes described above; and the electron-transfer activation within such a donor/acceptor pair can be achieved either via photoexcitation of charge-transfer absorption band (as described in the nitration of ESE with TNM) or via selective photoirradiation of either the separate donor or acceptor.[41] (The difference arising in the ion-pair dynamics from varied modes of photoactivation of donor/acceptor pairs will be discussed in detail in a later section.) Thus, actinic irradiation with $\lambda_{exc} > 380$ nm of a solution of chloranil and the prototypical cyclohexanone ESE leads to a mixture of cyclohexenone and/or an adduct depending on the reaction conditions summarized in Scheme 5.

Bhattacharya *et al.*[30b] have shown that the transformation depicted in Scheme 5 is also readily achieved via thermal activation by substituting chloranil ($E_{red} = 0.02$ V versus SCE) with a high-potential quinone such as

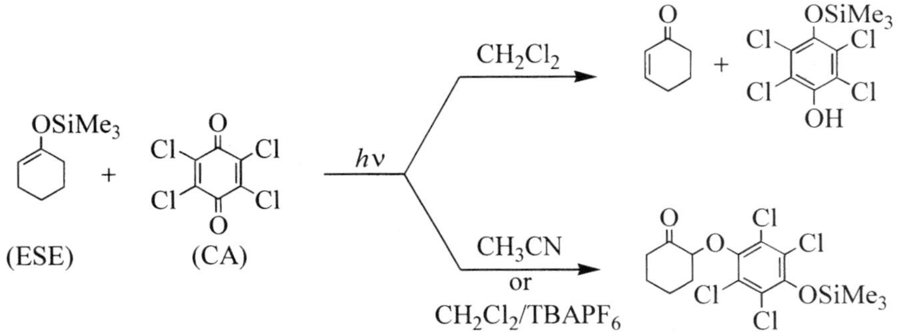

(ESE) $OSiMe_3$-cyclohexene + (CA) chloranil (Cl, Cl, Cl, Cl, O, O) $\xrightarrow{h\nu}$

- $\xrightarrow{CH_2Cl_2}$ cyclohexenone (O) + tetrachlorophenol bearing $OSiMe_3$ and OH (Cl, Cl, Cl, Cl)
- $\xrightarrow{CH_3CN \text{ or } CH_2Cl_2/TBAPF_6}$ 2-(aryloxy)cyclohexanone: O–aryl ring bearing Cl, Cl, Cl, Cl and $OSiMe_3$

Scheme 5

DDQ ($E_{red} = 0.52$ V). It is noteworthy that the strong medium effects (i.e., solvent polarity and added n-$Bu_4N^+PF_6^-$ salt) on product distribution (in Scheme 5) are observed both in thermal reaction with DDQ and photochemical reaction with chloranil. Moreover, the photochemical efficiencies for dehydrosilylation and oxidative addition in Scheme 5 are completely independent of the reaction media – as confirmed by the similar quantum yields ($\Phi = 0.85$ for the disappearance of cyclohexanone enol silyl ether) in nonpolar dichloromethane (with and without added salt) and in highly polar acetonitrile. Such observations strongly suggest the similarity of the reactive intermediates in thermal and photochemical transformation of the [ESE, quinone] complex despite changes in the reaction media.

Exploitation of time-resolved spectroscopy allows the direct observation of the reactive intermediates (i.e., ion-radical pair) involved in the oxidation of enol silyl ether (ESE) by photoactivated chloranil ($^3CA^*$), and their temporal evolution to the enone and adduct in the following way.[41c] Photoexcitation of chloranil (at $\lambda_{exc} = 355$ nm) produces excited chloranil triplet ($^3CA^*$) which is a powerful electron acceptor ($E_{red} = 2.70$ V versus SCE), and it readily oxidizes the electron-rich enol silyl ethers ($E_{ox} = 1.0$–1.5 V versus SCE) to the ion-radical pair with unit quantum yield, both in dichloromethane and in acetonitrile (equation 20).

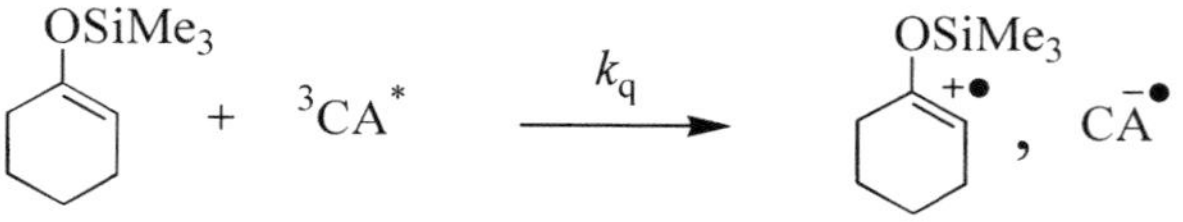

$$\text{OSiMe}_3\text{-cyclohexene} + {}^3\text{CA}^* \xrightarrow{k_q} [\text{OSiMe}_3\text{-cyclohexene}]^{+\bullet},\ \text{CA}^{-\bullet} \qquad (20)$$

In dichloromethane, the acidic ESE cation radical undergoes a rapid proton transfer ($k_f = 1.9 \times 10^9\ s^{-1}$) to the CA anion radical within the contact ion pair (CIP) to generate the uncharged radical pair (siloxycyclohexenyl radical and hydrochloranil radical) in Scheme 6. Based on the quantum yields of hydrochloranil radical ($HCA^\bullet$), we conclude that the oxidative elimination occurs by geminate combination of the radical pair within the cage as well as by diffusive separation and combination of the freely diffusing radicals to yield enone and hydrochloranil trimethylsilyl ether, as summarized in Scheme 6.

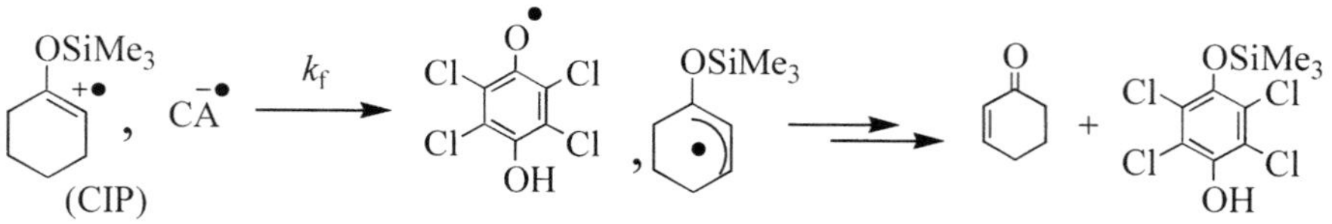

$$[\text{OSiMe}_3\text{-cyclohexene}]^{+\bullet},\ \text{CA}^{-\bullet}\ \text{(CIP)} \xrightarrow{k_f} \text{HCA}^\bullet,\ \text{OSiMe}_3\text{-cyclohexenyl}^\bullet \longrightarrow \text{cyclohexenone} + \text{Cl}_4\text{C}_6(\text{OH})(\text{OSiMe}_3)$$

Scheme 6

On the other hand, in highly polar acetonitrile the contact ion pair ($ESE^{+\bullet}$, $CA^{-\bullet}$) undergoes diffusive separation to the (long-lived) solvent-separated ion pair, which is accompanied by desilylation of the resulting $ESE^{+\bullet}$ to yield α-keto alkyl radical and solvated trimethylsilyl cation (see Scheme 7). Radical coupling with $CA^{-\bullet}$ followed by (re)silylation by trimethylsilyl cation (Scheme 7) affords the adduct described in Scheme 5.

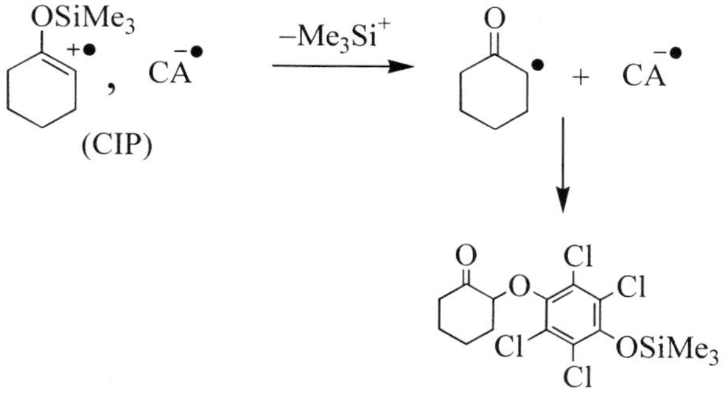

Scheme 7

Note that a similar separation of the CIP to free ion radicals can also be achieved by added $n\text{-}Bu_4N^+PF_6^-$ salt in dichloromethane to divert the course of reaction (via ion exchange) to favor adduct formation in Scheme 5.

Enol silyl ether cation radical as versatile synthetic intermediates

The one-electron oxidation of enol silyl ether donor (as described above) generates a paramagnetic cation radical of greatly enhanced homolytic and electrophilic reactivity. It is the unique dual reactivity of enol silyl ether cation radicals that provides the rich chemistry exploitable for organic synthesis. For example, Snider and coworkers[42] showed the facile homolytic capture of the cation radical moiety by a tethered olefinic group in a citronellal derivative to a novel multicyclic derivative from an acyclic precursor (Scheme 8).

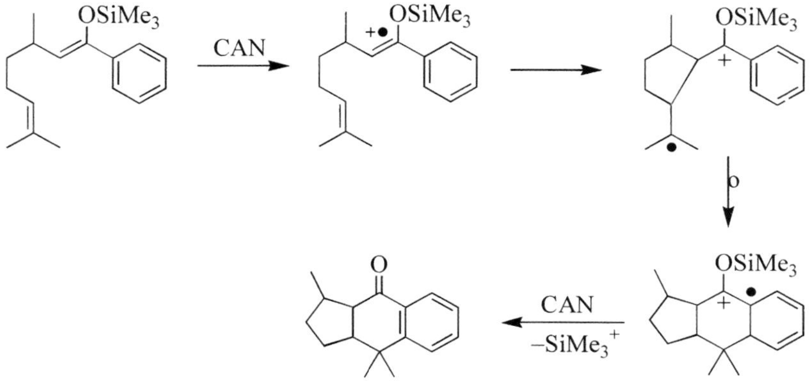

Scheme 8

KETONES AS ELECTRON ACCEPTORS

Ketones – in sharp contrast to enols – are electron acceptors in a wide variety of organic transformations that occur at the carbonyl carbon. For example, the familiar dark-blue benzophenone anion radical is produced via one-electron reduction of benzophenone with sodium in anhydrous THF (equation 21).

$$\mathrm{Ph_2C{=}O} \;+\; \mathrm{Na} \longrightarrow \mathrm{Ph_2C{=}O^{-\bullet}} \;+\; \mathrm{Na^+} \qquad (21)$$

(Acceptor) (Donor)

Another well-known transformation of carbonyl derivatives is their conversion to pinacols (1,2-diols) via an initial one-electron reduction with highly active metals (such as sodium, magnesium, aluminum, samarium iodide, cerium(III)/I_2, yttrium, low-valent titanium reagents (McMurry coupling), etc.), amines, and electron-rich olefins and aromatics as one-electron donors (D).[43] Ketyl formation is rapidly followed by dimerization[44] (equation 22).

$$\mathrm{RR'C{=}O} \;+\; \mathrm{D} \longrightarrow \mathrm{RR'C{=}\bar{O}^{\bullet}} \;+\; \mathrm{D^+} \longrightarrow {}^{-}\mathrm{O{-}C(R)(R'){-}C(R)(R'){-}O^{-}} \qquad (22)$$

The reduction of ketones to pinacols can also be achieved electrochemically via ketone anion radicals that are observable as transient intermediates.[45]

In the general context of donor/acceptor formulation, the carbonyl derivatives (especially ketones) are utilized as electron acceptors in a wide variety of reactions such as *additions* with Grignard reagents, alkyl metals, enolates (aldol condensation), hydroxide (Cannizzaro reaction), alkoxides (Meerwein–Pondorff–Verley reduction), thiolates, phenolates, etc.; reduction to alcohols with lithium aluminum hydride, sodium borohydride, trialkyltin hydrides, etc.; and cyloadditions with electron-rich olefins (Paterno–Büchi reaction), acetylenes, and dienes.[46]

Addition of Grignard reagents to carbonyl compounds

Grignard reagents are highly electron-rich donors that are among the most commonly utilized organometallics in a variety of nucleophilic addition reactions. In particular, the reactions of (electron-poor) carbonyl acceptors with (electron-rich) Grignard reagents constitute an interesting case for the electron-transfer paradigm to be examined as follows. As early as 1929, Blicke and Powers and others[47] noted that the reaction of diarylketones with alkylmagnesium halides (RMgX) results in a variety of byproducts (together with the simple 1,2-addition products), the formation of which could not be rationalized by simple concerted addition (Scheme 9).

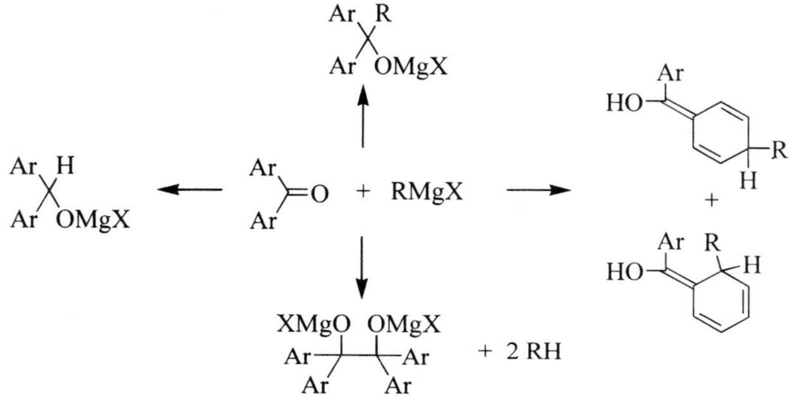

Scheme 9

The ESR detection of benzophenone-ketyl radical coupled with the formation of pinacols as byproducts (in Scheme 9) provides the basis for an electron-transfer mechanism between carbonyl acceptors and various Grignard reagents[48] (equation 23).

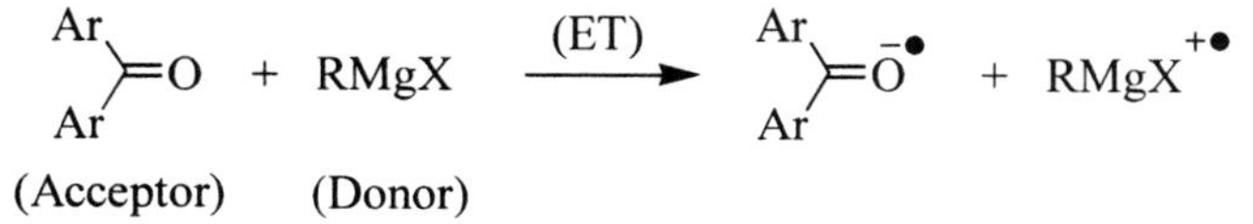

$$\mathrm{Ar_2C{=}O} + \mathrm{RMgX} \xrightarrow{\text{(ET)}} \mathrm{Ar_2C{=}O^{-\bullet}} + \mathrm{RMgX^{+\bullet}} \qquad (23)$$

(Acceptor) (Donor)

The electron-transfer formulation in equation (23) was further substantiated by Ashby *et al.*[49] by using cyclizable probes for the detection of alkyl radicals from the mesolytic scission of $RMgX^{+\bullet}$ (Scheme 10).

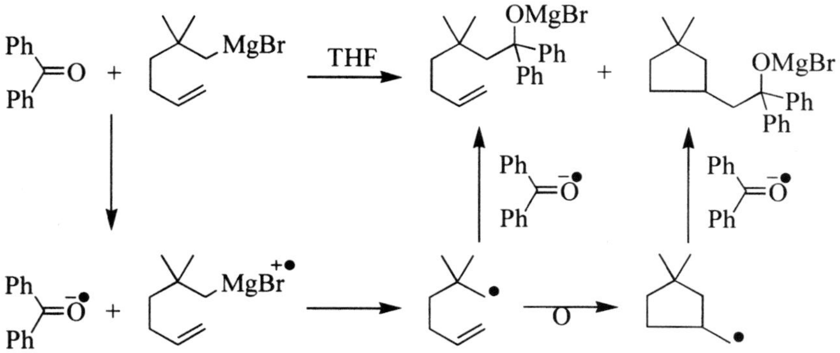

Scheme 10

A similar electron-transfer mechanism is readily applicable to the reaction between sterically hindered quinones and arylmagnesium bromides, which leads to biaryls as well as quinone anion radical as directly observable species,[50] (equation 24).

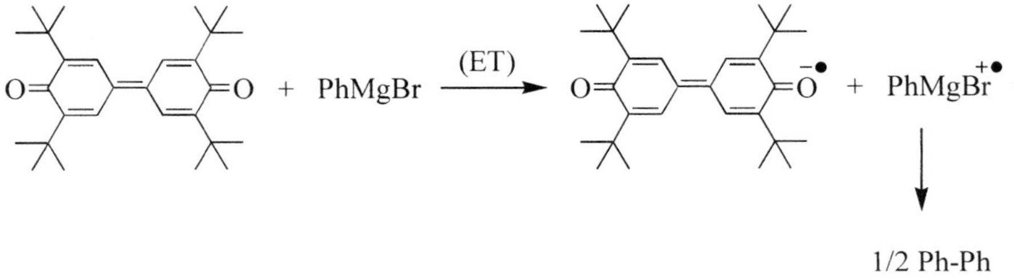

$$\text{Quinone} + \text{PhMgBr} \xrightarrow{\text{(ET)}} \text{Quinone}^{-\bullet} + \text{PhMgBr}^{+\bullet} \longrightarrow 1/2\ \text{Ph-Ph} \quad (24)$$

Holm and coworkers[51] provided a more quantitative basis for the electron-transfer mechanism in equation (23) by the detailed analyses of the kinetics and product distribution. For example, the reaction of *t*-butylmagnesium chloride with various substituted benzophenones followed the Hammett rate law for the consumption of diarylketones.[52] However, the product distribution in Scheme 9 depends heavily on the substitution pattern – the rates of the reaction of benzophenone showing a linear correlation with the electrode potentials of substituted alkylmagnesium bromides. The rates are not sensitive to steric hindrance since *t*-butylmagnesium bromide is ~1300 times more reactive than methylmagnesium bromide.[52] These observations suggest that the initial electron-transfer step in equation (23) is the rate-determining step, and the product distribution is determined in a series of fast consecutive steps, which are extremely sensitive to steric factors. The electron-transfer mechanism for the Grignard/ketone pair in equation (23) is similarly applicable to the reduction of electron-poor carbonyls to alcohols with aluminum hydride as well as a variety of other electron-rich reducing agents.

[2+2] Cycloadditions of olefins and carbonyl compounds (Paterno–Büchi reaction)

The photoinduced [2+2] cycloaddition of carbonyl acceptors with electron-rich olefins leads to oxetanes (Paterno-Büchi reaction) with high regio- and stereo-selectivities (equation 25).

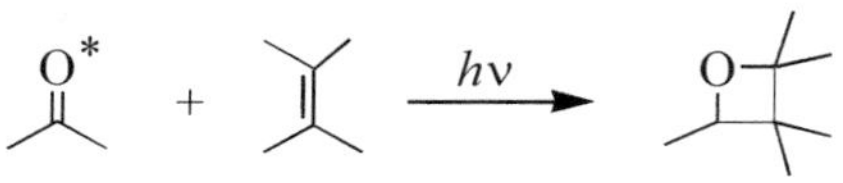

$$\text{Me}_2\text{C=O}^* + \text{Me}_2\text{C=CMe}_2 \xrightarrow{h\nu} \text{oxetane} \quad (25)$$

Mattay *et al.*[53] suggested from the photoreaction of biacetyl with highly electron-rich olefins that an initial electron transfer from an electron-rich olefin to photoexcited ketone is the key step in the oxetane formation via the ion-radical pair (equation 26).

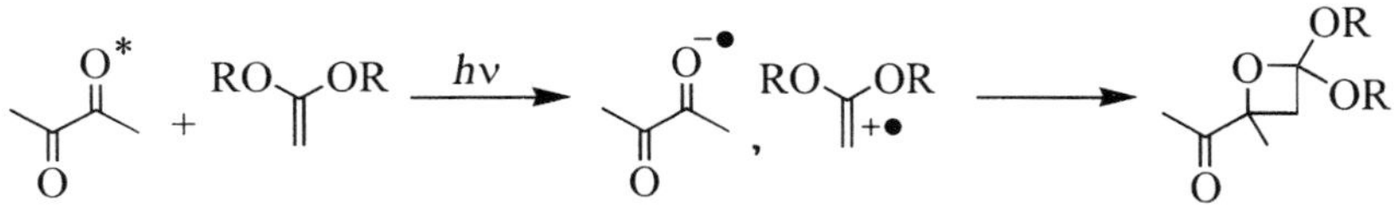

O* + RO OR $\xrightarrow{h\nu}$ $O^{-\bullet}$, RO OR $+\bullet$ ⟶ O OR OR

(26)

The scope of the Paterno–Büchi cycloaddition has been widely expanded for the oxetane synthesis from enone and quinone acceptors with a variety of olefins, stilbenes, acetylenes, etc. For example, an intense dark-red solution is obtained from an equimolar solution of tetrachlorobenzoquinone (CA) and stilbene owing to the spontaneous formation of 1:1 electron donor/acceptor complexes.[55] A selective photoirradiation of either the charge-transfer absorption band of the [D, A] complex or the specific irradiation of the carbonyl acceptor (i.e., CA) leads to the formation of the same oxetane regioisomers in identical molar ratios[56] (equation 27).

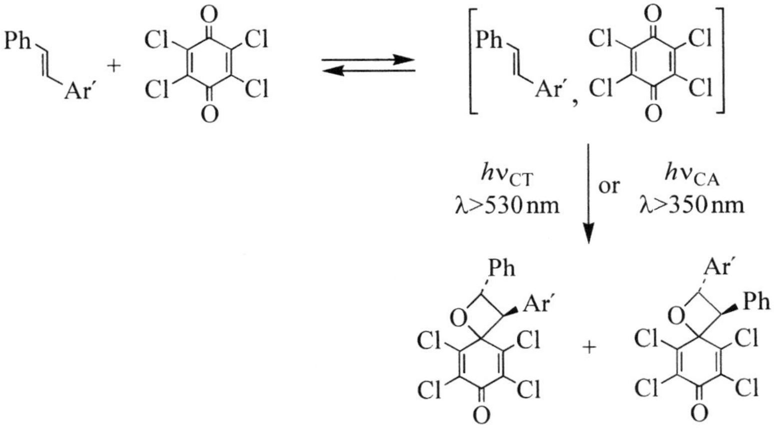

Ph Ar′ + Cl Cl Cl Cl O O ⇌ [Ph Ar′, Cl Cl Cl Cl O O]

$h\nu_{CT}$ λ>530 nm or $h\nu_{CA}$ λ>350 nm (27)

Ph Ar′ O Cl Cl Cl Cl O + Ar′ Ph O Cl Cl Cl Cl O

Time-resolved spectroscopy establishes the formation of an ion-radical pair as the critical reactive intermediate (both from direct excitation of the CT absorption band at 532 nm and from specific excitation of chloranil at 355 nm, see Fig. 3) which undergoes ion-pair collapse to the biradical adduct followed by the ring closure to oxetane, as summarized in Scheme 11.

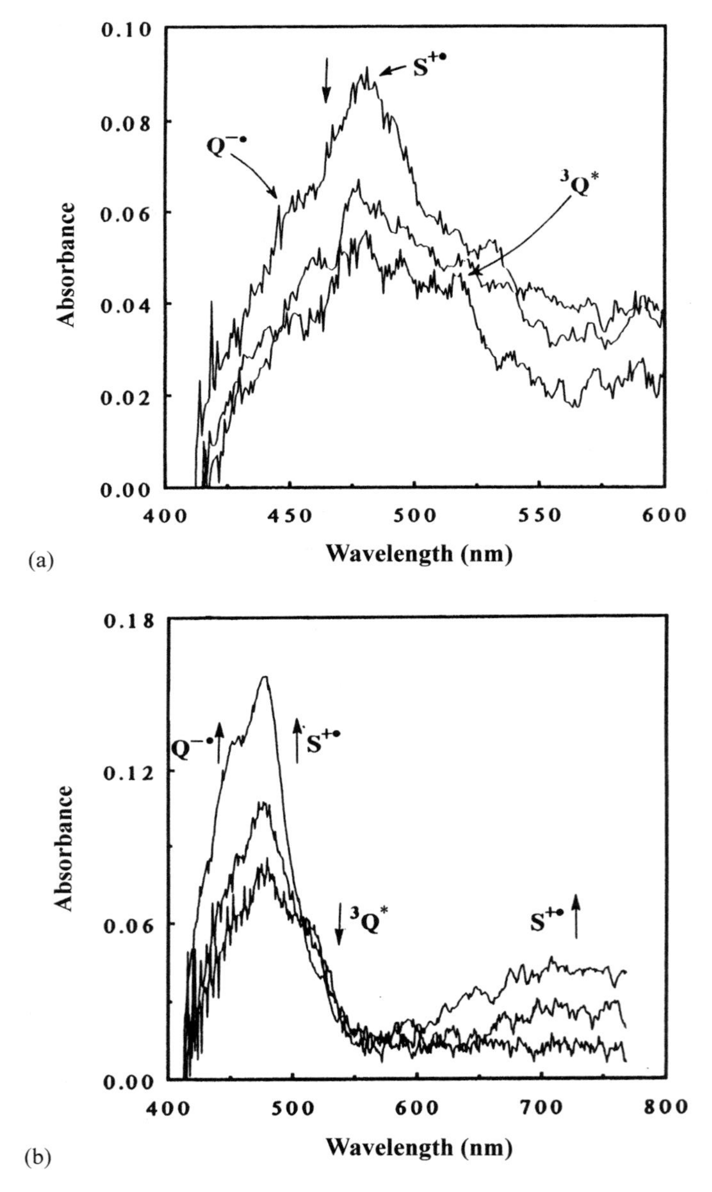

Fig. 3 Transient spectra obtained upon the application of a 200-fs laser pulse to a solution of stilbene (S) and chloranil (Q) in dioxane. (a) The fast decay ($\sim$20 ps) of the contact ion-radical pair $S^{+\bullet}$, $Q^{-\bullet}$ generated by direct charge-transfer excitation (CT path). (b) The slow growth ($\sim$1.6 ns) of the ion pair $S^{+\bullet}$, $Q^{-\bullet}$ due to the diffusional quenching of triplet chloranil (A* path) as described in Scheme 13. Reproduced with permission from Ref. 55.

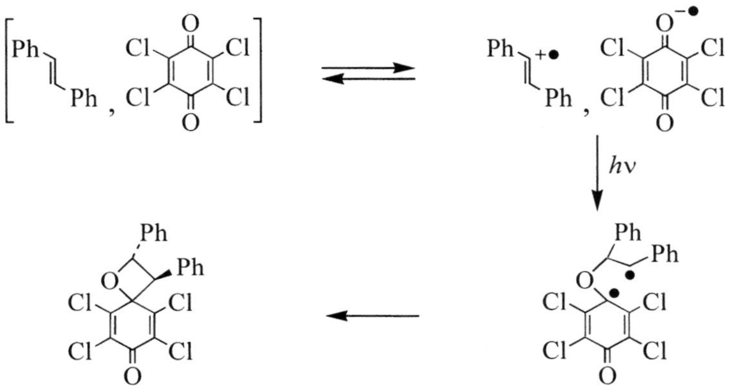

Scheme 11

We emphasize that the critical ion pair stilbene$^{+\bullet}$, CA$^{-\bullet}$ in the two photoactivation methodologies (i.e., charge-transfer activation as well as chloranil activation) is the same, and the different multiplicities of the ion pairs control only the timescale of reaction sequences.[14] Moreover, based on the detailed kinetic analysis of the time-resolved absorption spectra and the effect of solvent polarity (and added salt) on photochemical efficiencies for the oxetane formation, it is readily concluded that the initially formed ion pair undergoes a slow coupling ($k_c = 10^8\ s^{-1}$). Thus competition to form solvent-separated ion pairs as well as back electron transfer limits the quantum yields of oxetane production. Such ion-pair dynamics are readily modulated by choosing a solvent of low polarity for the efficient production of oxetane. Also note that a similar electron-transfer mechanism was demonstrated for the cycloaddition of a variety of diarylacetylenes with a quinone via the [D, A] complex[56] (Scheme 12).

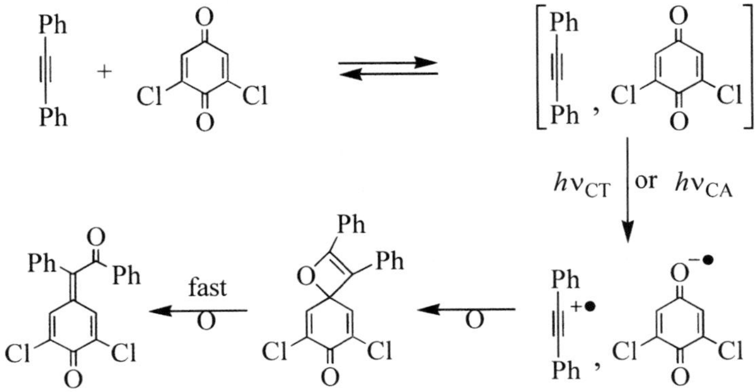

Scheme 12

ELECTRON TRANSFER AS THE UNIFYING THEME

The oxidative conversion of the enol silyl ether donor to its cation radical and the reductive conversion of the carbonyl acceptor to its anion radical, as illustrated in equations (16) and (21), respectively, represent the unifying theme in the electron-transfer paradigm of the keto–enol umpolung. As such, the α-substitutions in Chart 2 derive from the homolytic attack on (as in Scheme 3), the elimination of (as in Scheme 6), or the nucleophilic addition to (as in Scheme 7) the reactive cation radical ($ESE^{+\bullet}$) from the enol silyl ether donor. Likewise, an analogous set of rapid transformations involving the homolytic attack (as in Scheme 6), the cycloaddition (as in equation 27), or the electrophilic addition (as in Scheme 10) pertains to the anion radical ($>C{=}O^{-\bullet}$) from the carbonyl acceptor. The facility of each of these follow-up steps (k_f) relative to the reverse or back electron-transfer (k_{BET}) in equation (8) is thus critical for the successful completion of the ET paradigm in Scheme 1. Accordingly, we now direct our attention to the quantitative evaluation of the charge-transfer (pre-equilibrium) step in equation (8) by first presenting various measures of the electron-donor and electron-acceptor properties of some representative organic compounds.

7 Classification and quantitative evaluation of electron donors and electron acceptors

Electron donors and acceptors can be classified according to the symmetry of the highest occupied molecular orbital (HOMO) and lowest unoccupied molecular orbital (LUMO), respectively. For example, olefins are π-donors and carbonyl compounds are π-acceptors by virtue of the carbon–carbon and carbon–oxygen unsaturation.[57] Alcohols, amines, and sulfides are n- or σ-donors as a result of the presence of non-bonding (electron) pairs on oxygen, nitrogen, and sulfur.[58] Saturated hydrocarbons (especially cyclic ones) and coordinatively saturated organometallics (like alkylsilanes, etc.) are also σ-donors owing to the bonding electron pair in the carbon–carbon or carbon–metal single bond.[59] A heteroaromatic donor such as pyridine can be a π-donor toward some acceptors and a σ-donor toward others. Analogously, *N*-nitropyridinium cation can act as either a π- or σ-acceptor depending on the EDA interaction of the acceptor with the aromatic ring or the nitro group;[60] and the presence of the cationic charge enhances the overall acceptor strength relative to that of the neutral (*N*-nitroenamine) analog.[61]

Substituents modulate the electron-donor or acceptor strengths depending upon their electron-releasing or electron-withdrawing properties as evaluated by the Hammett σ^+ (or σ) constant(s).[62] For example, the multiple substitution of methyl groups on ethylene leads to the strong π-donor 2,3-dimethyl-2-

butene, whereas attachment of cyano groups affords the strong π-acceptor tetracyanoethylene.

ELECTRON DONORS

The donor strength is quantitatively evaluated by the energetics of the oxidative conversion of an organic donor (D) to its cation radical ($D^{+\bullet}$) as measured in solution by its reversible oxidation potential (E^0_{ox})[63] (equation 28).

$$D \underset{}{\overset{E^0_{ox}}{\rightleftharpoons}} D^{+\bullet} + e^- \tag{28}$$

Except for very electron-rich donors that yield stable, persistent radical cations, the E^0_{ox} values are not generally available.[64] Thus the cation radicals for most organic donors are too reactive to allow the measurement of their reversible oxidation potentials in either aqueous (or most organic) solvents by the standard techniques.[65] This problem is partially alleviated by the measurement of the irreversible anodic peak potentials E^a_p that are readily obtained from the linear sweep or cyclic voltammograms (CV). Since the values of E^a_p contain contributions from kinetic terms, comparison with the values of the thermodynamic E^0_{ox} is necessarily restricted to a series of structurally related donors,[66] i.e.,

$$E^0_{ox} = \beta E^a_p + \text{constant} \tag{29}$$

as illustrated in Fig. 4a with $\beta \cong 1.0$. It is important to emphasize this limitation when values of E^a_p (at a constant CV sweep rate) are employed as comparative measures of the electron-donor properties of various organic donors in solution.

An alternative measure of the electron-donor properties is obtained from the energetics of electron detachment in the gas phase; the ionization potentials (IP) of many organic donors have been experimentally determined from the photoelectron spectra obtained by their photoionization in the gas phase. Thus, the values of the ionization potential IP differ from the oxidation potential E^0_{ox} by solvation,[66] i.e.,

$$E^0_{ox} \cong \text{IP} + \Delta G_s \tag{30}$$

where ΔG_s is the solvation energy of the cation radical in comparison with the negligible contribution from the solvation of the uncharged donor. When the variations in ΔG_s are minor, the values of IP such as those listed in Table 3 can be adequate measures of the electron-donor properties of organic compounds applicable to a particular solvent. This generalization is especially tenable for a series of related compounds as illustrated in Fig. 4b.[66] Regardless of whether the electron-donor properties are evaluated in solution by such indirect measures as E^a_p or IP, cognizance must always be taken of the approximations that relate

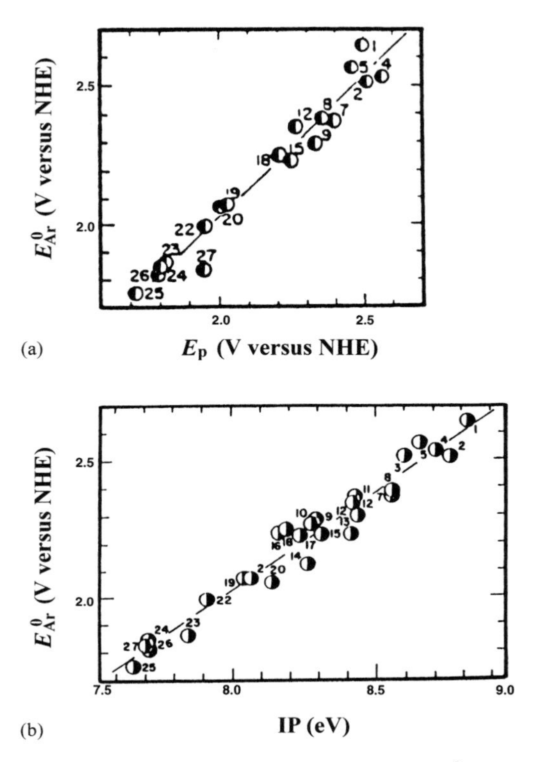

Fig. 4 (a) Correlation of the standard oxidation potentials E^0_{Ar} of various alkylbenzenes with the irreversible CV peak potentials E_p at scan rate $v = 100$ mV s^{-1}. (b) Correlation of the standard oxidation potentials E^0_{Ar} of various alkylbenzenes with the vertical ionization potentials IP. Numbers refer to the aromatic hydrocarbons identified in Tables I and III in Ref. 8. Reproduced with permission from Ref. 8.

them to the thermodynamic values of E^0_{ox}. None the less, the global plot shown in Fig. 5 of the extensive data in Table 3[67] indicates that the electron-donor property (i.e., HOMO energy) is consistently revealed in the trend of the gas-phase (ionization) potential relative to the solution-phase oxidation potentials, despite large differences in structural types. Such a correlation suggests that a similar conclusion pertains to the electron-donor property of organic anions in which gas-phase potentials are more difficult to measure consistently. However, the corresponding alkylmetal derivative such as the organolithium or Grignard reagent is amenable to electrochemical investigation.[59] Typically, the oxidation potentials decrease in the order: $CH_3MgCl > CH_3CH_2MgCl > (CH_3)_2CHMgCl > (CH_3)_3CMgCl$;[68] and the same trend applies to the series of dialkyl mercury donors to support the generalization that σ-donation increases as methyl < *n*-alkyl < *s*-alkyl < *t*-alkyl.[69]

Table 3 Oxidation potentials and ionization potentials of various classes of organic electron donors*.

No.	Donor (D)	E^0_{ox}	IP
Methylbenzenes			
1	Benzene	2.62 (a)†	9.24 (b)
2	Toluene	2.25 (a)	8.82 (b)
3	*o*-Xylene	2.16 (a)	8.56 (b)
4	*p*-Xylene	2.01 (a)	8.45 (b)
5	*m*-Xylene	2.11 (a)	8.56 (b)
6	Mesitylene	2.11 (a)	8.39 (b)
7	Durene	1.84 (a)	8.03 (b)
8	Pentamethylbenzene	1.71 (a)	7.92 (b)
9	Hexamethylbenzene	1.62 (a)	7.85 (b)
Methoxy(hydroxy)benzenes			
10	Phenol	1.04‡ (d)	8.50 (b)
11	Anisole	1.76 (c)	8.22 (b)
12	1,2-Dimethoxybenzene	1.45 (c)	8.08 (e)
13	1,3-Dimethoxybenzene	1.48 (c)	–
14	1,4- Dimethoxybenzene	1.34 (c)	7.90 (e)
15	1,3,5-Trimethoxybenzene	1.49 (c)	–
16	Hexamethoxybenzene	1.24 (c)	–
17	2,5-Dimethyl-1,4-dimethoxy-benzene	1.02‡ (f)	7.45 (f)
18	Tetramethyl-1,4-dimethoxy-benzene	1.46‡ (g)	8.06 (f)
Polycyclic aromatic hydrocarbons			
19	Biphenyl	1.92 (h)	7.95 (i)
20	Naphthalene	1.54 (h)	8.14 (i)
21	Phenanthrene	1.50 (h)	7.86 (i)
22	Triphenylene	1.55 (h)	7.84 (i)
23	Anthracene	1.09 (h)	7.45 (i)
24	Perylene	0.85 (h)	6.90 (i)
25	Chrysene	1.35 (h)	7.59 (i)
26	Tetracene	0.77 (h)	6.97 (i)
27	Pyrene	1.16 (h)	7.41 (i)
28	Biphenylene	–	7.56 (i)
Amines			
29	Aniline	0.98 (j)	7.72 (i)
30	*N*,*N*-Dimethylaniline	0.53 (k)	7.12 (i)
31	*N*,*N*,*N*,*N*-Tetramethyl-*p*-phenylenediamine	0.32 (k)	6.20 (i)
32	Ammonia	–	10.16 (i)
33	Methylamine	–	8.97 (i)
34	Dimethylamine	–	8.23 (i)
35	Trimethylamine	–	7.82 (i)
36	Triethylamine	1.15 (k)	7.50 (i)
37	Tributylamine	1.10 (k)	7.40 (i)
38	Hydrazine	0.07 (k)	9.91 (n)

(continued)

Table 3 *Continued*

No.	Donor (D)	E^0_{ox}	IP
Sulfides			
39	Methyl sulfide	1.35 (m)	8.68 (l)
40	Butyl sulfide	–	8.22 (l)
41	Tetrahydrothiophene	–	8.42 (l)
42	Diphenyl sulfide	–	8.07 (l)
43	Thioanisole	–	7.90 (l)
44	Thiophene	1.70 (m)	8.86 (i)
Olefins/dienes/alkynes			
45	Ethylene	3.20 (m)	10.51 (i)
46	Propene	3.08 (m)	10.03 (n)
47	*cis*-But-2-ene	2.51 (m)	9.36 (n)
48	*trans*-But-2-ene	2.51 (m)	9.37 (n)
49	Isobutene	–	9.45 (n)
50	Cyclopentene	–	9.18 (n)
51	Cyclohexene	1.98 (m)	8.94 (n)
52	Norbornylene	–	8.31 (i)
53	1,3-Butadiene	2.33 (m)	9.07 (i)
54	1,3-Butadiene, 2,3-dimethyl	–	8.71 (i)
55	Isoprene	–	8.84 (i)
56	1,3-Cyclohexadiene	–	8.25 (i)
57	1,4-Cyclohexadiene	1.90 (m)	8.82 (i)
58	Norbornadiene	–	8.35 (i)
59	Acetylene	–	11.40 (i)
60	Propyne	–	10.36 (i)
61	2-Butyne	–	9.56 (i)
62	1,2-Diphenylethylene	1.43 (o)	7.70 (i)
Electron-rich alkenes			
63	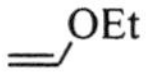 OEt	1.74 (p)	–
64	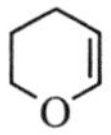 O	1.46 (p)	–
65	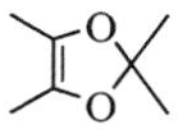 O O	0.72 (p)	–
66	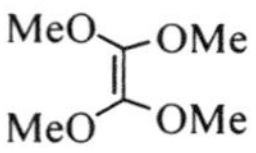 MeO OMe MeO OMe	0.62 (p)	–
67	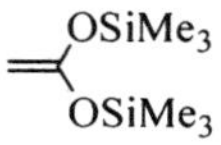 $OSiMe_3$ $OSiMe_3$	1.30 (q)	–
68	 Me_3SiO $OSiMe_3$	1.07 (p)	–

(*continued*)

Table 3 *Continued*

No.	Donor (D)	E^0_{ox}	IP
69	$Me_2C{=}C(OSiMe_3)(OMe)$	0.90 (q)	–
70	$Mes_2C{=}C(OH)Ph$	1.09 (r)	–
71	$Mes_2C{=}C(OMe)Ph$	1.15 (r)	–
72	$Mes_2C{=}C(OSiMe_3)Ph$	1.13 (r)	–
73	$Mes_2C{=}C(OPO(OEt)_2)Ph$	1.49 (r)	–
74	$Mes_2C{=}C(OAc)Ph$	1.68 (r)	–

*Oxidation potentials in V versus SCE in acetonitrile unless indicated otherwise. Ionization potential in eV.

†Literature cited is indicated by the letter in parentheses for Ref. 67.

‡In dichloromethane.

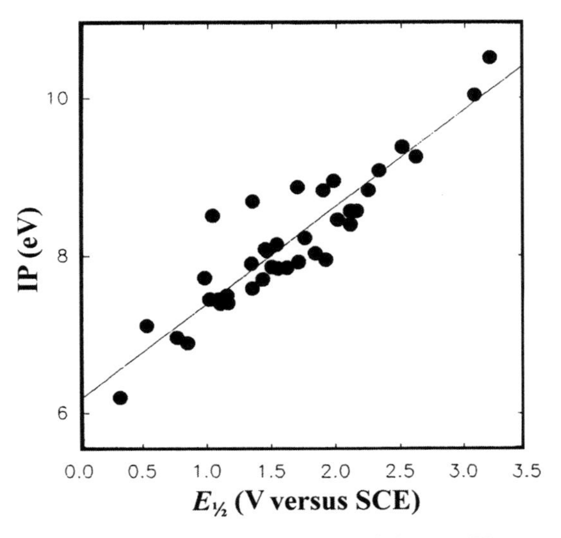

Fig. 5 Global correlation of the oxidation potentials E_{ox} (V versus SCE) with the vertical ionization potentials IP (eV) of various types of organic donors identified in Table 3.

Table 4 Reduction potential and electron affinity of various classes of organic electron acceptors*.

No.	Acceptor (A)	E^0_{red}	EA
Aromatic acceptors			
1	Nitrobenzene	−1.15 (a)†	1.01 (b)
2	1,4-Dinitrobenzene	−0.69 (a)	2.00 (b)
3	1,3,5-Trinitrobenzene	−0.42 (c)	1.8 (n)
4	Picric acid	−0.51 (c)	
5	4-Cyanonitrobenzene	−0.88 (a)	1.72 (b)
6	1,4-Dicyanobenzene	−1.6 (d)	1.10 (b)
7	1,2,4,5-Tetracyanobenzene	−0.65 (d)	1.6 (e)
8	1-Cyanonaphthalene	−1.98 (f)	0.68 (b)
9	1,4-Dicyanonaphthalene	−1.28 (f)	
10	9-Cyanoanthracene	−1.39 (g)	1.27 (b)
11	9,10-Dicyanoanthracene	−0.98 (g)	
12	2,6,9,10-Tetracyanoanthracene	−0.45 (h)	
Quinone acceptors			
13	Duroquinone	−0.84 (i)	1.59 (a)
14	*p*-Benzoquinone	−0.51 (d)	1.89 (b)
15	2,5-Dimethyl-3,6-dichloro-*p*-benzoquinone	−0.51 (j)	
16	2,6-Dimethyl-3,5-dichloro-*p*-benzoquinone	−0.51 (j)	
17	Tetrachloro-*p*-benzoquinone	0.02 (d)	2.76 (b)
18	Tetrabromo-*p*-benzoquinone	0.00 (d)	2.6 (n)
19	Tetraiodo-*p*-benzoquinone	−0.04 (d)	2.5 (n)
20	2,3-Dichloro-5,6-dicyano-*p*-benzoquinone	0.52 (d)	
21	Tetracyano-*p*-benzoquinone	0.90 (k)	
22	*o*-Benzoquinone	−0.31 (i)	
23	Tetrachloro-*o*-benzoquinone	0.32 (i)	
Olefinic acceptors			
24	1,1-Dicyanoethylene		1.54 (e)
25	(*E*)-1,2-Dicyanoethylene	−0.30 (l)	0.96 (e)
26	Tetracyanoethylene	0.24 (m)	2.9 (b)
27	Tetracyanoquinodimethane	0.19 (h)	2.84 (e)
28	Maleic anhydride	−0.85 (o)	1.44 (b)
29	Dichloromaleic anhydride	−1.51 (o)	1.90 (p)
30	Phthalic anhydride	−1.27 (o)	
31	Pyromellitic anhydride	−0.91 (o)	
Carbonyl acceptors			
32	Biacetyl	−1.03 (q)	0.75 (s)
33	9-Fluorenone	−1.74 (d)	1.19 (e)
34	Acetophenone	−2.14 (r)	
35	α,α,α-Trifluoroacetophenone	−1.38 (r)	
36	Benzophenone	−1.83 (r)	0.64 (b)
37	Decafluorobenzophenone	−1.61 (b)	

(*continued*)

Table 4 *Continued*

No.	Acceptor (A)	E^0_{red}	EA
Cationic acceptors			
38	Tropylium	−0.18 (t)	
39	2,4,6-Triphenylpyrilium	−0.29 (u)	
40	2,4,6-Triphenylthiapyrilium	−0.21 (u)	
41	4-Chlorobenzene diazonium	0.45 (v)	
42	4-Nitrobenzenediazonium	0.35 (v)	
43	Trityl cation	0.29 (w)	
44	Diphenyliodonium	−0.42 (x)	
45	*N*,*N*′-Dimethyl-4-bipyridinium	−0.45 (h)	
46	*N*-Methyl-4-cyanopyridinium	−0.67 (y)	
47	Nitrosonium (NO^+)	1.28 (z)	
48	Nitronium (NO_2^+)	1.27 (z))	
49	Dioxygenyl (OH_2^+)	5.3 (aa)	
Cation-radical acceptors			
50	Thianthrenium	1.28 (bb)	
51	Tris(4-bromophenyl)aminium	1.16 (cc)	
52	Tris(2,4,6-tribromophenyl)aminium	1.82 (cc)	
53	9,10-Dimethoxy-1,4:5,8-dimethanooctahydroanthracenium	1.11 (dd)	
54	9,10-Dimethoxy-1,4:5,8-diethanooctahydroanthracenium	1.30 (dd)	
Uncharged acceptors			
55	Tetranitromethane	~0.00 (ee)	1.8 (e)
56	Carbon tetrabromide	−0.30 (ff)	
57	2,4,4,6-Tetrabromocyclohexa-2,5-dienone	0.29 (gg)	
58	Iodine monochloride		2.84 (hh)
59	Chlorine	0.58 (x)	2.4 (e)
60	Bromine	0.47 (x)	2.6 (e)
61	Iodine	0.22 (x)	2.4 (e)
62	Dioxygen	−0.78 (d)	0.4 (n)
Inorganic oxidants			
63	Silver(I)	0.44 (ii)	
64	Cerium (IV) ammonium nitrate	1.61 (x)	
65	Osmium tetraoxide	0.18 (jj)	
66	Potassium permanganate	0.57 (jj)	
67	Tris(1,10-phenanthroline)iron(III)	0.99 (kk)	
68	Tris(1,10-Phenanthroline)ruthenium(III)	1.19 (kk)	
69	Tris(1.10-Phenanthroline)osmium(III)	0.74 (kk)	
70	Ferrocenium	0.46 (x)	
71	Decamethylferrocenium	−0.13 (x)	

*Reduction potential in V versus SCE in acetonitrile. Electron affinity in eV.
†Literature cited is indicated by the letter in parentheses for Ref. 71.

ELECTRON ACCEPTORS

The acceptor strength is quantitatively evaluated by the energetics of the reductive conversion of an organic acceptor (A) to its anion radical ($A^{-\bullet}$); and it is most readily evaluated in solution (Table 4) by the reversible potential E^0_{red} for one-electron reduction,[63] i.e.,

$$A \;+\; e^- \;\underset{}{\overset{E^0_{red}}{\rightleftharpoons}}\; A^{-\bullet} \tag{31}$$

Since the values of E^0_{red} for many organic acceptors are generally unobtainable (in organic solvents), an alternative measure of the electron-acceptor property is often based on the irreversible cathodic peak potential E^c_p (in cyclic voltammetry). Thus for a series of related compounds, Fig. 6 shows that the values of E^0_{red} are linearly related to gas-phase electron affinities (EA).[70]

$$E^0_{red} \cong \mathrm{EA} + \Delta G_s \tag{32}$$

Although the same limitations apply to the use of E^c_p as those described above for the anodic counterpart, the global trend in Fig. 7 shows that gas-phase electron affinities also generally reflect the trend in the reduction potentials measured in solution for the large variety of (uncharged) acceptor structures included in Table 4.[71]

Values of E^0_{red} for many types of cationic oxidants (particularly those based on transition metals[72]) have been tabulated,[73] and some of the more common

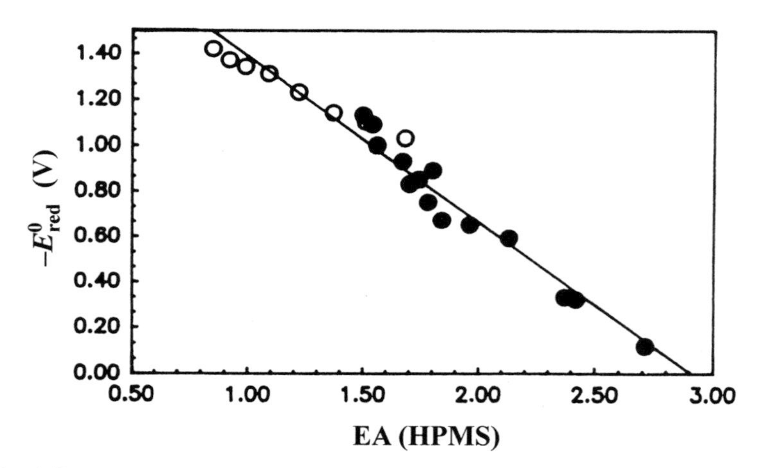

Fig. 6 A linear correlation of the reduction potentials E_{red} (V versus SCE) with the gas-phase electron affinities EA (eV) of various nitrobenzenes and quinones. Reproduced with permission from Ref. 70.

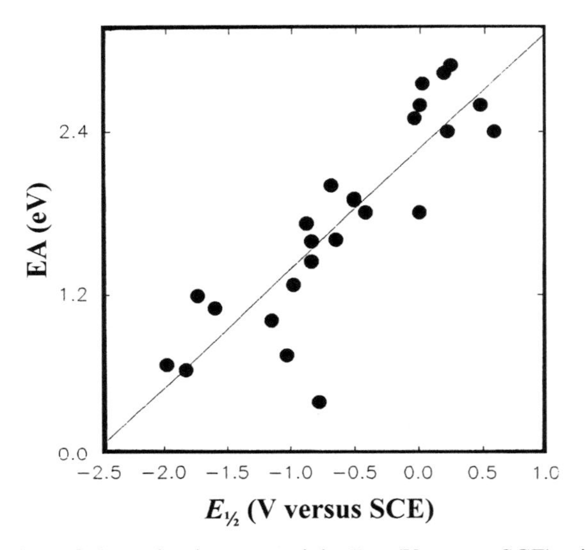

Fig. 7 Correlation of the reduction potentials E_{red} (V versus SCE) with the electron affinities EA (eV) of various types of organic acceptors (in Table 4).

ones are also included in Table 4. However, there are a number of useful oxidants that undergo multiple-electron change (e.g., $Cr^{V} \rightarrow Cr^{IV} \rightarrow Cr^{III}$), and E^{0}_{red} is known only for the overall change. With these oxidants, the one-electron potential of relevance to the electron-transfer paradigm must be evaluated separately by transient electrochemical techniques such as linear sweep micro-

Table 5 Reduction potentials of photoexcited (singlet and triplet) electron acceptors[a].

No.	Acceptor (A*)	E^{0}_{red}
1	2,4,6-Triphenylpyrillium (s)	2.8
2	2-Phenylpyrrolinium (s)	2.9
3	1,2,4,5-Tetracyanobenzene (s)	3.83
4	9,10-Dicyanoanthracene (s)	2.88
5	2,6,9,10-Tetracyanoanthracene (s)	2.82
6	1,4-Dicyanonaphthalene (s)	3.45
7	1-Cyanonaphthalene (s)	3.75
8	9-Cyanoanthracene (s)	2.96
9	1,4-Dicyanobenzene (s)	4.2
10	Tetrachlorobenzoquinone (t)	2.70
11	2,5-Dichloro-3,6-dimethyl-*p*-benzoquinone[b] (t)	1.76
12	*o*-Benzoquinone (t)	2.3
13	1,4-Dinitrobenzene (t)	2.6
14	*N*,*N*-Dimethyl-4-bipyridinium (t)	3.1

[a]From ref. 73. [b]From ref. 71j. (s) = singlet. (t) = triplet.

voltammetry.[66] Owing to their use as photochemical quenchers, the enhanced values of the reduction potential for the excited singlet and triplet acceptor species A* (see $h\nu_A$ in Scheme 13) are listed separately in Table 5.[73]

8 Follow-up reactions of ion radicals as critical (reactive) intermediates

Thermal or photochemical activation of the [D, A] pair leads to the contact-ion pair $D^{+\bullet}$, $A^{-\bullet}$, the fate of which is critical to the overall efficiency of donor/acceptor reactivity as described by the electron-transfer paradigm in Scheme 1 (equation 8). In photochemical reactions, the contact ion pair $D^{+\bullet}$, $A^{-\bullet}$ is generated either via direct excitation of the ground-state [D, A] complex (i.e., CT path via irradiation of the charge-transfer (CT) absorption band in Scheme 13) or by diffusional collision of either the locally excited acceptor with the donor (A* path) or the locally excited donor with the acceptor (D* path).

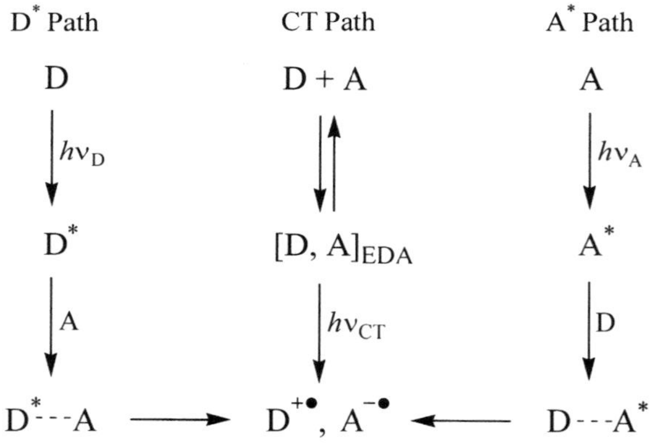

D^* Path	CT Path	A^* Path
D	D + A	A
$\downarrow h\nu_D$	$\rightleftarrows$	$\downarrow h\nu_A$
D^*	$[D, A]_{EDA}$	A^*
$\downarrow$ A	$\downarrow h\nu_{CT}$	$\downarrow$ D
$D^*\text{---}A \longrightarrow$	$D^{+\bullet}, A^{-\bullet}$	$\longleftarrow D\text{---}A^*$

Scheme 13

New synthetic transformations are highly dependent on the dynamics of the contact ion pair, as well as reactivity of the individual radical ions. For example, the electron-transfer paradigm is most efficient with those organic donors yielding highly unstable cation radicals that undergo rapid unimolecular reactions. Thus, the hexamethyl(Dewar)benzene cation radical that is generated either via CT activation of the [D, A] complex with tropylium cation,[74]

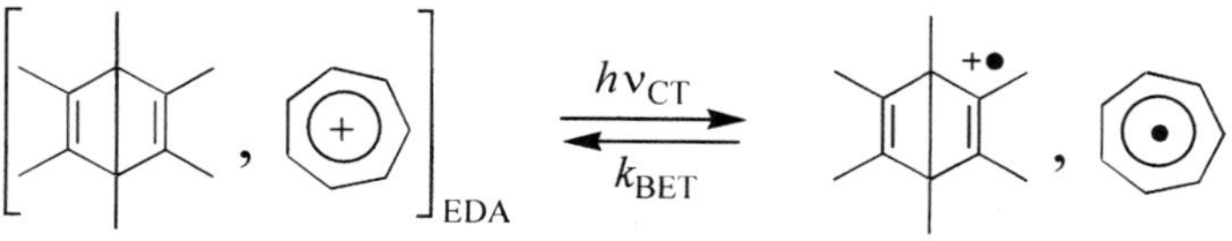

$$\left[\text{HMDB}, \text{C}_7\text{H}_7^+\right]_{EDA} \underset{k_{BET}}{\overset{h\nu_{CT}}{\rightleftarrows}} \text{HMDB}^{+\bullet}, \text{C}_7\text{H}_7^\bullet$$

or via diffusional collision with activated anthraquinone[75] as electron acceptor,

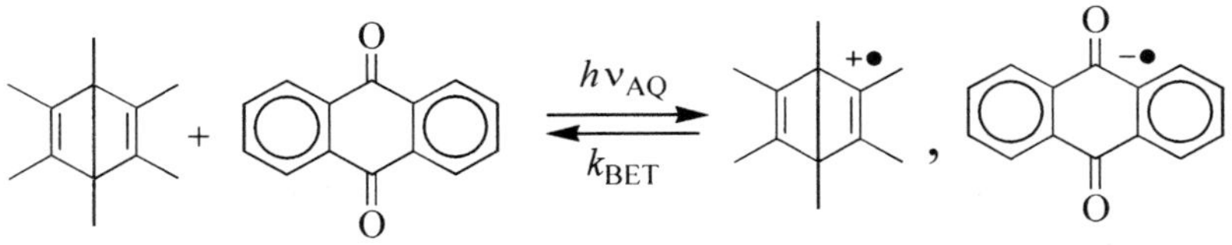

$h\nu_{AQ}$ ⇄ k_{BET}

undergoes rapid C—C bond cleavage and rearrangement to the hexamethylbenzene cation radical,[76]

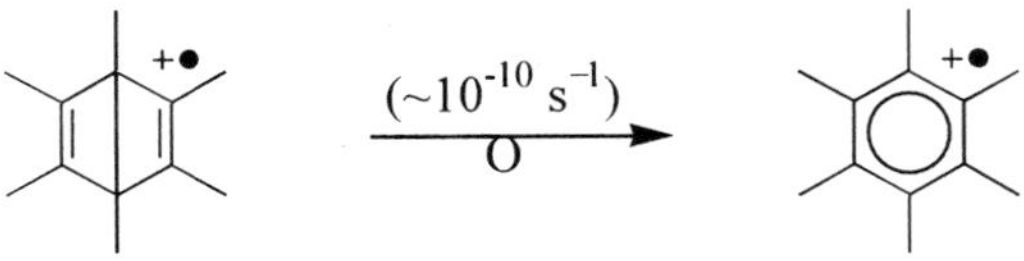

$$(\sim 10^{-10}\ s^{-1}) \qquad (33)$$

Similarly, the reaction of photoexcited 9,10-dicyanoanthracene (DCA) with a benzylstannane yields the contact ion pair in which the cation radical undergoes rapid mesolytic cleavage of the C—Sn bond to afford benzyl radical and tributyltin cation (which then adds to $DCA^{-\bullet}$)[77] (Scheme 14). When such unimolecular processes are faster than the energy-wasting back electron transfer (k_{BET}) within the contact ion pair, the D/A reactions occur rapidly despite unfavorable driving forces for electron transfer.

In a similar vein, various electron acceptors yielding anion radicals that undergo rapid unimolecular decomposition also facilitate the efficacy of Scheme 1 by effectively obviating the back-electron transfer. For example, the nitration of enol silyl ether with tetranitromethane (TNM) occurs rapidly (despite an unfavorable redox equilibrium)[78] owing to the fast mesolytic fragmentation of the TNM anion radical[79] (Scheme 15).

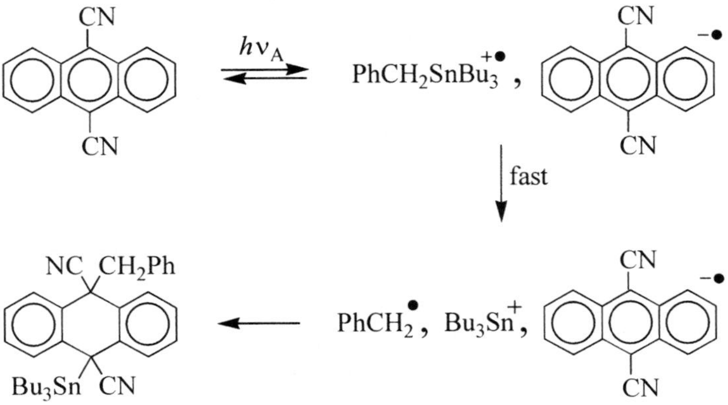

$h\nu_A$ ⇄ $PhCH_2SnBu_3^{+\bullet}$, $DCA^{-\bullet}$

fast

$PhCH_2^{\bullet}$, Bu_3Sn^+, $DCA^{-\bullet}$ ⟶ (NC, CH_2Ph, Bu_3Sn, CN adduct)

Scheme 14

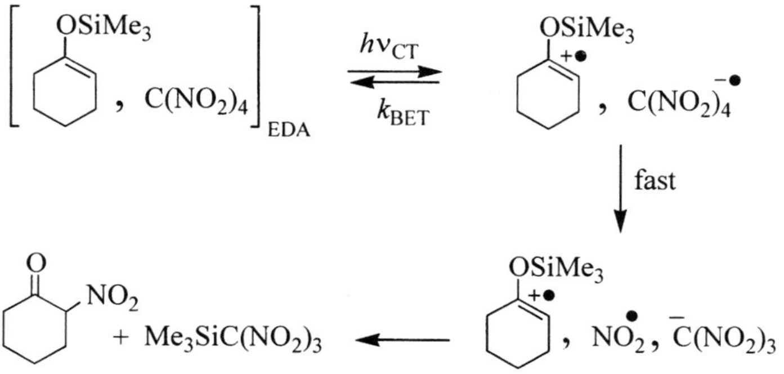

$$\left[\text{(1-trimethylsilyloxycyclohexene)}, C(NO_2)_4\right]_{EDA} \underset{k_{BET}}{\overset{h\nu_{CT}}{\rightleftharpoons}} \text{OSiMe}_3\text{-cyclohexene}^{+\bullet}, C(NO_2)_4^{-\bullet}$$

$$\xrightarrow{\text{fast}} \text{OSiMe}_3\text{-cyclohexene}^{+\bullet}, NO_2^{\bullet}, ^{-}C(NO_2)_3 \longrightarrow \text{2-nitrocyclohexanone} + Me_3SiC(NO_2)_3$$

Scheme 15

Productive bimolecular reactions of the ion radicals in the contact ion pair can effectively compete with the back electron transfer if either the cation radical or the anion radical undergoes a rapid reaction with an additive that is present during electron-transfer activation. For example, the [D, A] complex of an arene donor with nitrosonium cation exists in the equilibrium with a low steady-state concentration of the radical pair, which persists indefinitely. However, the introduction of oxygen rapidly oxidizes even small amounts of nitric oxide to compete with back electron transfer and thus successfully effects aromatic nitration[80] (Scheme 16).

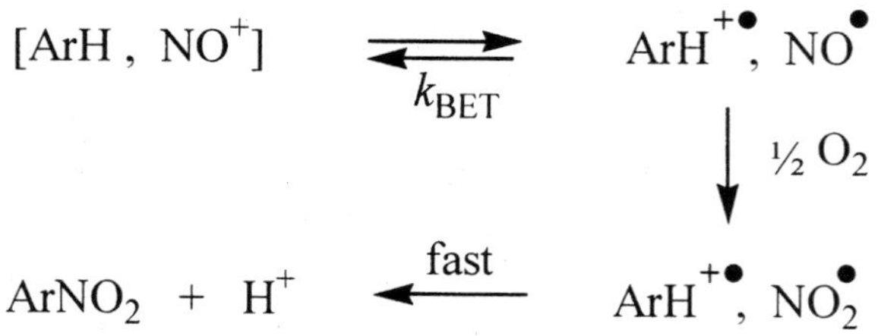

$$[ArH, NO^+] \underset{k_{BET}}{\rightleftharpoons} ArH^{+\bullet}, NO^{\bullet}$$

$$ArH^{+\bullet}, NO^{\bullet} \xrightarrow{\frac{1}{2} O_2} ArH^{+\bullet}, NO_2^{\bullet}$$

$$ArH^{+\bullet}, NO_2^{\bullet} \xrightarrow{\text{fast}} ArNO_2 + H^+$$

Scheme 16

In a related example, the [D, A] complex of hexamethylbenzene and maleic anhydride reaches a photostationary state with no productive reaction (Scheme 17). However, if the photoirradiation is carried out in the presence of an acid, the anion radical in the resulting contact ion pair[14] is readily protonated, and the redox equilibrium is driven toward the coupling (in competition with the back electron transfer) to yield the photoadduct.[81]

The examples in Schemes 14–17, which clearly demonstrate the critical role of ion radicals in D/A reactivity, suggest a thorough review of the diverse reaction pathways that are available to cation radicals and anion radicals as follows.

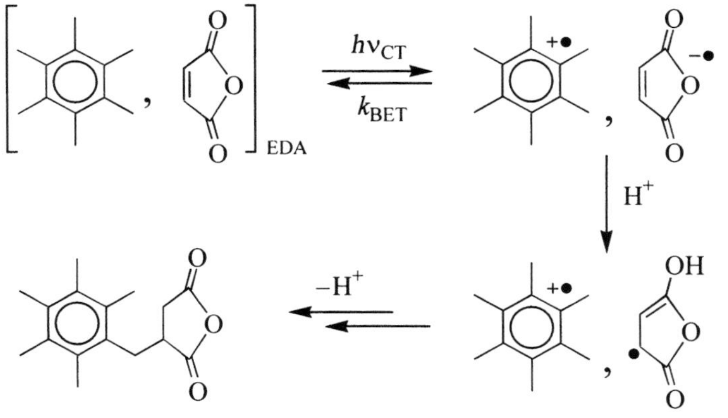

Scheme 17

CHEMISTRY OF CATION RADICALS FROM ELECTRON DONORS

Removal of an electron from a neutral organic donor leads to an "umpolung" cation radical with greatly enhanced electrophilic reactivity (due to the cationic charge) as well as homolytic reactivity (from the radical character). These highly reactive cation radicals undergo a variety of very fast unimolecular and bimolecular reactions.[82–85] Some examples of the unimolecular reactions are illustrated by the following examples, where Ar and R represent aryl and alkyl groups, respectively.

α-Fragmentation

$$ArCH_2CO_2H^{+\bullet} \longrightarrow ArCH_2^{\bullet} + CO_2 + H^+ \quad \text{(Ref. 86)}$$

$$RCH_2MR_3^{+\bullet} \longrightarrow RCH_2^{+} + R_3M^{\bullet} \quad \text{(Ref. 87)}$$

$$RCH_2MR_3^{+\bullet} \longrightarrow RCH_2^{\bullet} + R_3M^{+} \quad \text{(Ref. 88, 89)}$$

M = Sn, Si, Ge, Mg, B, etc.

$$R_3SiSiR_3^{+\bullet} \longrightarrow R_3Si^{\bullet} + R_3Si^{+} \quad \text{(Ref. 90)}$$

$\longrightarrow$ + N_2 (Ref. 91)

β-Fragmentation

$Ar_2C(OH)CO_2^{-}$ (radical cation) ⟶ $Ar_2\dot{C}OH + CO_2$ (Ref. 92)

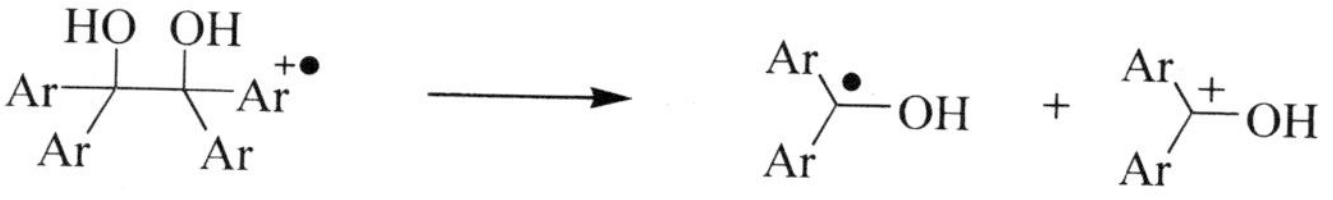

$Ar_2C(OH)-C(OH)Ar_2$ (radical cation) ⟶ $Ar_2\dot{C}OH + Ar_2C^{+}OH$ (Ref. 93)

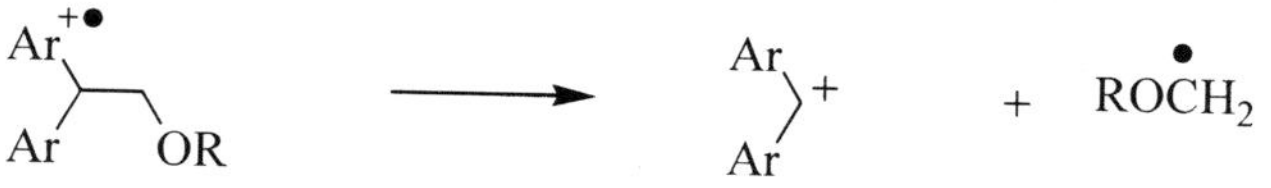

Ar_2CH-CH_2OR (radical cation) ⟶ $Ar_2CH^{+} + RO\dot{C}H_2$ (Ref. 94)

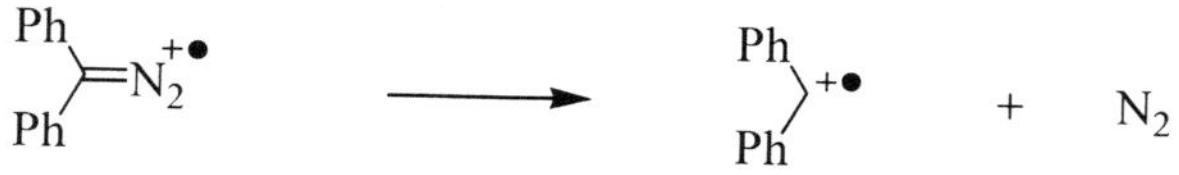

$Ph_2C{=}N_2^{+\bullet}$ ⟶ $Ph_2C^{+\bullet} + N_2$ (Ref. 95)

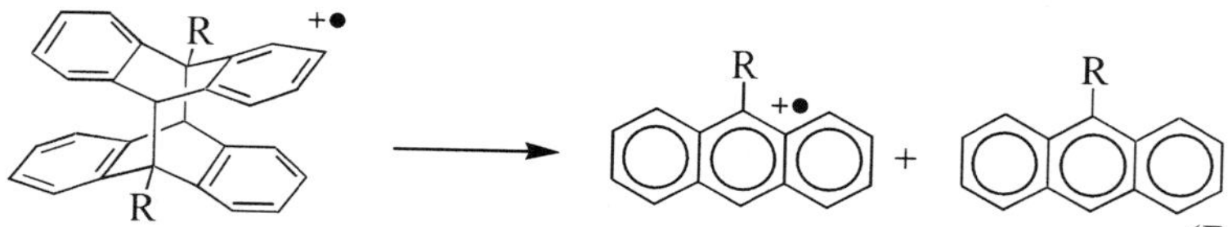

R, R (radical cation) ⟶ R (radical cation) + R (Ref. 96, 97)

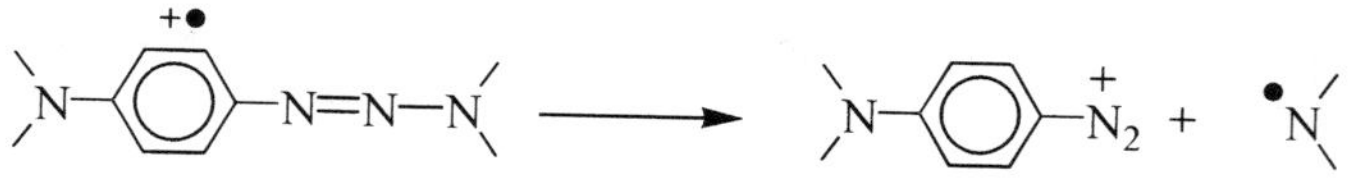

$Me_2N-C_6H_4-N{=}N-NMe_2$ (radical cation) ⟶ $Me_2N-C_6H_4-N_2^{+} + {}^{\bullet}NMe_2$ (Ref. 98)

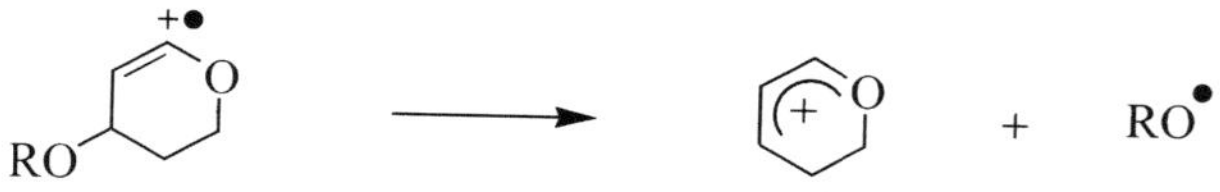

RO, O (radical cation) ⟶ O (+) + $RO^{\bullet}$ (Ref. 99)

Ring cleavage/rearrangement

O (radical cation), Ar, Ar ⟶ O, Ar, Ar (Ref. 100)

N (radical cation), Ar, Ar ⟶ N, Ar, Ar (Ref. 101)

S (radical cation), Ph, Ph ⟶ S, Ph, Ph (Ref. 102)

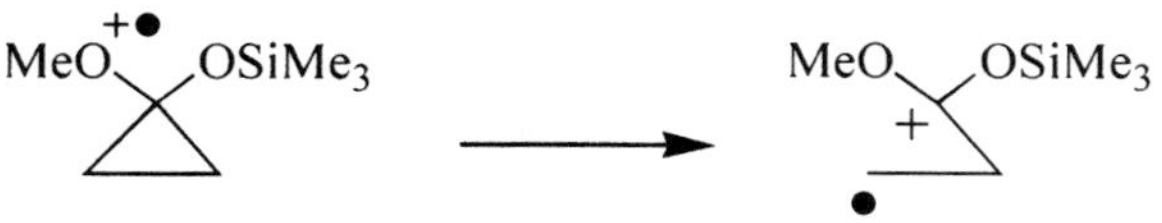

MeO OSiMe$_3$ (+•) → MeO OSiMe$_3$ (+, •) (Ref. 103)

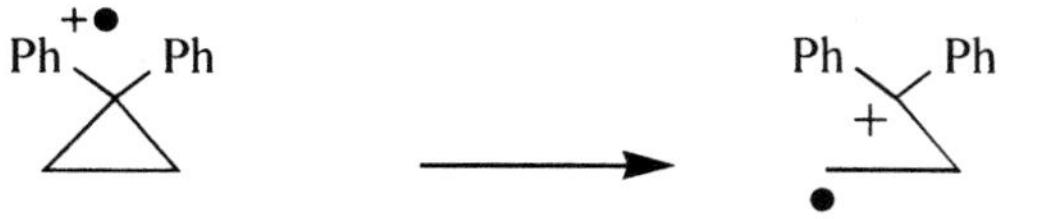

Ph Ph (+•) → Ph Ph (+, •) (Ref. 104)

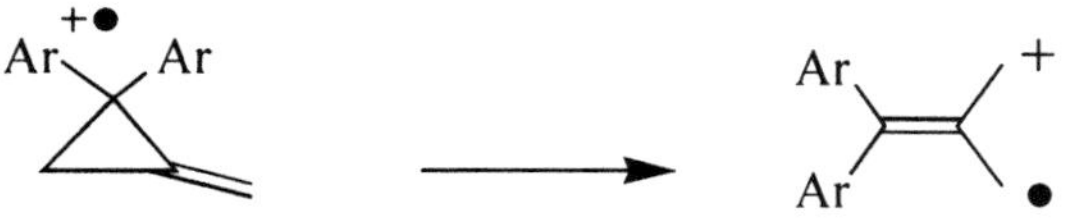

Ar Ar (+•) → Ar Ar (+, •) (Ref. 105)

Ph Ph (+•) → Ph Ph (+•) (Ref. 106)

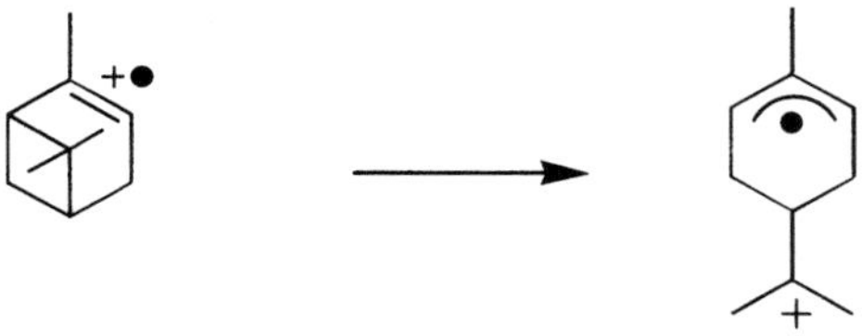

(+•) → (•, +) (Ref. 107)

Cyclization

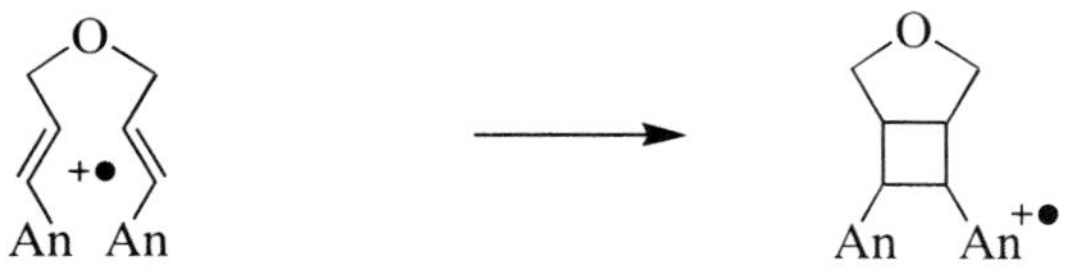

O An An (+•) → O An An (+•) (Ref. 108)

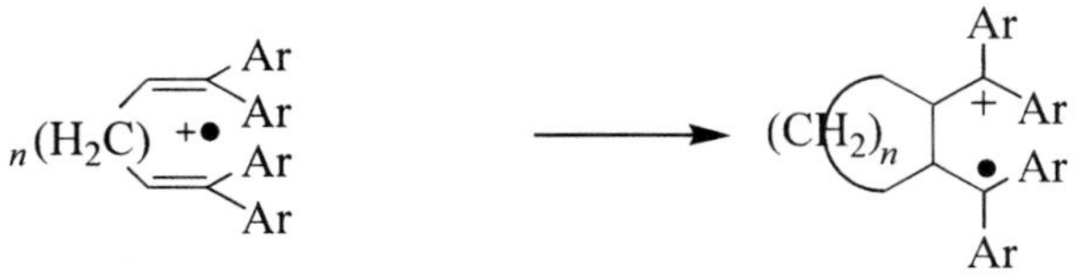

$_n(H_2C)$ Ar Ar Ar Ar (+•) → $(CH_2)_n$ Ar Ar Ar Ar (+, •) (Ref. 109)

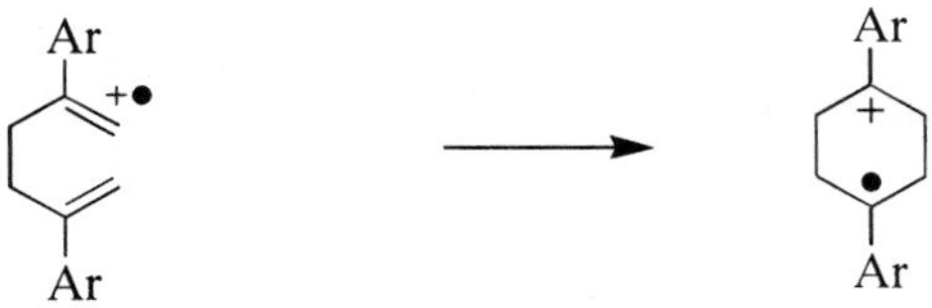

Ar Ar (+•) → Ar Ar (+, •) (Ref. 110)

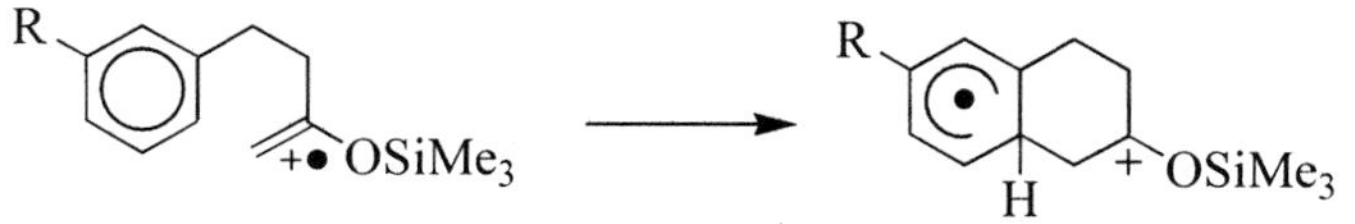

(Ref. 111)

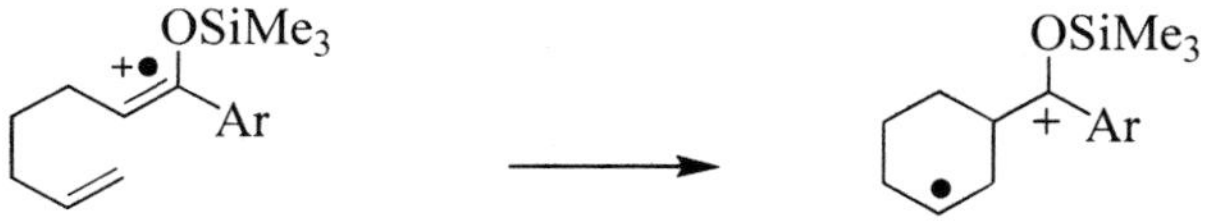

(Ref. 112)

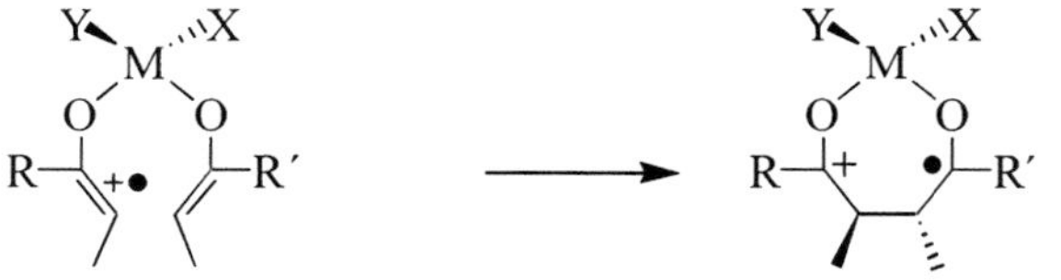

(Ref. 113)

Owing to their cationic and radical character, organic cation radicals also participate in a variety of bimolecular reactions with nucleophiles, bases, radicals, etc., as illustrated by the following examples.

Deprotonation

$$ArCH_3^{+\bullet} + Py \longrightarrow Ar\dot{C}H_2 + PyH^+ \quad \text{(Ref. 114)}$$

$$Me_3CH^{+\bullet} + H_2O \longrightarrow Me_3\dot{C} + H_3O^+ \quad \text{(Ref. 115)}$$

$$R_2\overset{+\bullet}{N}CH_2R' + MeOH \longrightarrow R_2N\dot{C}HR' + MeOH_2^+ \quad \text{(Ref. 116)}$$

$$HOC_6H_4OH^{+\bullet} + H_2O \longrightarrow {}^{\bullet}OC_6H_4OH + H_3O^+ \quad \text{(Ref. 117)}$$

Nucleophilic additions

$$RC_6H_5^{+\bullet} + Nu^- \longrightarrow \quad \text{(Ref. 118)}$$

Nu^- = amine, OAc^-, OH^-, CN^-, Cl^-, MeOH, F^-, etc.

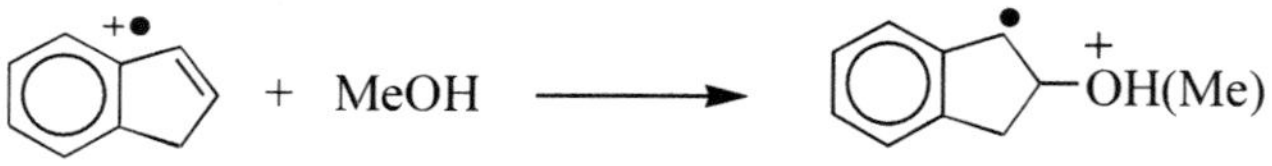

+ MeOH ⟶ $\overset{+}{O}H(Me)$ (Ref. 119)

(Ref. 120)

(Ref. 121)

Dimerization

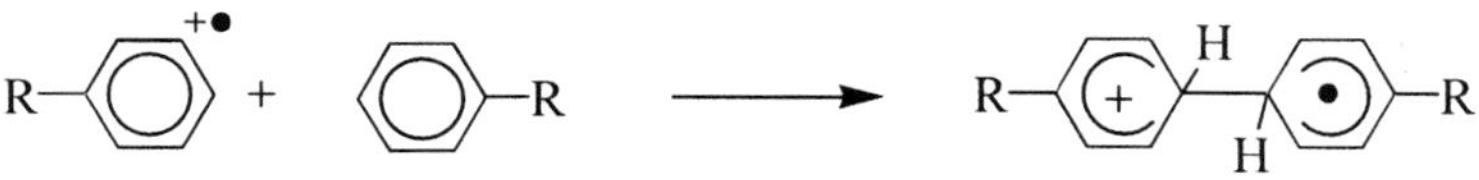

(Ref. 122)

(Ref. 123)

(Ref. 124)

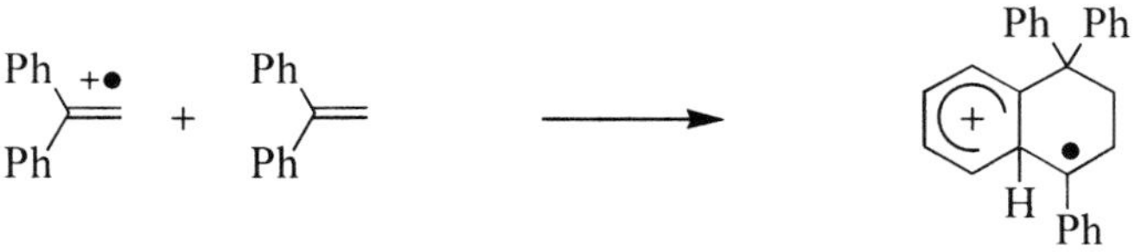

(Ref. 125)

Cycloaddition

(Ref. 126)

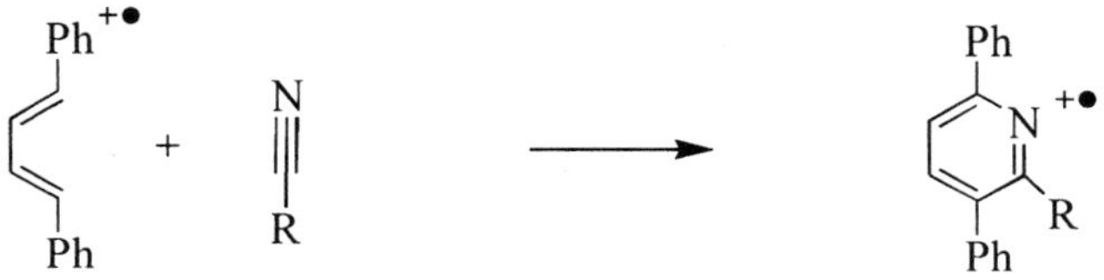

(Ref. 127)

(Ref. 124)

(Ref. 127)

Homolytic addition

R $^{+\bullet}$ + $^{\bullet}NO_2$ ⟶ R (+) H NO_2 (Ref. 128)

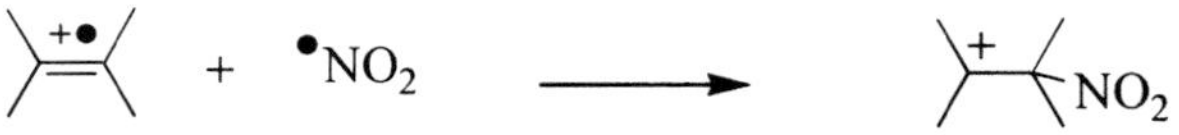

$^{+\bullet}$ + $^{\bullet}NO_2$ ⟶ + NO_2 (Ref. 129)

$^{+\bullet}$ + O_2 ⟶ + O-O$^{\bullet}$ (Ref. 130)

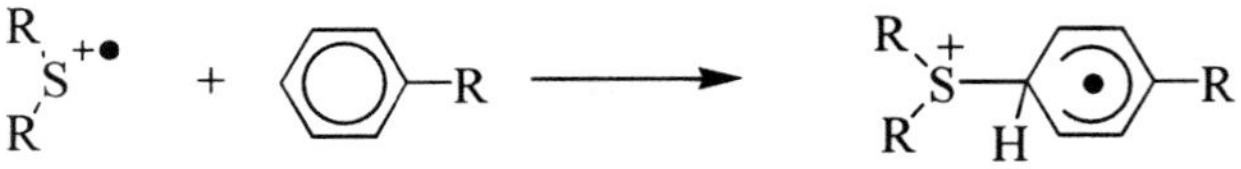

R R S $^{+\bullet}$ + R ⟶ R R S + H R (Ref. 131)

Electron transfer

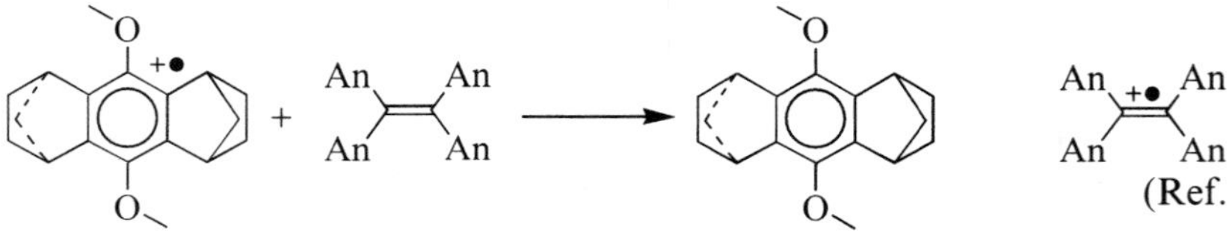

O O $^{+\bullet}$ + An An An An ⟶ O O An An $^{+\bullet}$ An An (Ref. 132)

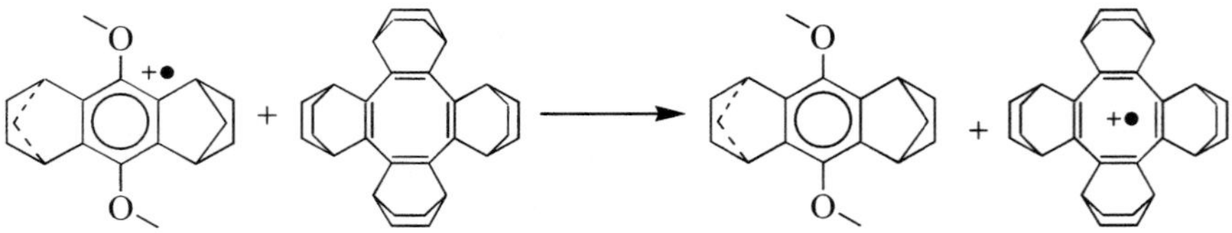

O O $^{+\bullet}$ + ⟶ O O + $^{+\bullet}$ (Ref. 132)

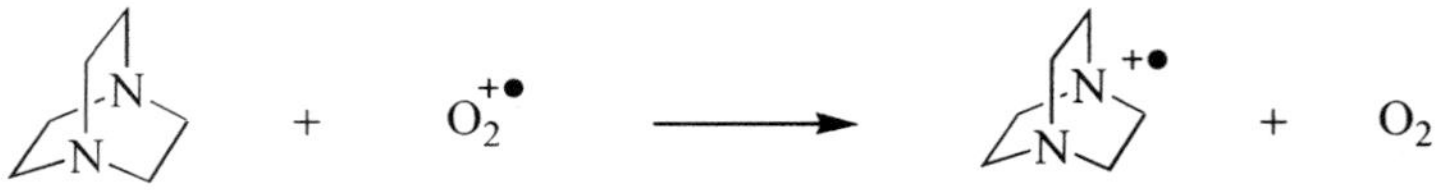

N N + $O_2^{+\bullet}$ ⟶ N N $^{+\bullet}$ + O_2 (Ref. 133)

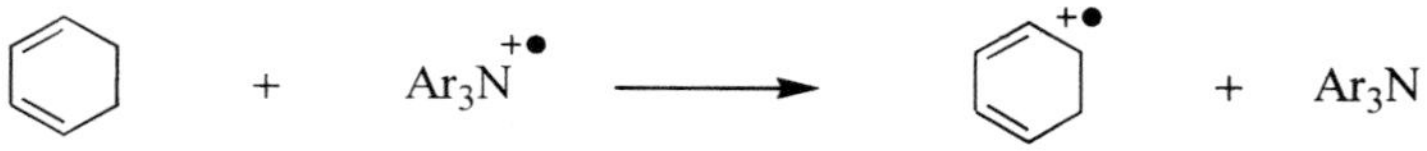

+ $Ar_3N^{+\bullet}$ ⟶ $^{+\bullet}$ + Ar_3N (Ref. 129)

Disproportionation

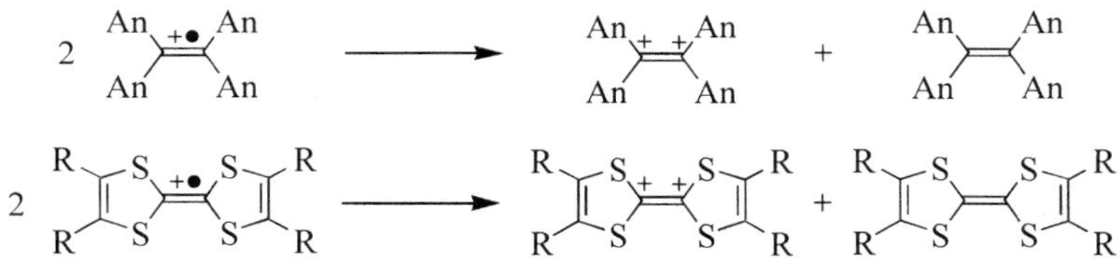

$$2\,An_2C{=}CAn_2^{+\bullet} \longrightarrow An_2C^{+}{-}C^{+}An_2 + An_2C{=}CAn_2 \qquad \text{(Ref. 132)}$$

$$2\,TTF(R_4)^{+\bullet} \longrightarrow TTF(R_4)^{2+} + TTF(R_4) \qquad \text{(Ref. 134)}$$

CHEMISTRY OF ANION RADICALS FROM ELECTRON ACCEPTORS

The attachment of an electron to an organic acceptor generates an "umpolung" anion radical that undergoes a variety of rapid unimolecular decompositions such as fragmentation, cyclization, rearrangement, etc., as well as bimolecular reactions with acids, electrophiles, electron acceptors, radicals, etc., as demonstrated by the following examples.[135–137]

Fragmentation

$$ArX^{-\bullet} \longrightarrow Ar^{\bullet} + X^{-} \qquad \text{(Ref. 138)}$$

$$RX^{-\bullet} \xrightarrow[X = Cl,\ Br,\ I]{} R^{\bullet} + X^{-} \qquad \text{(Ref. 139)}$$

$$Ar_2O^{-\bullet} \longrightarrow ArO^{\bullet} + Ar^{-} \qquad \text{(Ref. 140)}$$

$$RO{-}OR^{-\bullet} \longrightarrow RO^{\bullet} + RO^{-} \qquad \text{(Ref. 141)}$$

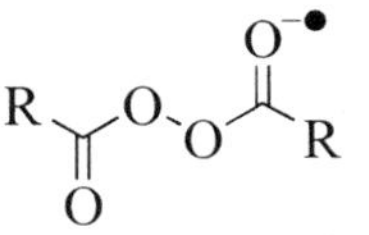

$$RC(O)O{-}OC(O^{-\bullet})R \longrightarrow RCO_2^{\bullet} + RCO_2^{-} \qquad \text{(Ref. 142)}$$

$$RS{-}RS^{-\bullet} \longrightarrow RS^{\bullet} + RS^{-} \qquad \text{(Ref. 143)}$$

Cyclization

CO_2Me / CO_2Me ⟶ CO_2Me / CO_2Me (Ref. 144)

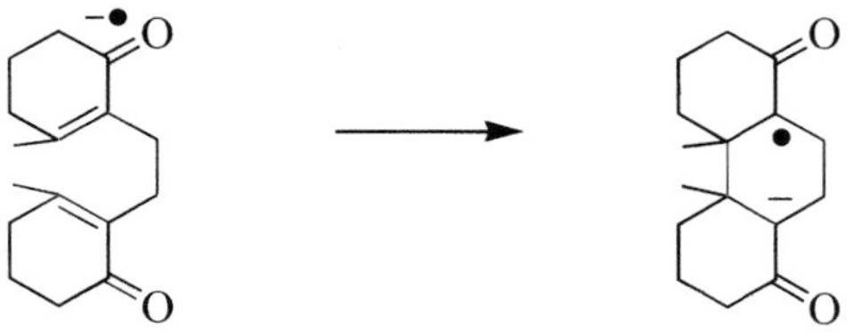

O / O ⟶ O / O (Ref. 145)

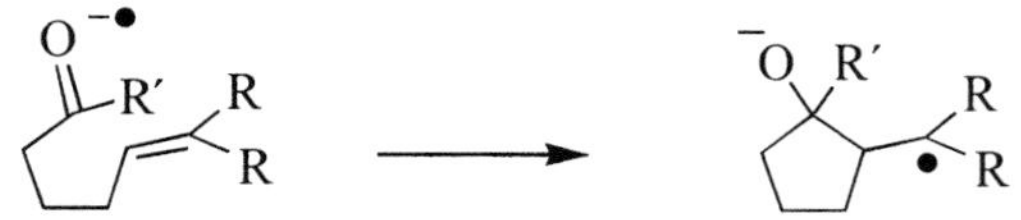

O, R′, R, R ⟶ $^-$O, R′, R, R (Ref. 146)

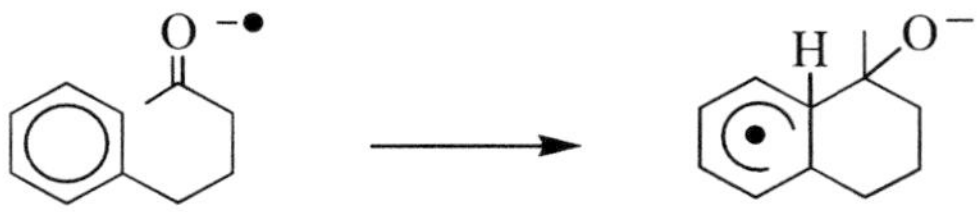

O ⟶ H, O^- (Ref. 147)

Protonation

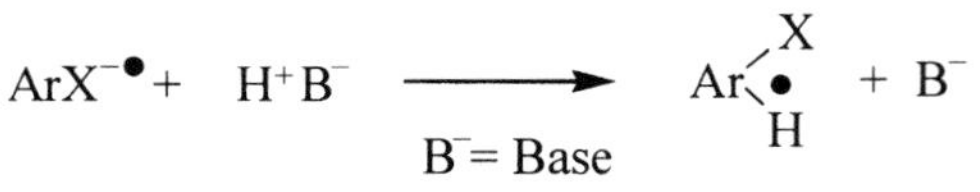

$$ArX^{-\bullet} + H^+B^- \xrightarrow[B^- = \text{Base}]{} Ar(X)(H)^{\bullet} + B^-$$ (Ref. 148)

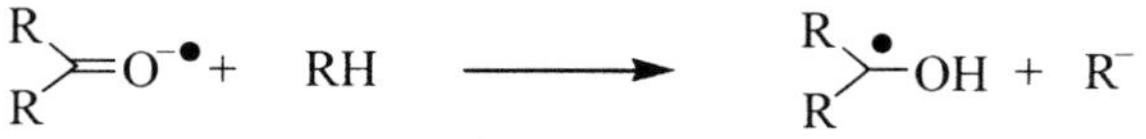

$$R_2C{=}O^{-\bullet} + RH \longrightarrow R_2\dot{C}{-}OH + R^-$$ (Ref. 149)

Electrophilic addition

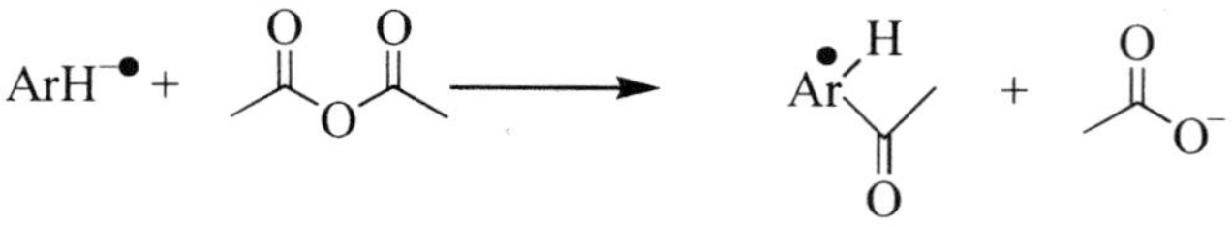

$$ArH^{-\bullet} + (CH_3CO)_2O \longrightarrow {}^{\bullet}Ar(H)COCH_3 + CH_3CO_2^-$$ (Ref. 149)

Dimerization

$$Ar_2C{=}CH_2^{-\bullet} + Ar_2C{=}CH_2^{-\bullet} \longrightarrow Ar_2\bar{C}{-}CH_2{-}CH_2{-}\bar{C}Ar_2$$ (Ref. 150)

(Ref. 151)

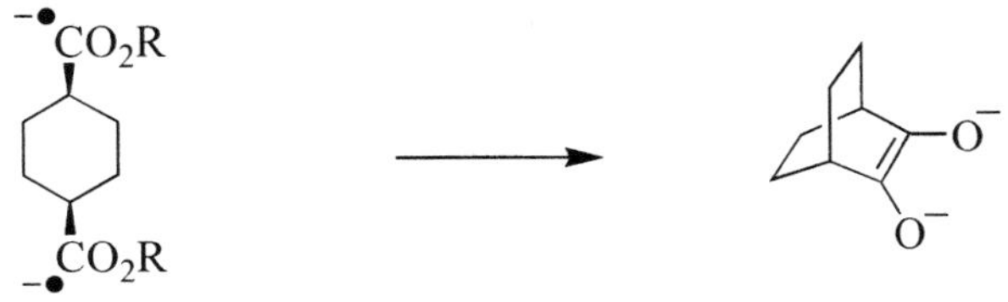

(Ref. 152)

Electron transfer

$ArH^{-\bullet}$ + RX ⟶ ArH + $RX^{-\bullet}$ (Ref. 153)

(Ref. 154)

$ArH^{-\bullet}$ + O_2 ⟶ ArH + $O_2^{-\bullet}$ (Ref. 155)

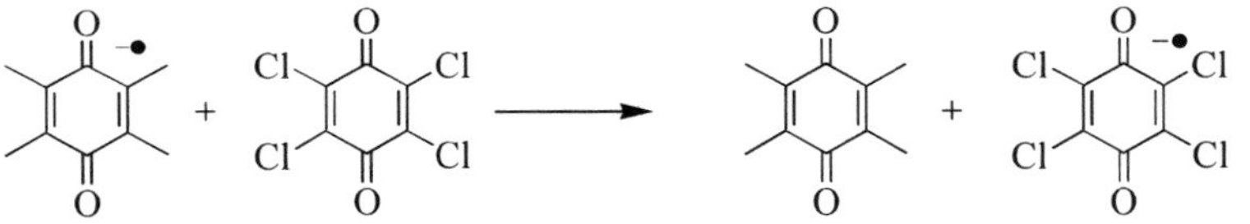

(Ref. 156)

Disproportionation

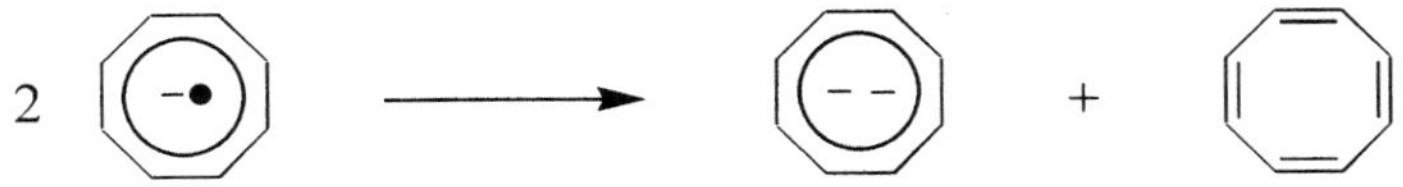

(Ref. 157)

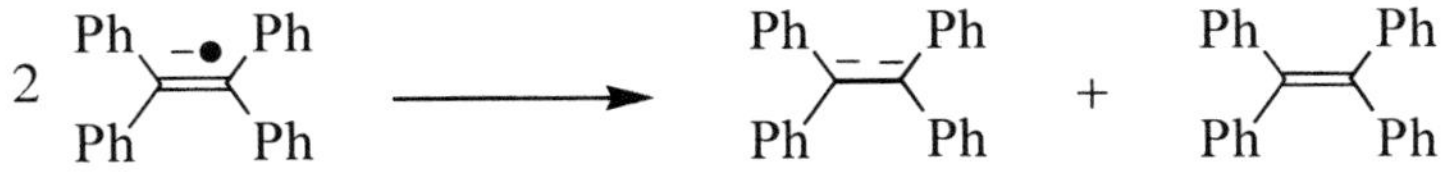

(Ref. 158)

(Ref. 159)

Another pathway, although less frequently encountered, is the ion-pair collapse of the contact ion pair $D^{+\bullet}$, $A^{-\bullet}$ by rapid formation of a bond between the cation radical and anion radical[14] (which competes effectively with the back electron transfer) as illustrated by the following examples.

C—C bond formation

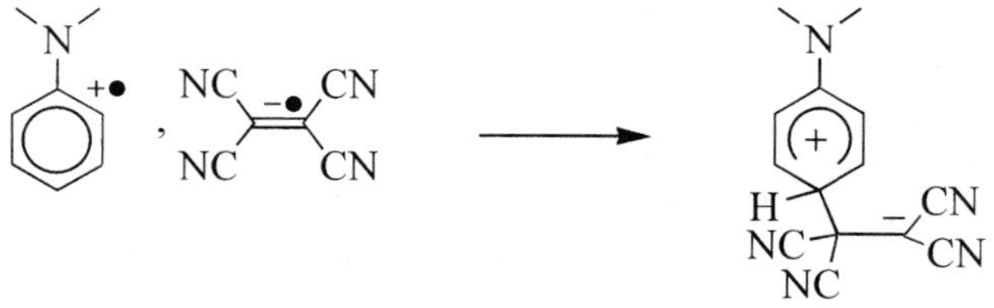

(Ref. 160)

(Ref. 161)

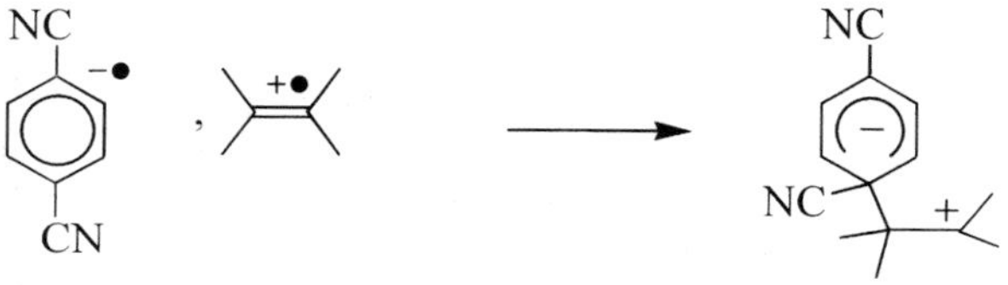

(Ref. 162)

Cycloaddition

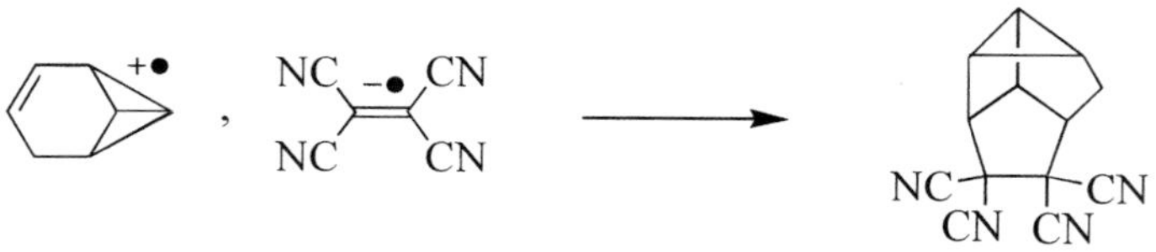

(Ref. 163)

(Ref. 164)

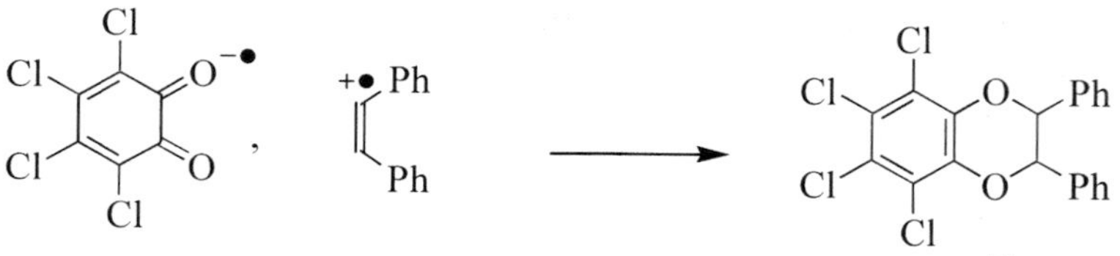

(Ref. 165)

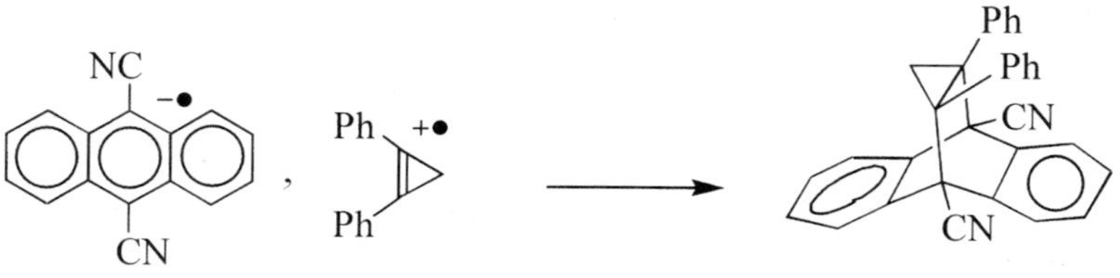

(Ref. 166)

PREPARATIVE ISOLATION OF ORGANIC CATION RADICALS

Despite the high reactivity of organic cation radicals, a large number of these intermediates have been characterized by spectroscopic (ESR and UV-vis) methods.[167,168] Several cation radicals have also been isolated as crystalline salts and the structures established by X-ray crystallography (see Table 6 for some representative examples).[169] The one-electron oxidation of organic donors to the corresponding cation radicals can be achieved with a variety of chemical oxidants, e.g., Brønsted acids (such as sulfuric acid, trifluoroacetic acid, methanesulfonic acid, etc.), thallium(III) trifluoroacetate/trifluoroacetic acid, chloranil/methanesulfonic acid, DDQ/trifluoroacetic acid, halogens (i.e., Cl_2, Br_2, I_2)/trifluoroacetic acid or silver tetrafluoroborate, aluminum trichloride/nitrobenzene, etc., in the absence of reactive nucleophiles and bases.[170] Organic cation radicals in solution can also be generated by electrochemical (anodic) oxidation[171] as well as by the pulse-radiolytic method.[172] Although all these methods are suitable for the spectroscopic identification of the paramagnetic cation radicals in solution, their use is not generally applicable to the isolation of pure crystalline salts for X-ray crystallographic characterization.

Crystalline cation-radical salts are generally isolated via three preparative procedures by using a nitrosonium (NO^+) salt, antimony pentachloride ($SbCl_5$), or triethyloxonium hexachloroantimonate ($Et_3O^+SbCl_6^-$) as a mild one-electron oxidant.

Nitrosonium (NO^+) salts

The BF_4^-, PF_6^- and $SbCl_6^-$ salts of cation radicals are readily prepared by oxidation of organic donors with the corresponding NO^+ salts in a relatively nonpolar solvent such as dichloromethane. For example, a solution of the hydroquinone ether MA in anhydrous (deaerated) dichloromethane turns purple upon the addition of crystalline $NO^+BF_4^-$ at low temperature ($\sim 50°C$).[173] The coloration is due to formation of the donor/acceptor complex [MA, NO^+] (equation 34).

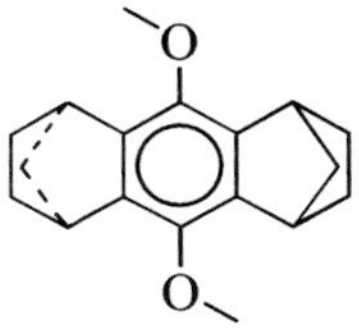

$$\text{(MA)} + NO^+BF_4^- \rightleftharpoons [\text{MA}, NO^+]BF_4^- \quad \text{(purple)} \tag{34}$$

When the solution is warmed to room temperature, the orange cation radical

Table 6 Representative examples of isolated and crystallographically characterized organic cation-radicals.

No.	Cation radical	E^0_{ox} (V versus SCE)	λ_{max} (nm)	Ref.[a]
1	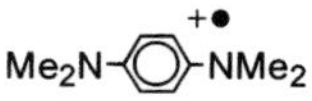 Me_2N–C₆H₄–NMe_2 +•	0.32	630	a
2	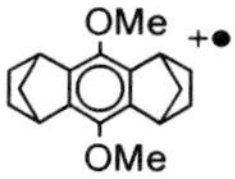 OMe, OMe +•	1.11	518	b
3	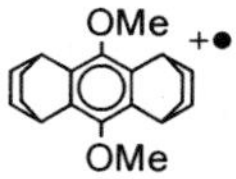 OMe, OMe +•	1.30	486	c
4	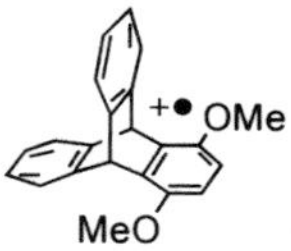 +• OMe, MeO	1.21	466	d
5	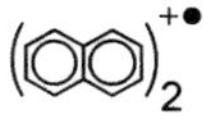 $( )_2$ +•	–	550	e
6	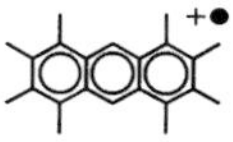 +•	1.05	476	f
7	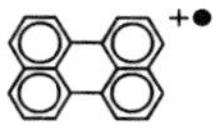 +•	0.98	543	g
8	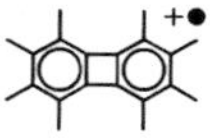 +•	0.80	600	d
9	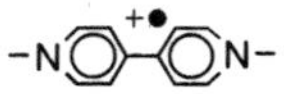 –N N– +•	0.40	608	h
10	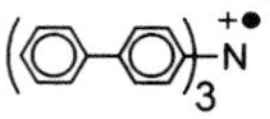 $( )_3$N +•	–	–	i
11	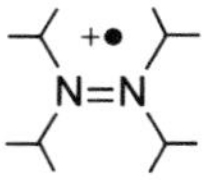 N=N +•	0.29	–	j
12	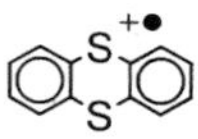 S, S +•	1.30	540	k
13	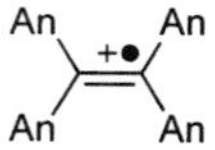 An, An, An, An +•	0.80	560	l
14	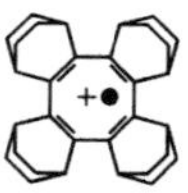 +•	0.75	745	m

[a]Literature cited is indicated by letter in Ref. 169.

$MA^{+\bullet}$ (λ_{max} = 518 nm) appears, and gaseous nitric oxide ($NO^{\bullet}$) is liberated (equation 35).

$$\underset{\text{(purple)}}{[MA, NO^{+}]} \underset{-50^{\circ}C}{\overset{25^{\circ}C}{\rightleftharpoons}} \underset{\text{(orange)}}{MA^{+\bullet}} + NO^{\bullet}\uparrow \qquad (35)$$

Importantly, the purple color is completely restored upon recooling the solution. Thus, the thermal electron-transfer equilibrium depicted in equation (35) is completely reversible over multiple cooling/warming cycles. On the other hand, the isolation of the pure cation-radical salt in quantitative yield is readily achieved by *in vacuo* removal of the gaseous nitric oxide and precipitation of the $MA^{+\bullet}BF_4^-$ salt with diethyl ether. This methodology has been employed for the isolation of a variety of organic cation radicals from aromatic, olefinic and heteroatom-centered donors.[174] However, competitive donor/acceptor complexation complicates the isolation process in some cases.[175]

Antimony pentachloride ($SbCl_5$)

The strongly oxidizing $SbCl_5$ is an effective oxidant for the preparation of cation-radical salts of hexachloroantimonate ($SbCl_6^-$) from a variety of organic donors, such as *para*-substituted triarylamines, fully-substituted hydroquinone ethers, tetraarylethylenes, etc.[176] For example, the treatment of the hydroquinone ether EA (2 mmol) with $SbCl_5$ (3 mmol) in anhydrous dichloromethane at −78°C immediately results in an orange-red solution from which the crystalline cation radical salt readily precipitates in quantitative yield upon the slow addition of anhydrous diethyl ether (or hexane)[173] (equation 36).

$$2\ \underset{\text{(EA)}}{\text{EA}} + 3\ SbCl_5 \rightleftharpoons 2\ EA^{+\bullet}\ SbCl_6^- + SbCl_3 \qquad (36)$$

The ready separation of the hexachloroantimonate salts of various cation radicals is possible owing to their insolubility in diethyl ether (or hexane) under conditions in which the reduced antimony(III) chloride is highly soluble. In the case of $EA^{+\bullet}SbCl_6^-$, the isolated product is quite pure as determined by iodometric titration. However in many other cases, the Lewis acid $SbCl_5$ effects electrophilic chlorination of substitution-labile aromatic and olefinic donors in competition with one-electron oxidation[177] (e.g. equation 37).

$$\text{1,4-dimethoxybenzene} + SbCl_5 \longrightarrow \text{1,4-dimethoxybenzene}^{+\bullet}\, SbCl_6^- + \text{2-chloro-1,4-dimethoxybenzene} + \text{etc.} \qquad (37)$$

Triethyloxonium hexachloroantimonate ($Et_3O^+ SbCl_6^-$)

Triethyloxonium hexachloroantimonate is a selective and mild one-electron oxidant for the facile preparation and isolation of crystalline cation-radical salts from a variety of aromatic and olefinic donors.[177] Thus, in a general procedure, a slurry of $Et_3O^+SbCl_6^-$ (3 mmol) and dimethoxytriptycene, DMT (2 mmol) is stirred in dichloromethane at 0°C. The heterogeneous mixture immediately takes on a bright yellow-green coloration (λ_{max} = 466 nm), and on continued stirring (for about 1 h) it is transformed to yield a dark solution of $DMT^{+\bullet}$. The highly pure $DMT^{+\bullet}SbCl_6^-$ salt is isolated in quantitative yield by precipitation with diethyl ether (equation 38).

$$2\ \text{(DMT)} + 3\ Et_3O^+\ SbCl_6^- \longrightarrow 2\ DMT^{+\bullet}\ SbCl_6^- + 3\ EtCl + 3\ Et_2O + SbCl_3 \qquad (38)$$

(Note that the competitive aromatic chlorination of the aromatic donors effected by the strongly oxidizing antimony pentachloride (in equation 37) is circumvented by the use of the mild $Et_3O^+SbCl_6^-$ as the one-electron oxidant.[177])

PREPARATION OF ORGANIC ANION RADICALS

A variety of organic anion radicals have been prepared in solution and characterized by spectroscopic (ESR and UV-vis) methods. Most often, the anion radicals are formed by reaction of electron-poor molecules with highly reducing alkali metal mirrors (such as sodium, potassium, and lithium) in anhydrous (oxygen-free) ethereal solvents (such as dimethoxyethane, tetrahydrofuran, diethyl ether, etc.).[178] Alternatively, the highly stable 2,6-diphenylpyridine (DPP) anion radical (formed by reaction of sodium metal with DPP in tetrahydrofuran) is often employed for the preparation of anion radicals of ketones by electron exchange[179] (equation 39).

$$\text{DPP}^{-\bullet} + Ph_2C{=}O \longrightarrow \text{DPP} + Ph_2C{=}O^{-\bullet} \qquad (39)$$

(DPP)

Although organic anion radicals are oxygen sensitive, they have been isolated as crystalline salts from a variety of electron acceptors (e.g., chloranil, tetracyanoethylene, tetracyanoquinodimethane, perylene, naphthalene, anthracene, tetraphenylethylene, etc.) and their structures have been established by X-ray crystallography.[180]

9 Typical donor/acceptor transformations using the electron-transfer paradigm

Different organic transformations are critically dependent on the donor/acceptor organization in bringing the reactants together via charge-transfer forces. The thermal and/or photochemical activation of the D/A pair (either via direct excitation of [D, A] complex or by diffusional quenching of an excited acceptor with a donor, see Scheme 13) generates the reactive $D^{+\bullet}$, $A^{-\bullet}$ in which the subsequent ion-pair dynamics ultimately governs the overall efficiency. The following examples selected for a detailed analysis of the donor/acceptor organization and the electron-transfer paradigm fall into several categories of conventional reactions such as nucleophilic addition, electrophilic aromatic substitution, cycloaddition, epoxidation and oxidation.

ORGANOMETALS AND METAL HYDRIDES AS ELECTRON DONORS IN ADDITION REACTIONS

Reactions of highly electron-rich organometalate salts (organocuprates, organoborates, Grignard reagents, etc.) and metal hydrides (trialkyltin hydride, triethylsilane, borohydrides, etc.) with cyano-substituted olefins, enones, ketones, carbocations, pyridinium cations, etc. are conventionally formulated as nucleophilic addition reactions. We illustrate the utility of donor/acceptor association and electron-transfer below.

Addition of organocuprates to α,β-unsaturated ketones

The stereoselective 1,4-addition of lithium diorganocuprates (R_2CuLi) to α,β-unsaturated carbonyl acceptors is a valuable synthetic tool for creating a new C—C bond.[181] As early as in 1972, House and Umen noted that the reactivity of diorganocuprates directly correlates with the reduction potentials of a series of α,β-unsaturated carbonyl compounds.[182] Moreover, the ESR detection of 9-fluorenone anion radical in the reaction with Me_2CuLi, coupled with the observation of pinacols as byproducts in equation (40) provides the experimental evidence for an electron-transfer mechanism of the reaction between carbonyl acceptors and organocuprates.[183]

+ Me_2CuLi —THF, -78°C→ O −• ; HO, OH + OH (40)

Most importantly, Vellekoop and Smith have recently reported the formation of colored donor/acceptor complexes between Me_2CuLi and enones at low temperatures[184] (equation 41).

$$(Me_2CuLi)_n \; + \; \text{O} \rightleftharpoons \left[(Me_2CuLi)_n, \; \text{O} \right]_{EDA} \quad (41)$$

The D/A complexation in equation (41) is further substantiated by infrared and NMR studies. These observations suggest that an initial thermal electron transfer within the D/A charge-transfer complex generates an ion-radical pair, and a rapid methyl transfer subsequently completes the 1,4-addition (equation 42).

$$(\mathrm{Me_2CuLi})_n \;+\; \text{enone} \longrightarrow (\mathrm{Me_2CuLi})_n^{+\bullet},\ \text{enone}^{-\bullet} \longrightarrow \text{enolate } (\mathrm{O^-}) \qquad (42)$$

Despite considerable efforts, the formulation in equation (42) remains incomplete owing to the high reactivity of organocuprates as well as their oligomeric nature. Accordingly, we select organoborates as stable electron donors to study alkyl additions to various pyridinium acceptors (by thermal and photoinduced electron transfer) via charge-transfer salts as follows.

Alkylation of pyridinium acceptors with organoborates as electron donors

Addition of lithium tetramethylborate to an aqueous solution of *N*-methylisoquinolinium (iQ^+) triflate yields a bright yellow precipitate immediately[185] (equation 43).

$$iQ^+\ \mathrm{OTf^-} + \mathrm{Li^+\,BMe_4^-} \longrightarrow \left[iQ^+,\ \mathrm{BMe_4^-}\right]_{CIP} + \mathrm{LiOTf} \qquad (43)$$

(CT Salt)

The structure of the resulting 1:1 charge-transfer salt [iQ^+, BMe_4^-] is established by X-ray crystallography. The origin of the charge-transfer color in [iQ^+, BMe_4^-] arises from an interaction of BMe_4^- with the cationic iQ^+ acceptor (see Fig. 8). (It is important to note that the HOMO in organometals is generally centered on the carbon–metal bond.[69]) Similarly, various substituted pyridinium cations (with varying acceptor strengths in Table 7) yield highly colored charge-transfer salts with BMe_4^-, $BMePh_3^-$ and BPh_4^- as electron donors. It is noteworthy that all the pyridinium triflate salts and alkali metal organoborates (in Table 7) are colorless, whereas the resulting pyridinium borates display colors that range from pale yellow to dark orange. The quantitative (UV-vis) spectral analysis of various pyridinium tetramethylborates shows a consistent red shift of the charge-transfer absorption bands (λ_{CT}) with increasing reduction potential of the pyridinium acceptor (see Table 7), in accord with the Mulliken theory. Moreover, a monotonic blue shift of the λ_{CT} is observed with increasing oxidation potential of the organoborates in the order $BMe_4^- < BMePh_3^- < BPh_4^-$ when iQ^+ is used as electron acceptor.

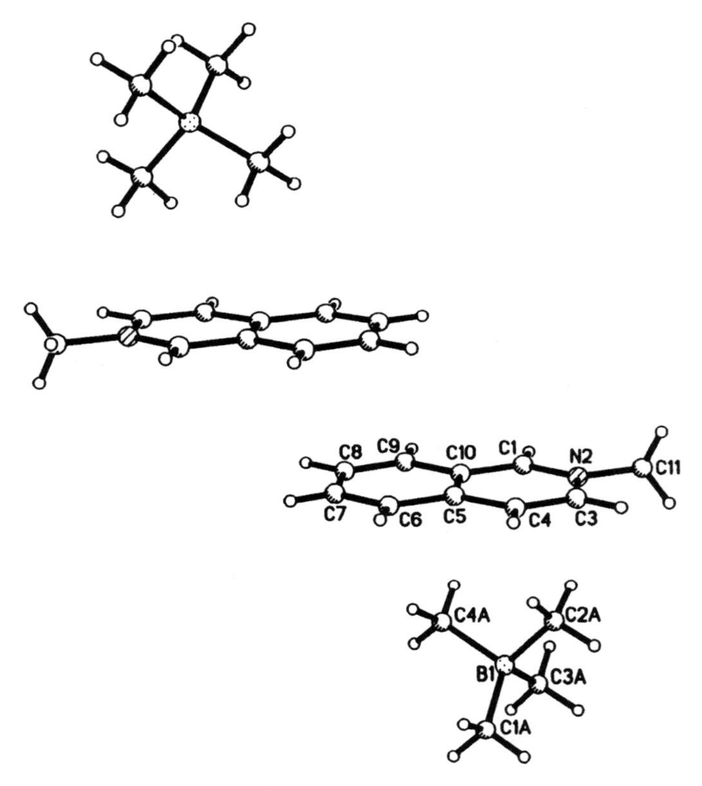

Fig. 8 X-ray structure of the [iQ$^+$, BMe_4^-] showing the charge-transfer interaction of the tetramethylborate anion with the cationic *N*-methylisoquinolinium acceptor. Reproduced with permission from Ref. 185.

Thermal methyl transfer. An orange solution of [iQ$^+$, BMe_4^-] in tetrahydrofuran loses its color in 1 h at room temperature to afford the adduct 1,2-dihydro-1,2-dimethylisoquinoline together with BMe_3 in quantitative yield (equation 44).

$$\left[\text{iQ}^+ , \; BMe_4^- \right] \xrightarrow[\text{THF}]{\text{RT}} \text{1,2-dihydro-1,2-dimethylisoquinoline} + BMe_3 \qquad (44)$$

Interestingly, the methyl-transfer reaction between BMe_4^- and 3-cyano-*N*-methylpyridinium (with the highest reduction potential, see Table 7) occurs instantaneously and thus precludes the isolation of the charge-transfer salt.

Table 7 Absorption spectra of [Py^+, BR_4^-] charge-transfer salts[a].

No.	Pyridinium acceptor (Py^+)		E^0_{red} (V versus SCE)	Charge-transfer band (nm)		
				BMe_4^-	$BMePh_3^-$	BPh_4^-
1	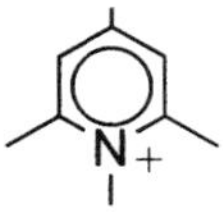	(mC^+)	1.63	371		
2	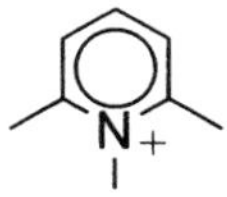	(mL^+)	1.45	387		
3	 Ph	(PP^+)	1.27	470		
4		(iQ^+)	1.08	490	437	406
5	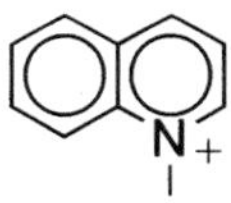	(Q^+)	0.90		417	418
6	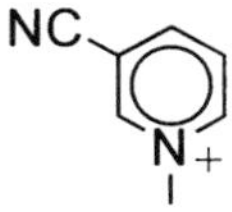 NC	(NCP^+)	0.83		485	485
7	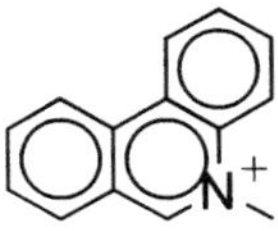	(mP^+)	0.78		472	450
8	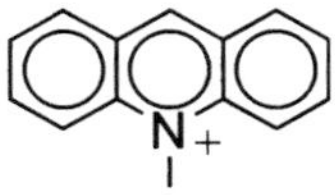	(mAc^+)	0.43		580	512

[a] From Ref. 185.

Photoinduced methyl transfer. The yellow mixture of 4-phenylpyridinium cation (with relatively low reduction potential) and tetramethylborate anion in tetrahydrofuran persists for 24 h without any reaction (in the dark). However, the deliberate irradiation of the charge-transfer band (at $\lambda_{exc} = 370$ nm) bleaches the yellow color and affords the methyl-transfer product in excellent yield[185] (equation 45).

$$\left[\text{Ph-C}_5\text{H}_4\text{N}^+\text{-Me},\ \text{BMe}_4^-\right]_{\text{CIP}} \xrightarrow{h\nu_{\text{CT}}} \text{Ph-(2-methyl-1-methyl-1,2-dihydropyridine)} + \text{BMe}_3 \quad (45)$$

Note that the photolysis of various pyridinium borate salts at $-78°C$ (to prevent thermal reaction) affords the same methyl-transfer products as obtained in thermal reactions (i.e., equation 44).

Electron-transfer mechanism for nucleophilic addition. In accord with Mulliken theory, irradiation of the charge-transfer band of $[Py^+, BMe_4^-]$ directly affords the radical pair via one-electron transfer (equation 46).

$$\left[\text{Py}^+,\ \text{BMe}_4^-\right]_{\text{CIP}} \underset{k_{\text{BET}}}{\overset{h\nu_{\text{CT}}}{\rightleftharpoons}} \text{Py}^\bullet,\ \text{BMe}_4^\bullet \quad (46)$$

A rapid scission of the Me—B bond (on the picosecond timescale) generates the methyl radical ($CH_3^\bullet$) and BMe_3. The coupling of pyridine and methyl radicals within the solvent cage completes the methyl transfer (equation 47).

$$\text{Py}^\bullet,\ \text{BMe}_4^\bullet \longrightarrow \text{Py}^\bullet,\ \text{Me}^\bullet,\ \text{BMe}_3 \longrightarrow \text{Py-Me} + \text{BMe}_3 \quad (47)$$

Comparison of thermal and photochemical activation. The identical color changes that accompany the thermal and photochemical methyl transfer in various $[Py^+, BMe_4^-]$ salts suggests that pre-equilibrium charge-transfer complexation is common to both processes. Moreover, the methyl transfer either by charge-transfer photolysis or by thermal activation of $[Py^+, BMe_4^-]$ leads to the same products, which strongly suggests common reactive intermediates (i.e., the radical pair in equation (46)) for both thermal and photochemical processes.

In this context, tetrabutylammonium borohydride – a strong reducing agent ($E_{ox} \sim -1.5$ V versus SCE) – reacts thermally via electron transfer with a variety of electron acceptors (such as TCNE, dicyanobenzene, duroquinone, *p*-benzoquinone, 9-fluorenone, benzil, etc.) in tetrahydrofuran to afford the corresponding anion radicals (observed by ESR spectroscopy) and borane[186] (e.g., equation 48).

$$\text{BH}_4^- + \text{PhC(O)Ph} \xrightarrow{\text{THF}} \text{PhC(O}^{-\bullet}\text{)Ph} + \tfrac{1}{2}\,\text{H}_2 + \text{BH}_3 \quad (48)$$

Hydrogen transfer from electron-rich metal hydrides to electron acceptors

The highly electron-rich trialkylmetal hydrides of tin, germanium, and silicon (e.g., tributyltin hydride, tributylgermanium hydride, and triethylsilicon hydride or triethylsilane) are the common reagents for the reduction of a wide variety of organic functional groups.[187] Reductions with these metal hydrides are generally described as hydride transfer; however, the involvement of radical intermediates is often noted. Donor/acceptor association and the electron-transfer paradigm can be applied to these metal-hydride reductions of electron-poor functional groups (such as olefins, quinones, ketones, etc.) as follows.

Charge-transfer complexes as intermediates in metal hydride additions to tetracyanoethylene (TCNE). Strong charge-transfer colors are observed when a colorless solution of TCNE is exposed to various metal hydrides owing to the formation of the [D, A] complex[188] (equation 49).

$$R_3MH + TCNE \rightleftharpoons [R_3MH,\ TCNE]_{EDA} \qquad (49)$$

The UV-vis spectral analysis confirms the appearance of a new charge-transfer absorption band of the complexes of colorless σ-donors (R_3MH) and the π-acceptor (TCNE). In accord with Mulliken theory, the absorption maxima (λ_{CT}) of the [R_3MH, TCNE] complexes shift toward blue with increasing ionization potential of the metal hydrides (i.e., tin > germanium > silicon) as listed in Table 8.

Thermal reaction. The charge-transfer absorption (or color) of the [R_3MH, TCNE] complex fades in the dark, and a concomitant addition of the metal hydride to the TCNE double bond is observed[188] (equation 50).

$$[R_3MH,\ TCNE]_{EDA} \xrightarrow{\text{dark}} H{-}C(CN)_2{-}C(CN)_2{-}MR_3 \qquad (50)$$

Table 8 Donor–acceptor complexes of metal hydrides with tetracyanoethylene[a].

Metal hydride	IP	λ_{CT} (cm^{-1})	Relative[b] rate (k_t)
n-Bu_3SnH	8.72	15 400	10^7
n-Bu_3GeH	9.62	25 600	10^3
Et_3SiH	9.95	34 500	10^0

[a]From Ref. 188. [b]Relative insertion rate of metal hydrides into TCNE.

Careful kinetic analysis of this thermal reaction shows that the rate of disappearance of the CT band is identical to that of the adduct formation in equation (50). Most importantly, the relative reactivity of the metal hydrides in Table 8 decreases with the increasing ionization potential in the order $Bu_3SnH < Bu_3GeH < Et_3SiH$.

Photochemical reaction. The selective photoactivation of the CT absorption band of the [R_3MH, TCNE] complex at low temperature (to obviate a thermal reaction) leads to the same insertion product as obtained in the thermal reaction (equation 51).

$$[\,R_3MH\,,\ TCNE\,] \xrightarrow{h\nu_{CT}} H-\overset{NC}{\underset{NC}{C}}-\overset{CN}{\underset{CN}{C}}-MR_3 \tag{51}$$

Electron transfer mechanism. In accord with the Mulliken theory, photoexcitation of the charge-transfer band effects the formation of a contact ion radical pair, which undergoes a rapid mesolytic scission of the M—H bond (equation 52).

$$[R_3MH\,,\ TCNE]_{EDA} \longrightarrow R_3MH^{+\bullet},\ TCNE^{-\bullet} \xrightarrow{fast} R_3M^+,\ H^\bullet,\ TCNE^{-\bullet},\ etc. \tag{52}$$

Note that the identity of the radical pair in equation (52) is confirmed by the observation of both the TCNE anion radical and hydrogen atom ($H^\bullet$) by ESR spectroscopy in a frozen matrix.[188]

The fact that both the thermal and the photochemical insertion reactions yield the same products via formation of charge-transfer complexes leads to the conclusion that the reactive ion-radical pair in equation (52) is the common intermediate for both activation processes. Such a conclusion is further verified by the direct observation of anion-radical intermediates from the thermal reaction of TCNE and DDQ with various metal hydrides.[188]

Tributyltin hydride reduction of carbonyl compounds. The reduction of carbonyl compounds with metal hydrides can also proceed via an electron-transfer activation in analogy to the metal hydride insertion into TCNE.[188] Such a notion is further supported by the following observations: (a) the reaction rates are enhanced by light as well as heat;[189] (b) the rate of the reduction depends strongly on the reduction potentials of ketones. For example, trifluoroacetophenone ($E_{red} = -1.38$ V versus SCE) is quantitatively reduced by Bu_3SnH in propionitrile within 5 min, whereas the reduction of cyclohexanone ($E_{red} \sim -2.4$ V versus SCE) to cyclohexanol (under identical reaction conditions) occurs only in 5% yield during the course of 5 h[190] (equation 53).

$$\text{PhCH(OH)CF}_3\ (100\%) \xleftarrow[\text{PhC(O)CF}_3\ (5\ \text{min})]{\text{CH}_3\text{CH}_2\text{CN}} \text{Bu}_3\text{SnH} \xrightarrow[\text{cyclohexanone}\ (5\ \text{h})]{\text{CH}_3\text{CH}_2\text{CN}} \text{cyclohexanol}\ (5\%) \tag{53}$$

In summary, the electron transfer from cuprates, metal hydrides, or organoborates (as electron donors) to various electron acceptors such as enones, olefins, pyridinium cations and ketones generates the ion-radical pair as the primary reactive intermediate via electron-transfer activation. The efficiency of the transformations in equations (42), (44), (45), (50), (51) and (53) is a result of the rapid unimolecular (mesolytic) fragmentation of the organometallic cation radical to effectively block back electron transfer. The resulting radical fragments then undergo addition to the anion radical (or radical) within the solvent cage. Let us now examine organic donors that undergo facile (unimolecular) scission of C—C and C—H bonds upon oxidation with different organic and inorganic oxidants.

OXIDATIVE CLEAVAGE OF CARBON–CARBON AND CARBON–HYDROGEN BONDS

Donor/acceptor association and the electron-transfer paradigm form the unifying theme for the C—C bond cleavage of various benzpinacols and diarylethane-like donors in the presence of different electron acceptors (such as chloranil (CA), dichlorodicyanobenzoquinone (DDQ), tetracyanobenzene (TCNB), triphenylpyrylium (TPP^+), methyl viologen, nitrosonium cation, etc.). Scheme 13 reminds us how this is achieved by either CT photolysis of the D/A pair or via diffusional quenching of the excited electron acceptor A* by the electron donor D.

Cleavage of benzpinacols

Benzpinacols (or their trimethylsilyl ethers) are effective electron donors and readily form vividly colored charge-transfer complexes with common electron acceptors such as chloranil (CA), dichlorodicyanobenzoquinone (DDQ), tetracyanobenzene (TCNB), methyl viologen (MV^{2+}), and nitrosonium (NO^+) cation.[191–194] For example, the exposure of a silylated benzpinacol to chloranil affords a bright orange solution ($\lambda_{max} = 468$ nm) owing to the formation of the intermolecular [D, A] complex[191] (equation 54).

$$\mathrm{An(RO)C{-}C(OR)An} + \text{CA} \rightleftharpoons [\mathrm{An(RO)C{-}C(OR)An},\ \text{CA}]_{EDA} \qquad (54)$$

(CA) = 2,3,5,6-tetrachloro-1,4-benzoquinone (chloranil)

R = H or $SiMe_3$; An = *p*-Anisyl

Since the same orange color ($\lambda_{max} = 472$ nm) is observed upon mixing the parent *p*-methoxytoluene and chloranil (see Fig. 9), the aromatic moiety in the benzpinacols must be the donor site and thus responsible for the charge-transfer interactions with the chloranil acceptor.

The bright-orange solution of the quinone/pinacol complex is stable at 23°C if protected from light. However, the deliberate irradiation ($h\nu_{CT}$) of the charge-

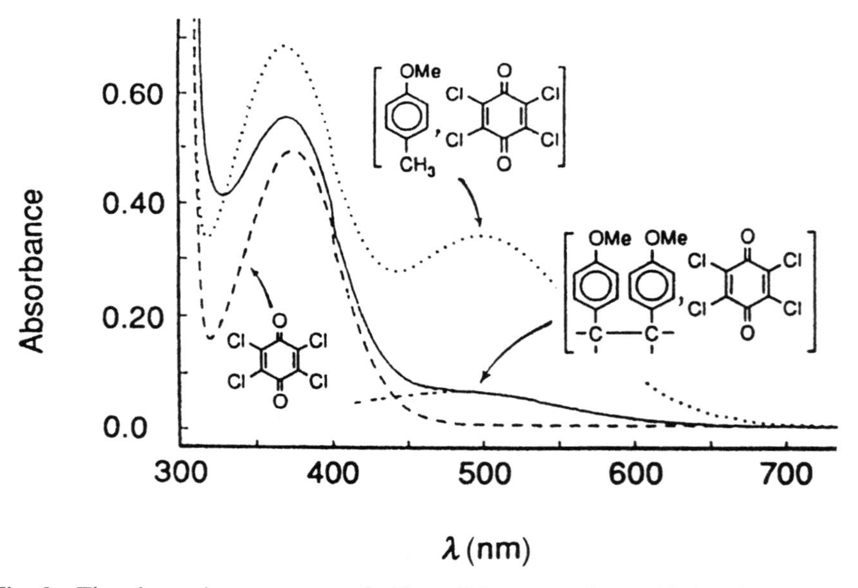

Fig. 9 The absorption spectrum of chloranil in comparison with its charge-transfer spectrum with methoxytoluene and with anisyl(TMS)pinacol (as indicated). Reproduced with permission from Ref. 191.

transfer band ($\lambda_{exc} > 500$ nm) leads to a slow bleaching of the color with concomitant formation of the retropinacol product (equation 55).

$$\left[\mathrm{An(RO)C{-}C(OR)An_2}\ ,\ \mathrm{chloranil}\right]_{EDA} \xrightarrow[\sim 24\ h]{h\nu_{CT}} 2\ \mathrm{An_2C{=}O} + \mathrm{C_6Cl_4(OR)_2} \tag{55}$$

Interestingly, the irradiation of the same solution at the (local) absorption band of chloranil ($h\nu_Q$) results in a much faster and complete bleaching of the solution to yield identical retropinacol products (equation 56).

$$\mathrm{An_2(RO)C{-}C(OR)An_2} + \mathrm{chloranil} \xrightarrow[\sim 45\ min]{h\nu_{Q}} 2\ \mathrm{An_2C{=}O} + \mathrm{C_6Cl_4(OR)_2} \tag{56}$$

Despite the difference in the photoefficiencies in benzpinacol fragmentation via charge-transfer excitation (equation 55) and photosensitization (equation 56), the observed stoichiometries are identical for both activation modes.

Electron-transfer activation. Both charge-transfer (CT) photolysis as well as the diffusional quenching of photoexcited chloranil with pinacol donors occur via a reactive ion pair as the common intermediate (equation 57).

$$\left[\mathrm{An_2(RO)C{-}C(OR)An_2}\ ,\ \mathrm{CA}\right]_{EDA} \overset{h\nu_{CT}}{\rightleftharpoons} \mathrm{An_2(RO)C{-}C(OR)An_2^{+\bullet}},\ \mathrm{CA^{-\bullet}} \overset{h\nu_{Q}}{\rightleftharpoons} \mathrm{An_2(RO)C{-}C(OR)An_2} + \mathrm{CA} \tag{57}$$

Subsequently, the rapid unimolecular (mesolytic) fragmentation of the resulting pinacol cation radical followed by proton (or trimethylsilyl cation) transfer to quinone anion radical (within the solvent cage) yields the retropinacol products in equations (55) and (56) (equation 58).

$$\mathrm{An}\!\!-\!\!\overset{\mathrm{RO}}{\underset{}{\mid}}\!\!-\!\!\underset{\mathrm{OR}}{\mid}\!\!-\!\!\mathrm{An}^{+\bullet};\ \mathrm{CA}^{-\bullet} \longrightarrow \mathrm{An}\!-\!\overset{\mathrm{RO}}{\mid}\bullet,\ +\!\underset{\mathrm{OR}}{\mid}\!-\!\mathrm{An},\ \mathrm{CA}^{-\bullet} \longrightarrow 2\ \mathrm{An_2C{=}O} + \mathrm{C_6Cl_4(OR)_2} \quad (58)$$

The significantly enhanced quantum efficiency for the sensitized photofragmentation ($\Phi_{sens} = 0.75$) compared to that for the charge-transfer activation ($\Phi_{CT} = 0.1$) is related to the lifetimes of the ion pairs involved. In fact, the difference in the lifetimes arises from the different spin multiplicities of the ion pairs. Thus, the singlet ion pair generated by CT excitation decays very rapidly by back electron transfer (k_{BET}) so that the fragmentation ($k_{C\text{–}C}$) of the pinacol cation radical is inefficient (as judged by the low quantum yields). In contrast, the triplet ion pair generated by electron-transfer quenching of triplet chloranil lives much longer owing to the spin-forbidden character of the back electron transfer, and allows the pinacol cation radical to undergo efficient fragmentation. The two competitive reaction pathways (i.e. back-electron transfer in the ion pair and C—C fragmentation of pinacol cation radical) have been experimentally verified by time-resolved spectroscopy[193] (equation 59).

$$\left[\mathrm{An}\!-\!\overset{\mathrm{RO}}{\mid}\!-\!\underset{\mathrm{OR}}{\mid}\!-\!\mathrm{An},\ \mathrm{CA}\right] \underset{k_{BET}}{\overset{h\nu_{CT}}{\rightleftharpoons}} \mathrm{An}\!-\!\overset{\mathrm{RO}}{\mid}\!-\!\underset{\mathrm{OR}}{\mid}\!-\!\mathrm{An}^{+\bullet},\ \mathrm{CA}^{-\bullet} \xrightarrow{k_{C\text{-}C}} \mathrm{An}\!-\!\overset{\mathrm{RO}}{\mid}\bullet,\ +\!\underset{\mathrm{OR}}{\mid}\!-\!\mathrm{An},\ \mathrm{CA}^{-\bullet} \quad (59)$$

Indeed, the (200-fs) laser excitation of the EDA complexes of various benzpinacols with methyl viologen (MV^{2+}) confirms the formation of all the transient species in equation (59). A careful kinetic analysis of the decay rates of pinacol cation radical and reduced methyl viologen leads to the conclusion that the ultrafast C—C bond cleavage ($k_{C\text{–}C} = 10^{10}$ to $10^{11}\ s^{-1}$) of the various pinacol cation radicals competes effectively with the back electron transfer in the reactive ion pair.

Thermal activation. Cleavage of benzpinacols can also be achieved by thermal reactions, in which powerful electron acceptors such as DDQ ($E^0_{red} = 0.6$ V versus SCE) effectively oxidize electron-rich pinacols. For example, the blue-green color of the EDA complex formed upon mixing of tetraanisylpinacol and DDQ in dichloromethane bleaches within minutes (in the dark) to afford the retropinacol products in quantitative yield[191] (equation 60).

$$\left[\text{An}-\underset{\text{OH}}{\overset{\text{HO}}{\text{C}}}-\text{C}-\text{An} \,,\, \text{DDQ} \right]_{\text{EDA}} \xrightarrow{\text{dark}} 2\ \text{An}(\text{Me})\text{C=O} + \text{NC}_2\text{Cl}_2\text{C}_6(\text{OR})_2 \tag{60}$$

As such, the thermal process in equation (60) proceeds via the same reactive intermediates (arising from an adiabatic electron transfer) as that observed in the photochemical processes in equations (57) and (58). The proposed electron-transfer activation for the thermal retropinacol reaction is further confirmed by the efficient cleavage of benzpinacol with tris-phenanthroline iron(III), which is a prototypical outer-sphere one-electron oxidant[195] (equation 61).

$$\text{An}-\underset{\text{OH}}{\overset{\text{HO}}{\text{C}}}-\text{C}-\text{An} + 2\,(\text{Phen})_3\text{Fe}^{3+} \longrightarrow 2\ \text{An}(\text{Me})\text{C=O} + 2\,(\text{Phen})_3\text{Fe}^{2+} + 2\,\text{H}^+ \tag{61}$$

Cleavage of bicumene

Bicumene readily forms a highly colored charge-transfer complex with nitrosonium (NO^+) cation that can be isolated as a crystalline salt.[194] Thus, a solution of bicumene and $NO^+SbCl_6^-$ shows a pair of characteristic charge-transfer bands with λ_{max} = 338 and 420 nm that are indistinguishable from those of the nitrosonium complexes with other methylbenzenes (equation 62).

$$\text{PhC(Me)}_2\text{C(Me)}_2\text{Ph} + \text{NO}^+ \rightleftharpoons [\text{PhC(Me)}_2\text{C(Me)}_2\text{Ph}\,,\ \text{NO}^+]_{\text{EDA}} \tag{62}$$

X-ray crystallographic analysis of the crystalline [bicumene, NO^+] charge-transfer salt confirms that the charge-transfer color arises from a close approach of NO^+ to the centroid of the phenyl moiety (see Fig. 10) with a non-bonded contact to an aromatic carbon of 2.63 Å.[194] The orange solution of bicumene bleaches slowly over a long period in a thermal reaction at room temperature (in the dark) or rapidly via irradiation of the CT band at low temperature. In both cases, 1,1,3-trimethyl-3-phenylindane is obtained as the principal organic product (equation 63).

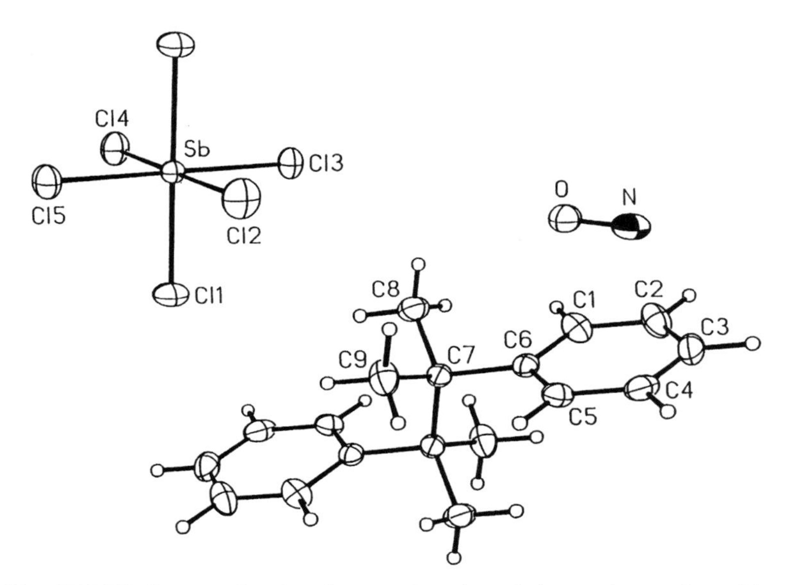

Fig. 10 ORTEP diagram showing the complexation of nitrosonium cation with a single phenyl moiety of bicumene. Reproduced with permission from Ref. 194.

$$[\ldots,\ NO^{+}]_{EDA} \xrightarrow[\text{or dark}]{h\nu_{CT}} \ldots + \text{etc.} \qquad (63)$$

Charge-transfer activation of the EDA complex leads to the ion-radical pair, in which the bicumene cation radical undergoes a unimolecular fragmentation (equation 64).

$$[\ldots,\ NO^{+}] \underset{k_{BET}}{\overset{h\nu_{CT}}{\rightleftharpoons}} \ldots^{+\bullet},\ NO \xrightarrow{k_{C\text{-}C}} \ldots^{\bullet},\ ^{+}\ldots,\ NO \qquad (64)$$

The cumyl radical and the cation that are co-generated by the mesolytic fragmentation in equation (64) are then converted into the photoproduct (in equation 64), as summarized in equations (65a,b).

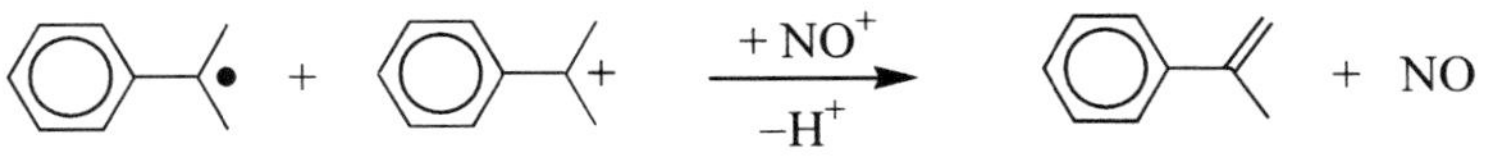

$$\text{PhC(CH}_3)_2^{\bullet} + \text{PhC(CH}_3)_2^{+} \xrightarrow[-H^+]{+NO^+} \text{PhC(CH}_3)\text{=CH}_2 + \text{NO} \quad (65a)$$

$$\text{PhC(CH}_3)\text{=CH}_2 + \text{PhC(CH}_3)_2^{+} \longrightarrow \text{PhC(CH}_3)_2\text{CH}_2\text{C}^{+}\text{(CH}_3)\text{Ph} \xrightarrow{-H^+} \text{1,1,3-trimethyl-3-phenylindane} \quad (65b)$$

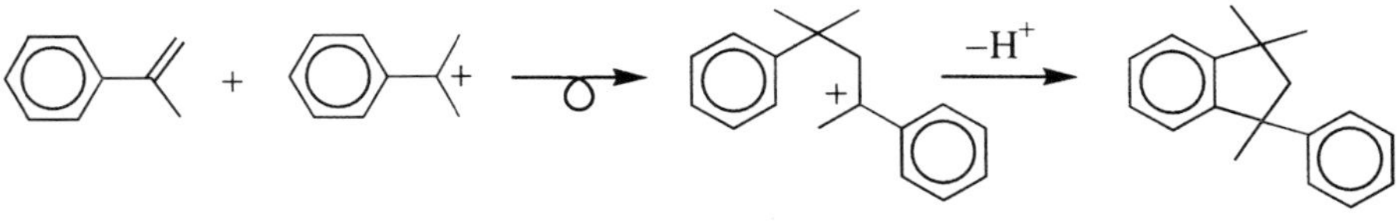

Importantly, the overall quantum yield ($\Phi = 0.2$) in equation (63) indicates that the fragmentation of bicumene cation radical occurs at a slightly slower rate ($k_{C\text{-}C} \sim 10^8\ s^{-1}$) than the back electron transfer ($k_{BET} \sim 4 \times 10^8\ s^{-1}$) within the ion-radical pair in equation (64).

In a similar vein, the cleavage of the photodimer of anthracene occurs in the dark in the presence of nitrosonium cation to afford the anthracene cation radical as its π-dimer via a thermal electron transfer[196] (equation 66).

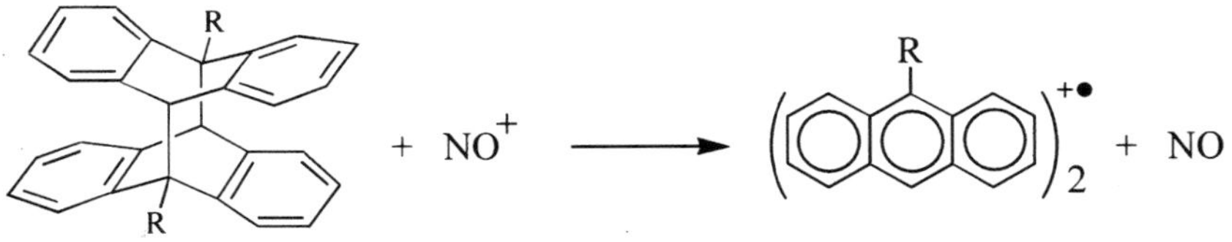

$$\text{(anthracene photodimer, R)} + NO^+ \longrightarrow (\text{9-R-anthracene})_2^{+\bullet} + NO \quad (66)$$

Oxidative decarboxylation

Anodic decarboxylation proceeds via a C—C bond scission of carboxylate anions to afford the Kolbe dimer,[197] i.e.,

$$RCO_2^- \xrightarrow{-e^-} RCO_2^{\bullet} \xrightarrow{k_{C\text{-}C}} R^{\bullet} + CO_2 \longrightarrow \tfrac{1}{2}\,R\text{—}R$$

or the Brown–Walker modification[198] for the synthesis of long-chain dicarboxylic acids, i.e.,

$$HO_2C(CH_2)_nCO_2^- \xrightarrow[-CO_2]{-e^-} HO_2C(CH_2)_{n-1}\dot{C}H_2 \longrightarrow \tfrac{1}{2}\ HO_2C(CH_2)_{2n}CO_2H$$

The decarboxylation of carboxylate anions is carried out chemically by a variety of one-electron oxidants such as lead tetraacetate, uranyl nitrate, peroxides, quinones, pyridinium cations, etc.[199] Importantly, the carboxylate anion (as electron donor) and the pyridinium cation (as electron acceptor) readily form colored charge-transfer salts[200] (equation 67).

$$Ar_2C(OH)CO_2^- + MV^{2+} \rightleftharpoons [Ar_2C(OH)CO_2^-,\ MV^{2+}]_{CIP} \quad (67)$$

(MV^{2+})

The steady-state charge-transfer photolysis of the yellow solution of $[MV^{2+}, Ar_2C(OH)CO_2^-]$ leads to a quantitative yield of diarylketone[200] (equation 68).

$$[Ar_2C(OH)CO_2^-,\ MV^{2+}]_{CIP} \xrightarrow{h\nu_{CT}} Ar_2C{=}O + CO_2,\ \text{etc.} \quad (68)$$

Such an efficient decarboxylation via charge-transfer activation of the CT salts of a benzilate anion and methyl viologen (MV^{2+}) in equation (68) provides a unique opportunity to examine the C—C bond cleavage directly by time-resolved (ps) spectroscopy.[200] For example, the 200-fs laser excitation of benzilate salt $[Ph_2C(OH)CO_2^-, MV^{2+}]$ reveals the simultaneous formation of the ketyl radical ($Ph_2COH^\bullet$) and reduced methyl viologen ($MV^{+\bullet}$), as shown in the transient spectrum in Fig. 11. The fact that both the ketyl radical and $MV^{+\bullet}$ are formed on the early picosecond time scale leads to the conclusions that electron transfer from benzilate anion to MV^{2+} generates the ion-radical pair $Ph_2C(OH)CO_2^\bullet$, $MV^{+\bullet}$, in which the acyloxy radical expels CO_2 by an ultrafast C—CO_2 bond scission to effectively compete with the back electron transfer (equation 69).

$$[Ph_2C(OH)CO_2^-,\ MV^{2+}] \underset{k_{BET}}{\overset{h\nu_{CT}}{\rightleftharpoons}} Ph_2C(OH)CO_2^\bullet,\ MV^{+\bullet} \xrightarrow[-CO_2]{k_{C\text{-}C}} Ar_2\dot{C}{-}OH \quad (69)$$

Most importantly, the careful kinetic analysis of the rise and decay of the transient species in equation (69) shows that the decarboxylation of $Ph_2C(OH)CO_2^\bullet$ occurs within a few picoseconds ($k_{C-C} = (2\text{–}8) \times 10^{11}\ s^{-1}$). The observation of such ultrafast (decarboxylation) rate constants, which nearly approach those of barrier-free unimolecular reactions, suggests that the advances in time-resolved spectroscopy can be exploited to probe the transition state for C—C bond cleavages via charge-transfer photolysis.

C—H bond activation of hydrocarbons

Facile C—H bond activation is critical to such synthetically useful transfor-

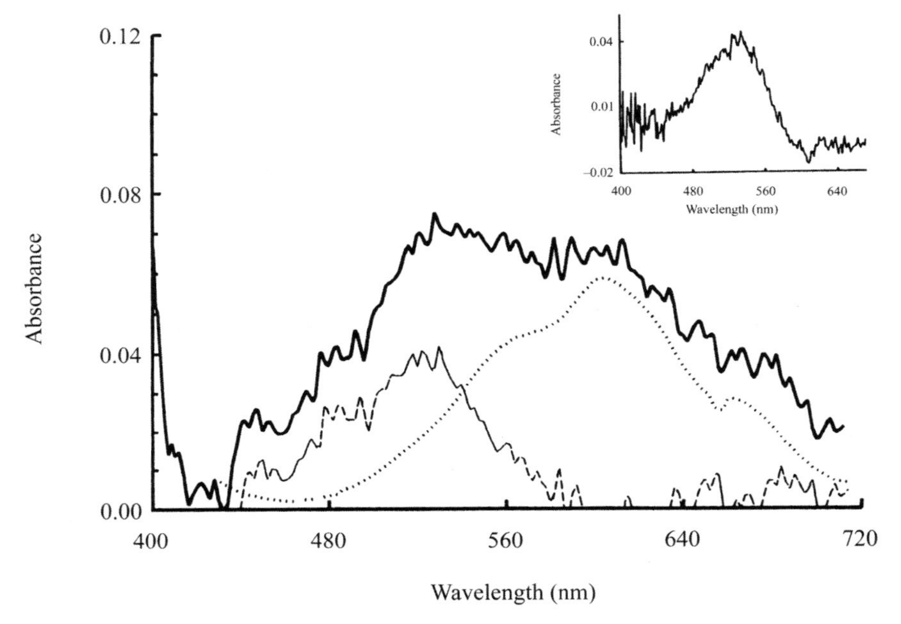

Fig. 11 Transient spectra obtained upon the application of a 200-fs laser pulse to a solution of $[Ph_2C(OH)CO_2^-, MV^{2+}]$ charge-transfer salt showing the simultaneous formation of benzophenone ketyl radical (dashed line) and the reduced methyl viologen (dotted line). The inset is the authentic spectrum of ketyl radical. Reproduced with permission from Ref. 92a.

mations as benzylic and allylic oxidations, dehydrogenation, aromatization, etc., which are carried out by a variety of organic and inorganic oxidants. Chart 4 gives some typical examples of organic transformations that employ quinones (such as DDQ) as oxidants.[201] Generally, the transformations in Chart 4 are rationalized by invoking either a hydride or a hydrogen-atom abstraction mechanism,[201] i.e.,

$$DH_2 + Q \longrightarrow \begin{cases} \xrightarrow{(H^-)} DH^+ + QH^- \\ \xrightarrow[(H^\bullet)]{} DH^\bullet + QH^\bullet \end{cases}$$

Interestingly, the common feature in each of the transformations in Chart 4 is

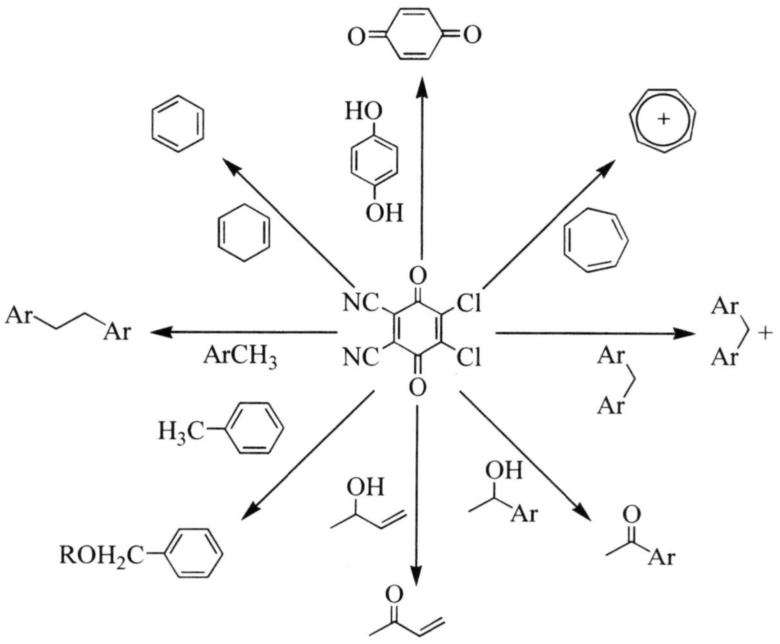

Chart 4

reported to be the observation of transient charge-transfer colorations that arise from the formation of donor/acceptor complexes,[202] e.g.,

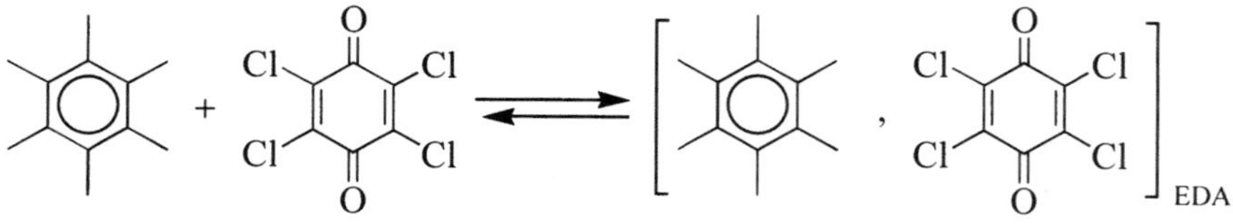

Owing to its relatively high reduction potential, DDQ undergoes rapid thermal reactions, which hamper the direct examination of the electron-transfer paradigm in C—H activations. However, low-potential quinones such as chloranil (CA) and dichloroxyloquinone (CX) form thermally stable donor/acceptor complexes with methylbenzenes, which have been extensively characterized by spectroscopic as well as X-ray crystallographic methods.[202] For example, the mixing of different methylbenzenes with chloranil spontaneously results in brightly colored solutions, which progressively change from pale yellow (*p*-xylene) to yellow (mesitylene) to orange (durene) to orange-red (pentamethylbenzene) to purple (hexamethylbenzene) in line with the decreasing ionization potentials of the aromatic donors.

Electron-transfer activation. Time-resolved spectroscopy has established that the irradiation of the CT bands ($h\nu_{CT}$) of [ArMe, CA] complexes results in direct electron transfer to form the contact ion pair instantaneously,[203] i.e.,

$$[\mathrm{ArMe,\ CA}]_{\mathrm{EDA}} \underset{k_{\mathrm{BET}}}{\overset{h\nu_{\mathrm{CT}}}{\rightleftarrows}} \mathrm{ArMe}^{+\bullet},\ \mathrm{CA}^{-\bullet}$$

However, the short lifetimes (~ 50 ps) of the ion-radical pair $ArMe^{+\bullet}$, $CA^{-\bullet}$ owing to rapid back electron transfer (k_{BET}) does not allow other reactions to compete effectively.[203] In contrast, the diffusional quenching of the photoexcited chloranil with methylbenzenes leads to (spectrally) indistinguishable ion-radical pairs with greatly enhanced lifetimes,[204] i.e.,

$$\mathrm{ArMe} + \mathrm{CA}^{*} \underset{k_{\mathrm{BET}}}{\overset{h\nu_{\mathrm{CA}}}{\rightleftarrows}} \mathrm{ArMe}^{+\bullet},\ \mathrm{CA}^{-\bullet}$$

Such variation in the lifetimes of the ion pairs, which depends on the mode of activation, primarily arises from the difference in the spin multiplicities (see above). None the less, the long-lived ion-radical pair allows the in-cage proton transfer from the cation radical $ArMe^{+\bullet}$ to the $CA^{-\bullet}$ anion radical to effectively compete with the back electron transfer,[205] i.e.,

$$\mathrm{ArMe} + \mathrm{CA} \xleftarrow{k_{\mathrm{BET}}} \mathrm{ArMe}^{+\bullet},\ \mathrm{CA}^{-\bullet} \xrightarrow{k_{\mathrm{C\text{-}H}}} \mathrm{ArCH}_2^{\bullet} + \mathrm{CAH}^{\bullet}$$

The modulation of the ion-pair dynamics by salt and solvent effects as well as the observation of significant kinetic isotope effects unambiguously establishes that benzylic C—H activation proceeds via a two-step sequence involving reversible electron transfer followed by proton transfer within the contact ion pair,[41c,205] (Scheme 18).

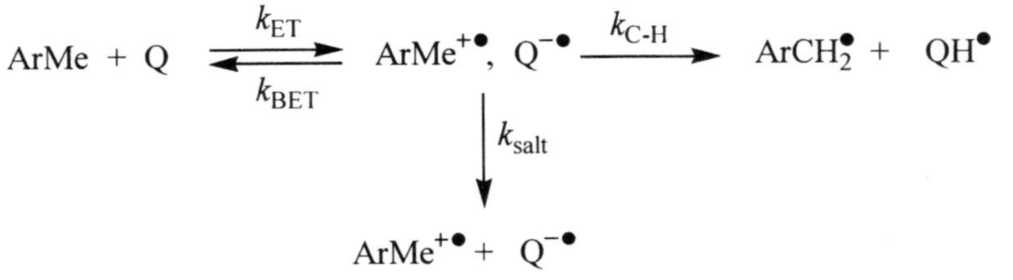

$$\mathrm{ArMe} + \mathrm{Q} \underset{k_{\mathrm{BET}}}{\overset{k_{\mathrm{ET}}}{\rightleftarrows}} \mathrm{ArMe}^{+\bullet},\ \mathrm{Q}^{-\bullet} \xrightarrow{k_{\mathrm{C\text{-}H}}} \mathrm{ArCH}_2^{\bullet} + \mathrm{QH}^{\bullet}$$

$$\mathrm{ArMe}^{+\bullet},\ \mathrm{Q}^{-\bullet} \xrightarrow{k_{\mathrm{salt}}} \mathrm{ArMe}^{+\bullet} + \mathrm{Q}^{-\bullet}$$

Scheme 18

Analogously, ion-radical pairs[14] such as those in Scheme 18 are invoked as the critical reactive intermediates in the other donor/acceptor transformations depicted in Chart 4.

ELECTRON-TRANSFER ACTIVATION IN CYCLOADDITION REACTIONS

Diels–Alder cycloaddition reactions of electron-poor dienophiles to electron-rich dienes, which are generally carried out thermally, afford widespread applications for C—C bond formation. On the basis of their electronic properties, numerous dienes can be characterized as electron donors and dienophiles as electron acceptors. Despite the early suggestions by Woodward,[206] the donor/acceptor association and electron-transfer paradigm are usually not considered as the simplest mechanistic formulation for the Diels–Alder reaction. However, the examples of cycloaddition reactions described below will show that photo-irradiation of various D/A pairs leads to efficient cycloaddition reactions via electron-transfer activation.

Cycloaddition of benzocyclobutene and various electron acceptors

The mixing of *cis*-1,2-diphenylbenzocyclobutene (*c*-DBC) with tetracyanoethylene (TCNE) produces a yellow-red coloration due to charge-transfer complexation,[207] i.e.,

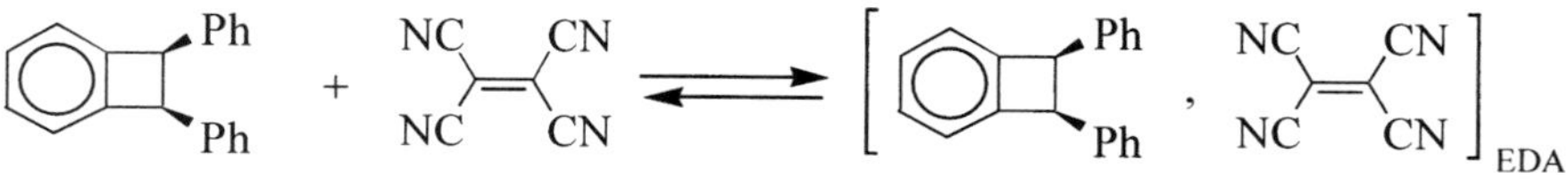

Ph Ph + NC CN NC CN ⇌ [Ph Ph , NC CN NC CN]$_{EDA}$

Charge-transfer activation by irradiation at the CT absorption band ($h\nu_{CT}$) bleaches the yellow color within 20 min to afford a *trans*-tetralin adduct (*t*-DTT) in quantitative yields[207] (equation 70).

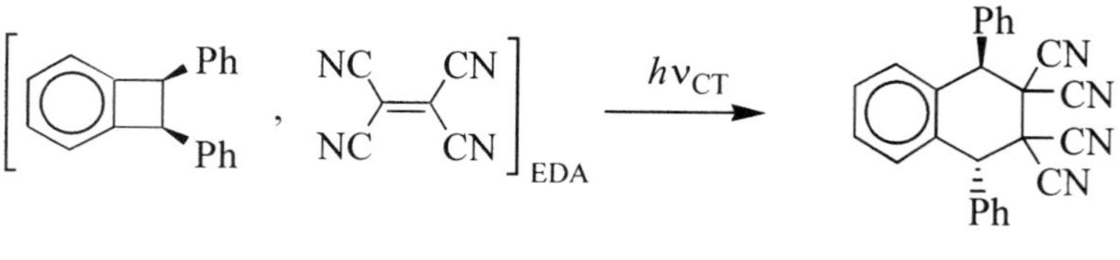

$$[c\text{-DBC}, \text{TCNE}]_{EDA} \xrightarrow{h\nu_{CT}} t\text{-DTT} \quad (70)$$

Similarly, the CT irradiation of the isomeric [*t*-DBC, TCNE] charge-transfer complex leads to the *cis*-tetralin adduct (*c*-DTT) as the sole product[207] (equation 71).

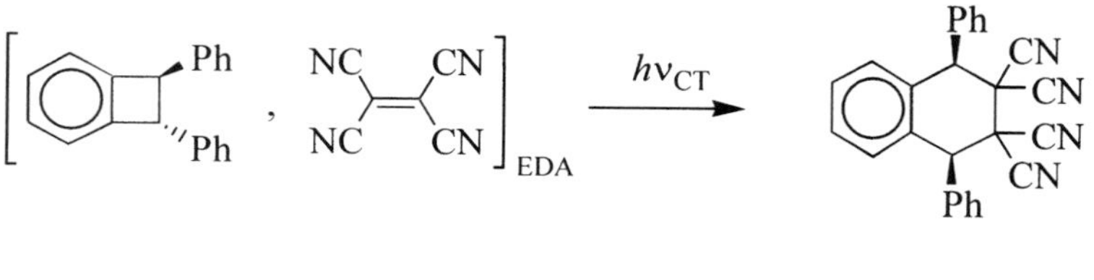

$$[t\text{-DBC}, \text{TCNE}]_{EDA} \xrightarrow{h\nu_{CT}} c\text{-DTT} \quad (71)$$

Note that the cycloaddition reactions in equations (70) and (71) can also be induced thermally, although the completion of the reactions requires several days.

Electron-transfer activation. The photoinduced electron-transfer activation of the [DBC, TCNE] complex generates the $DBC^{+\bullet}$ cation radical, which undergoes a fast electrocyclic ring opening, i.e.,

$$[\text{DBC}, \text{TCNE}]_{EDA} \xrightarrow{h\nu_{CT}} \text{DBC}^{+\bullet}, \text{TCNE}^{-\bullet} \longrightarrow (\text{ring-opened DBC})^{+\bullet}, \text{TCNE}^{-\bullet}$$

The new (rearranged) ion-radical pair then collapses with extreme rapidity (as indicated by the high quantum yields) to a stereospecific cycloadduct, i.e.,

$$(\text{ring-opened DBC})^{+\bullet}, \text{TCNE}^{-\bullet} \xrightarrow{k_{C\text{-}C}} \text{cycloadduct}$$

Note that the observed high stereospecificity of the reactions in equations (70) and (71) also points to a rapid coupling of ion pair after the electrocyclic rearrangement of the DBC cation radicals in the contact ion pairs.

Interestingly, the electron-transfer activation of *cis*- and *trans*-DBC cycloadditions with chloranil as a sensitizer leads to a mixture of *cis*- and *trans*-decalin adducts (equation 72).

$$\text{DBC} + \text{chloranil} \xrightarrow{h\nu_{CA}} \text{adduct} + \text{adduct} \quad (72)$$

The lack of stereospecificity in equation (72) (as compared to equations (70) and (71)) is readily accommodated by the long lifetime of the triplet ion pair produced by the sensitized irradiation method. The latter allows the facile ring opening of $DBC^{+\bullet}$ cation radical followed by isomerization of the resulting xylylene cation radical, i.e.,

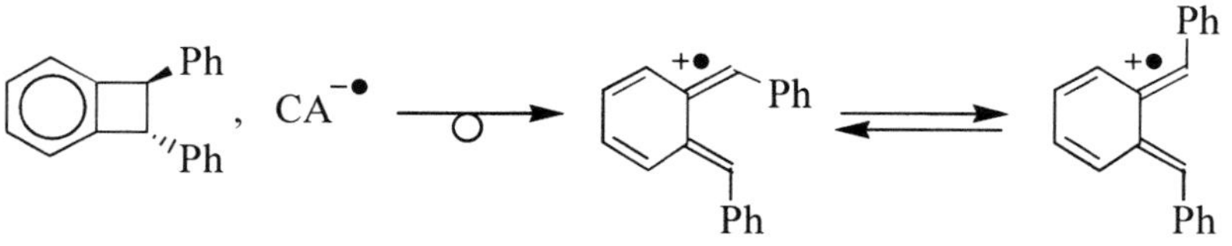

The long-lived isomeric xylylene cation radical then undergoes either coupling to the adducts in equation (72) or back electron transfer followed by Diels–Alder reaction of the resulting neutral xylylenes and chloranil.

Cycloaddition of homobenzvalene with TCNE

Homobenzvalene (HB) is an electron-rich donor (IP = 8.02 eV) owing to the presence of a strained ring system, and thus readily forms a charge-transfer complex with TCNE. Charge-transfer irradiation of the [HB, TCNE] complex leads to rapid bleaching of the yellow color, and the formation of a mixture of isomeric cycloadducts[208] (equation 73).

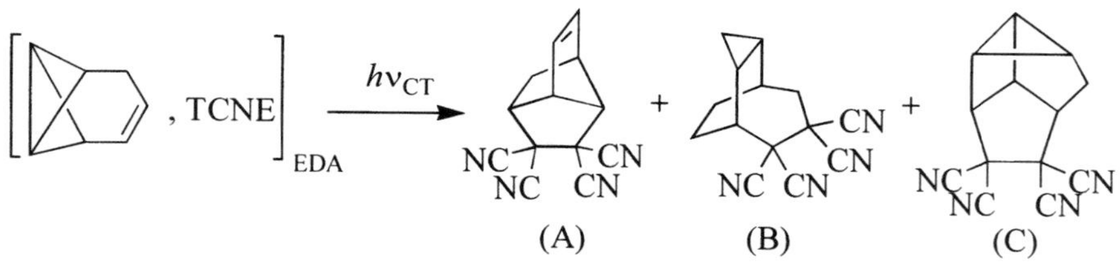

(73)

As described above, the charge-transfer irradiation of the [HB, TCNE] complex produces the contact ion pair, in which the strained $HB^{+\bullet}$ undergoes multiple bond cleavages to afford three isomeric cation radicals depicted in Scheme 19, which then undergo coupling with $TCNE^{-\bullet}$ to form a mixture of isomeric cycloadducts (A, B, and C) in equation (73).[208]

The thermal reaction of the [HB, TCNE] complex leads to a different mixture of isomeric adducts compared to those obtained by charge-transfer cycloaddition[208] (equation 74).

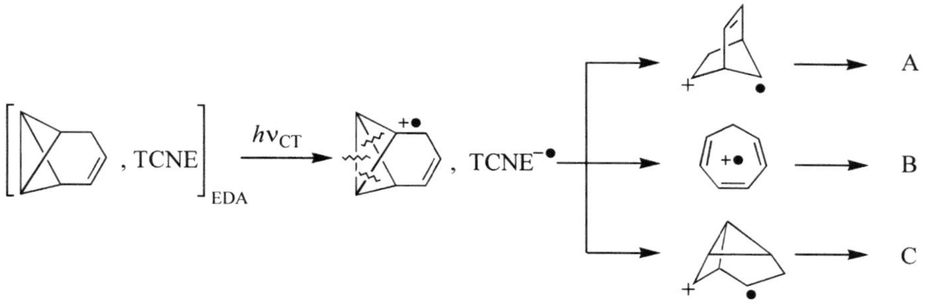

[, TCNE]$_{EDA}$ $\xrightarrow{h\nu_{CT}}$ $^{+\bullet}$, TCNE$^{-\bullet}$ → A, B, C

Scheme 19

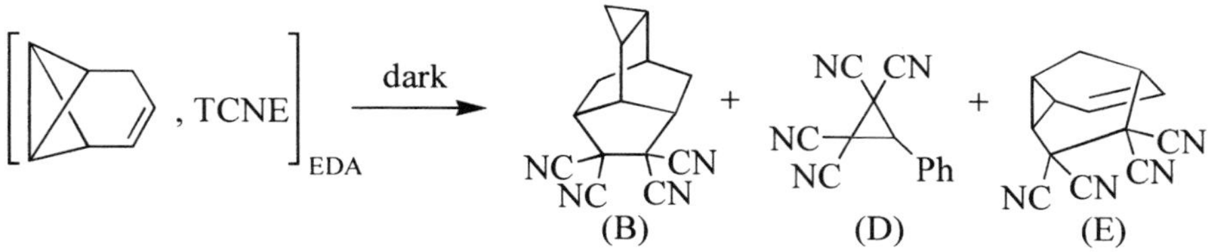

[, TCNE]$_{EDA}$ $\xrightarrow{\text{dark}}$ (B) + (D) + (E)

NC CN NC CN (B); NC CN NC Ph (D); CN CN NC CN (E)

(74)

Interestingly, the common appearance of product (B) in the thermal and the charge-transfer activated reactions suggests that both activation processes are strongly coupled. However, differences in the energetics and the nature of the $HB^{+\bullet}$, $TCNE^{-\bullet}$ ion pairs (obtained from thermal and photoactivation) are probably responsible for the variation in the products in equations (73) and (74).

A variety of other highly-strained electron-rich donors also form colored complexes (similar to homobenzvalene) with various electron acceptors, which readily undergo thermal cycloadditions (with concomitant bleaching of the color).[209] For example, Tsuji *et al.*[210] reported that dispiro[2.2.2.2]deca-4,9-diene (DDD), with an unusually low ionization potential of 7.5 eV,[211] readily forms a colored charge-transfer complex with tetracyanoquinodimethane (TCNQ). The [DDD, TCNQ] charge-transfer complex undergoes a thermal cycloaddition to [3,3]paracyclophane in excellent yield, i.e.,

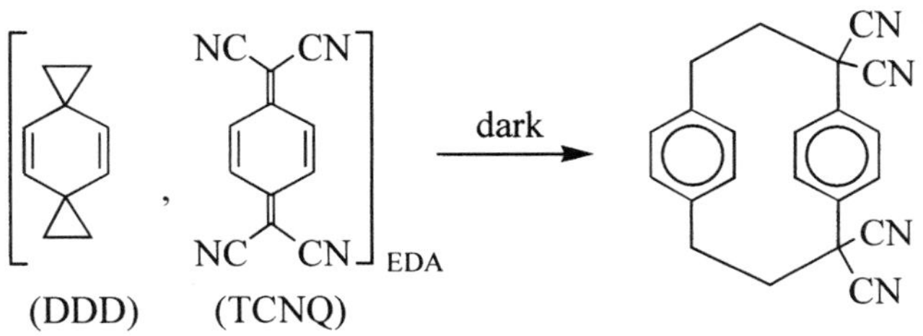

[(DDD), (TCNQ)]$_{EDA}$ $\xrightarrow{\text{dark}}$ [3,3]paracyclophane (CN, CN, CN, CN)

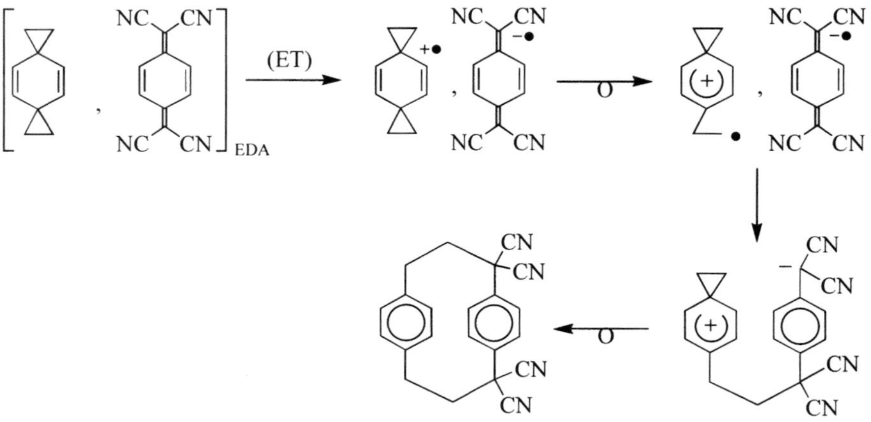

Scheme 20

The efficient formation of the cyclophane suggests that the [DDD, TCNQ] charge-transfer complex is converted to the ion-radical pair via a thermal electron transfer. Subsequent fragmentation of the resulting $DDD^{+\bullet}$ and a homolytic coupling with $TCNQ^{-\bullet}$ leads then to a zwitterionic intermediate which collapses to the cyclophane within the ion pair, as summarized in Scheme 20.

It is important to note that the efficiency of the various cycloaddition reactions presented above arises from a rapid cleavage of the resulting cation radical intermediates, which renders the back electron transfer process ineffective.

Cycloaddition of anthracene to maleic anhydride

The thermal Diels–Alder reactions of anthracene with electron-poor olefinic acceptors such as tetracyanoethylene, maleic anhydride, maleimides, etc. have been studied extensively. It is noteworthy that these reactions are often accelerated in the presence of light. Since photoinduced [4 + 2] cycloadditions are symmetry-forbidden according to the Woodward–Hoffman rules, an electron-transfer mechanism has been suggested to reconcile experiment and theory.[212] For example, photocycloaddition of anthracene to maleic anhydride and various maleimides occurs in high yield (> 90%) under conditions in which the thermal reaction is completely suppressed (equation 75).

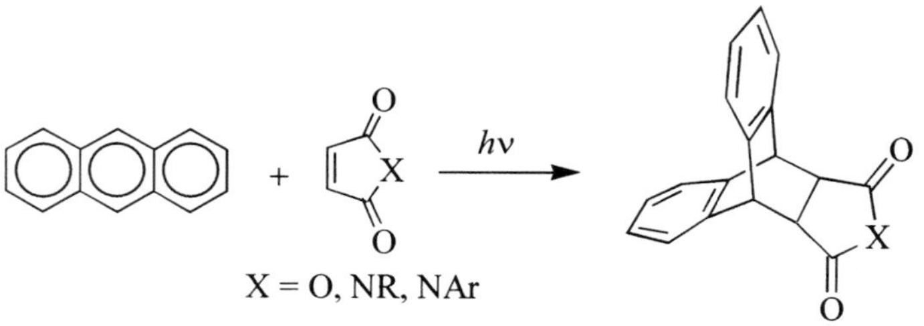

X = O, NR, NAr (75)

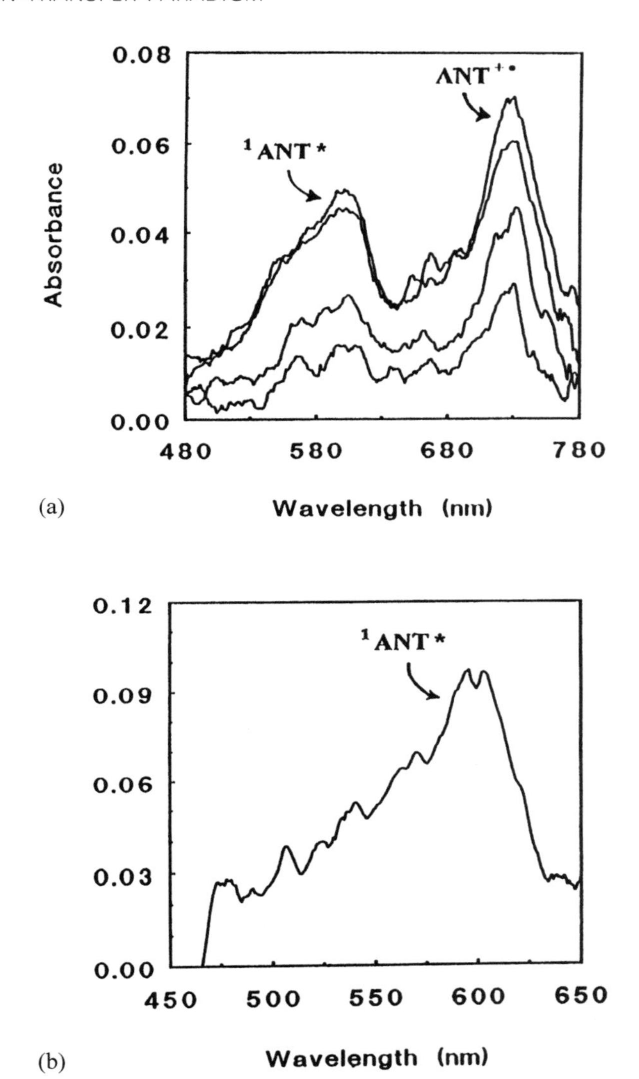

Fig. 12 (a) The absorption spectra of singlet excited anthracene (^{1}ANT*) and anthracene cation radical (ANT$^{+\bullet}$) obtained upon 25-ps laser excitation of anthracene in the presence of excess maleic anhydride (MA). (b) The authentic spectrum of singlet excited anthracene (^{1}ANT*). Reproduced with permission from Ref. 212.

Time-resolved spectroscopy establishes that the fluorescence of the excited (singlet) anthracene (^{1}ANT*) is readily quenched by maleic anhydride (MA), which leads to the formation of the ion pair ANT$^{+\bullet}$, MA$^{-\bullet}$ via diffusional electron transfer (see Fig. 12), i.e.,

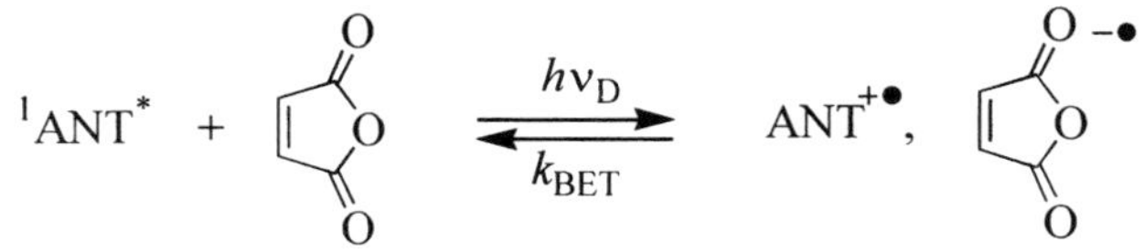

$$ {}^{1}\text{ANT}^{*} + \text{MA} \underset{k_{\text{BET}}}{\overset{h\nu_{\text{D}}}{\rightleftharpoons}} \text{ANT}^{+\bullet}, \text{MA}^{-\bullet} $$

The critical role of the ion-radical pair in the cycloaddition reactions in equation (75) is demonstrated by a careful measurement of the quantum yields as a function of the dienophile concentration and by a study of the effect of solvent and salt on the dynamics of the ion pair $ANT^{+\bullet}$, $MA^{-\bullet}$.[212] However, in the reported cases, back electron transfer effectively competes with the coupling within the ion-radical pair and thus limits the quantum yields for the formation of the Diels–Alder adduct.[212]

OSMYLATION OF ARENE DONORS

Osmium(VIII) tetraoxide (OsO_4) is an effective reagent for the *cis* hydroxylation of olefins under stoichiometric conditions as well as in a variety of catalytic variants.[213] Under both catalytic and stoichiometric conditions, the critical step is the formation of an osmium(VI) cycloadduct, the formation of which is dramatically accelerated in the presence of amine bases such as pyridine,[214] i.e.,

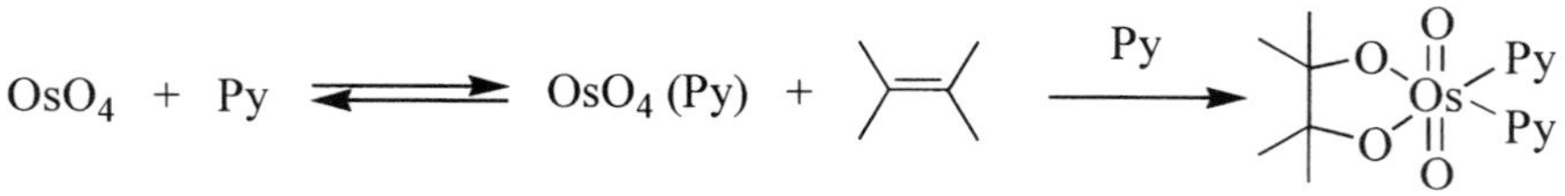

$$ OsO_4 + Py \rightleftharpoons OsO_4(Py) + \text{olefin} \xrightarrow{Py} \text{cyclic osmate ester } (Py)_2OsO_2(O_2C_2R_4) $$

The 1,2-diol is liberated easily from cyclic osmate ester by either reductive or oxidative hydrolysis.[213] Importantly, the ligand acceleration has been utilized extensively for the production of chiral 1,2-diols from (achiral) olefins using optically active amine bases (such as L = dihydroquinidine, dihydroquinine and various chiral diamine ligands).[215]

Despite the extensive utility of osmylation reactions, the mechanism remains controversial – generally focusing on either a concerted [3 + 2] cycloaddition to directly yield the cyclic osmate ester or a two-step mechanism where [2 + 2] cycloaddition forms a 4-membered osmaoxetane which then undergoes a ligand-assisted ring expansion to yield the cyclic osmate ester[216] (Scheme 21).

An initial osmium tetraoxide/olefin complex (in Scheme 21) is not generally disputed and is considered a donor/acceptor complex between an alkene as electron donor and OsO_4 as an electron acceptor (with a reversible reduction potential $E^{0}_{\text{red}} = -0.06$ V versus SEC),[217] e.g.,

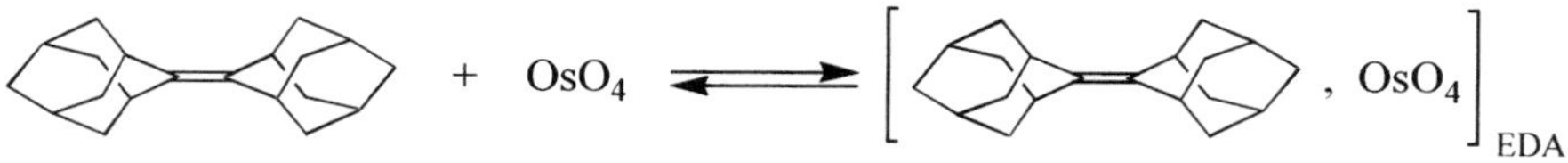

$$ \text{Ad=Ad} + OsO_4 \rightleftharpoons [\text{Ad=Ad}, OsO_4]_{\text{EDA}} $$

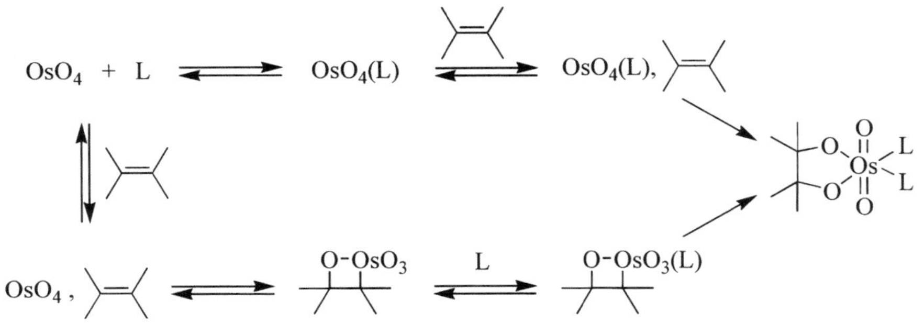

$$OsO_4 + L \rightleftharpoons OsO_4(L) \rightleftharpoons OsO_4(L), \text{alkene}$$

Scheme 21

Unfortunately, the fast rates of OsO_4 addition to most alkenes preclude the observation of D/A complexes, and they are not readily characterized. However, a variety of aromatic electron donors form similar (colored) D/A complexes with OsO_4 that are more persistent; and the observation of ArH/OsO_4 complexes forms the basis for examining the electron-transfer paradigm in osmylation reactions.

Donor–acceptor complexes. A colorless hexane (or dichloromethane) solution of osmium tetraoxide upon exposure to benzene turns yellow instantaneously owing to the formation of a donor/acceptor complex,[218] i.e.,

$$C_6H_6 + OsO_4 \rightleftharpoons [C_6H_6, OsO_4]_{EDA}$$

Similar vivid colorations are observed when other aromatic donors (such as methylbenzenes, naphthalenes and anthracenes) are exposed to OsO_4.[218] The quantitative effect of such dramatic colorations is illustrated in Fig. 13 by the systematic spectral shift in the new electronic absorption bands that parallels the decrease in the arene ionization potentials in the order benzene 9.23 eV, naphthalene 8.12 eV, anthracene 7.55 eV. The progressive bathochromic shift in the charge-transfer transitions ($h\nu_{CT}$) in Fig. 13 is in accord with the Mulliken theory for a related series of [D, A] complexes.

Thermal osmylation. Upon standing in the dark, the purple CT color of anthracene complex [ANT, OsO_4] slowly diminishes to afford the 1:2 (insoluble) osmium adduct as the sole product, i.e.,

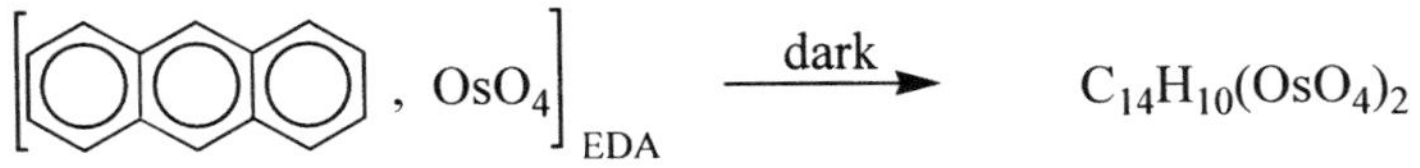

$$[\text{anthracene}, OsO_4]_{EDA} \xrightarrow{\text{dark}} C_{14}H_{10}(OsO_4)_2$$

Naphthalene and phenanthrene also yield similar osmium adducts that are

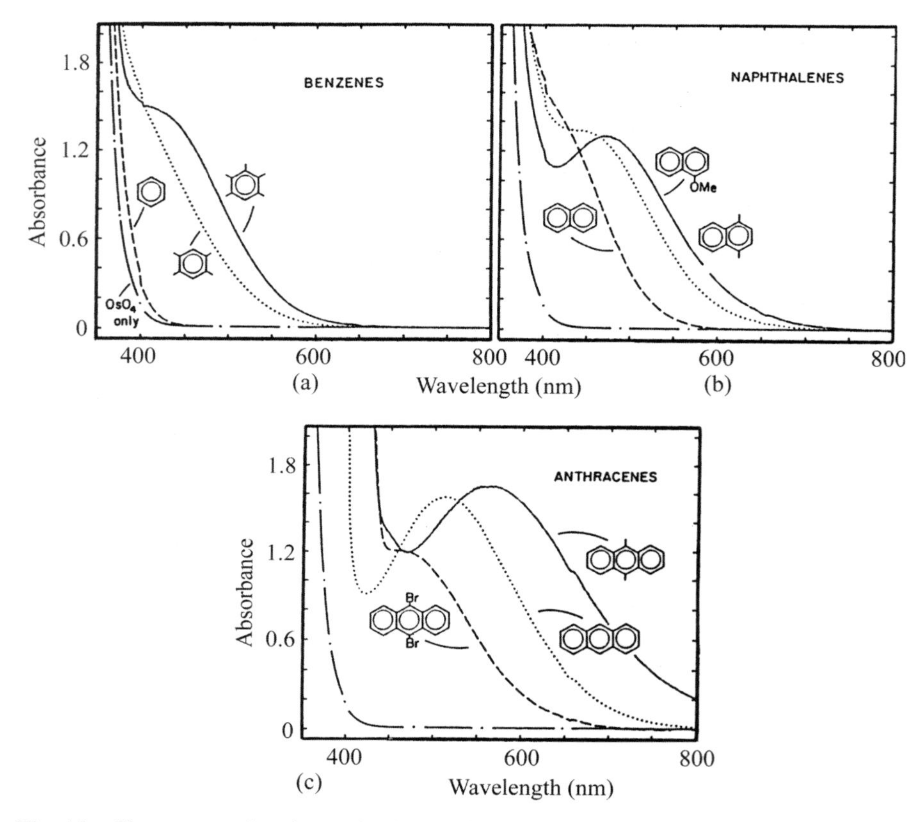

Fig. 13 Charge-transfer absorption bands from dichloromethane solutions containing OsO_4 and various (a) benzene, (b) naphthalene, and (c) anthracene donors (as indicated) showing the progressive bathochromic shift with aromatic donor strength. Reproduced with permission from Ref. 96b.

isolable as pyridine complexes. The structures of osmium adducts A, B, C are established by NMR spectroscopy and by X-ray crystallography.

(A) (B) (C)

Photochemical osmylation. The irradiation of the charge-transfer bands (Fig. 13) of the EDA complex of OsO_4 with various benzenes, naphthalenes, anthracenes, and phenanthrene yields the same osmylated adducts as obtained in the thermal reactions. For example, irradiation of the purple solution of anthracene and OsO_4 in dichloromethane at $\lambda > 480$ nm yields the same 2:1 adduct (B) together with its syn isomer as the sole products, i.e.,

$$[\text{anthracene}, OsO_4]_{EDA} \xrightarrow{h\nu_{CT}} \text{anthracene}(O_2OsO_2)_2$$

(syn and anti)

Note that the charge-transfer irradiation of the same anthracene/OsO_4 complex in hexane solution yields only a small amount of the adduct (B) together with considerable amounts of anthraquinone. The latter probably arises from osmylation at the (9,10) positions followed by decomposition of the unstable osmium adduct.[218]

Electron-transfer activation. Time-resolved spectroscopy establishes that irradiation of the charge-transfer band ($h\nu_{CT}$) of various arene/OsO_4 complexes directly leads to the contact ion pair. For example, 25-ps laser excitation of the [anthracene, OsO_4] charge-transfer complex results in the ion-radical pair instantaneously, as shown in Fig. 14[218] (equation 76).

$$[\text{anthracene}, OsO_4]_{EDA} \underset{}{\overset{h\nu_{CT}}{\rightleftharpoons}} \text{anthracene}^{+\bullet}, OsO_4^{-\bullet} \qquad (76)$$

Furthermore, kinetic analysis of the decay rate of anthracene cation radical, together with quantum yield measurements, establishes that the ion-radical pair in equation (76) is the critical reactive intermediate in osmylation reaction. Subsequent rapid ion-pair collapse then leads to the osmium adduct with a rate constant $k \sim 10^9\ s^{-1}$ in competition with back electron-transfer, i.e.,

$$[Ar, OsO_4]_{EDA} \underset{k_{BET}}{\overset{h\nu_{CT}}{\rightleftharpoons}} Ar^{+\bullet}, OsO_4^{-\bullet} \xrightarrow[\text{O}]{\text{fast}} Ar(OsO_4)$$

Such a charge-transfer osmylation of aromatic donors constitutes the critical

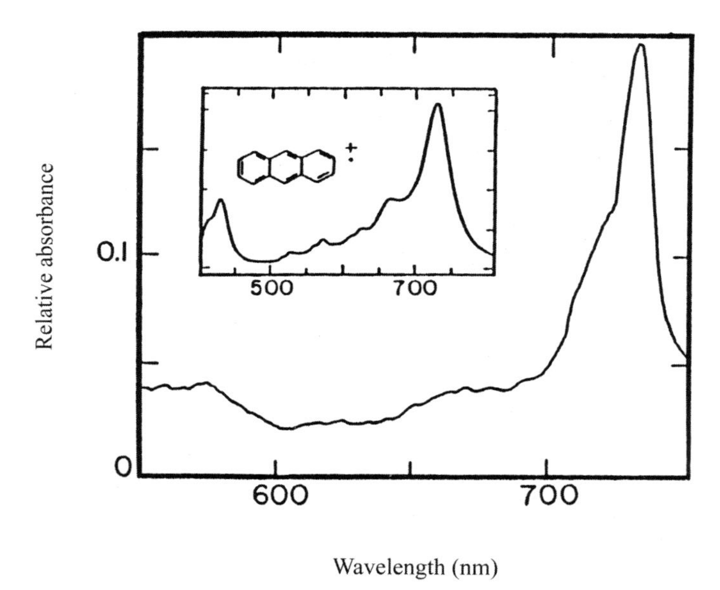

Fig. 14 Transient absorption spectrum of anthracene cation radical ($ANT^{+\bullet}$) obtained upon 30-ps laser excitation of the [ANT, OsO_4] charge-transfer complex in dichloromethane. The inset shows the authentic spectrum of $ANT^{+\bullet}$ obtained by an independent (electrochemical) method. Reproduced with permission from Ref. 96b.

step in stereospecific tetrakis-hydroxylation to various cyclitol derivatives with catalytic amounts of osmium tetraoxide,[219] e.g.,

$$C_6H_6 + Ba(ClO_3)_2 \xrightarrow[h\nu_{CT}]{[OsO_4]} \text{(cyclohexene-tetrol: OH, OH, HO, OH)} + \text{etc.}$$

The identical stoichiometries and the color changes that are observed in thermal and photochemical aromatic osmylations point to the ion-radical pair $Ar^{+\bullet}$, $OsO_4^{-\bullet}$ as the seminal intermediate in both activation processes. It is similarly possible that the osmylation of olefinic donors may proceed via the same types of reactive intermediates as delineated for the aromatic osmylation.

ELECTRON-TRANSFER ACTIVATION OF ELECTROPHILIC AROMATIC SUBSTITUTION

Electrophilic aromatic substitution is generally considered to proceed via a π-complex (or an encounter complex) between the electrophile E^+ and the aromatic substrate ArH which collapses in a single rate-limiting step to form the Wheland intermediate (or σ-complex)[220] (equation 77).

$$\mathrm{ArH} + \mathrm{E}^{+} \rightleftharpoons [\mathrm{ArH}, \mathrm{E}^{+}] \longrightarrow \overset{+}{\mathrm{Ar}}\genfrac{}{}{0pt}{}{\mathrm{H}}{\mathrm{E}} \tag{77}$$

The focal point in the mechanism of electrophilic aromatic substitution is the activation process that leads to the well-established Wheland intermediate.[221] In order to address the mechanism of the activation of π-complexes (in equation (77)), we first recognize that most electrophiles (e.g., nitrosonium and nitronium cations, various nitrating agents, halogens, carbocations, diazonium cations, sulfur trioxide, lead(IV), mercury(II), and thallium(III) salts, etc.[222]) are excellent electron acceptors (or oxidants) as judged by their electrochemical reduction potentials (see Table 4) as well as by their ready ability to oxidize a variety of organic and organometallic electron donors to the corresponding cation radicals. Furthermore, various aromatic compounds are electron donors, as reflected by their low ionization potentials and easily accessible anodic oxidations in solution (see Table 3). As a consequence, the π-complex in equation (77) is more accurately described as an electron donor/acceptor (EDA) complex between the electrophile and the aromatic nucleophile. Most importantly, charge-transfer absorption bands (or colors) of such EDA complexes are visible indications of the strong molecular orbital interaction that precedes the reaction of the electrophile with the aromatic nucleophile.[223]

The application of the electron-transfer paradigm to electrophilic aromatic substitution is based on (i) the observation of EDA interactions between electrophiles and aromatic donors, and (ii) the quantitative correlations of the second-order rate constants with the charge-transfer transition energies ($h\nu_{CT}$) for halogenation, mercuration, thallation, etc.[224] Accordingly, we shall take halogenation, nitration, and nitrosation with various electrophilic agents as typical examples to elucidate the importance of donor/acceptor organization and electron-transfer activation as the fundamental steps that lead to the Wheland intermediate.

Aromatic halogenation with iodine monochloride

The exposure of a colorless solution of anisole to iodine monochloride (ICl) in dichloromethane leads to the appearance of a new electronic absorption band ($\lambda_{max} = 350$ nm) arising from the formation of a donor/acceptor complex,[225] i.e.,

$$\mathrm{ArOMe} + \mathrm{ICl} \rightleftharpoons [\mathrm{ArOMe}, \mathrm{ICl}]_{\mathrm{EDA}}$$

The molecular structure of the [anisole, ICl] complex is analogous to the [benzene, Cl_2] complex – the earliest known example of a charge-transfer complex for which the molecular structure has been established by X-ray crystallography[226] and the electronic structure theoretically predicted.[227]

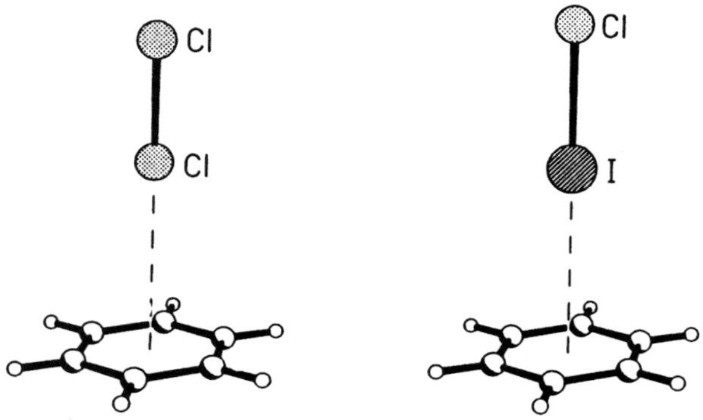

Chart 5
Reproduced with permission from Ref. 226

Iodination versus chlorination. Three distinct classes of halogenation reactions are observed with various substituted methoxybenzenes and ICl when carried out under an identical set of conditions, i.e., exclusive iodination, exclusive chlorination, and mixed chlorination and iodination. For example, equimolar mixtures of anisole, 2,5-dimethyl-1,4-dimethoxybenzene and 1,4-dimethoxybenzene and iodine monochloride (kept in the dark) yield *p*-iodoanisole, chloro-2,5-dimethyl-1,4-dimethoxybenzene, and a (4:6) mixture of chloro- and iododimethoxybenzene, respectively, in nearly quantitative yields,[225] i.e.,

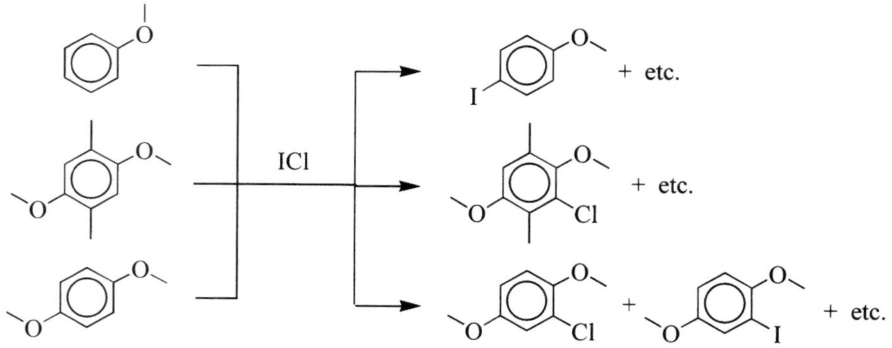

A careful analysis of the reaction mixtures establishes the stoichiometries for exclusive chlorination and iodination in equations (78a) and (78b).

$$ArH + 2\,ICl \longrightarrow ArCl + HCl + I_2 \qquad (78a)$$

$$ArH + ICl \longrightarrow ArI + HCl \qquad (78b)$$

Electron-transfer activation. UV-vis spectroscopic studies at low temperatures provide direct evidence for the electron-transfer activation in aromatic halogenation. For example, immediately upon mixing dimethoxybenzene and iodine monochloride at $-78°C$, the formation of dimethoxybenzene cation radical is noted (see Fig. 15a) (equation 79).

$$MeO\text{-}C_6H_4\text{-}OMe + 2\,ICl \longrightarrow [MeO\text{-}C_6H_4\text{-}OMe]^{+\bullet}\ ICl_2^- + \tfrac{1}{2}\,I_2 \qquad (79)$$

The stoichiometry in equation (79) is confirmed by the spectroscopic quantification of the dimethoxybenzene cation radical as well as by the earlier observation that the reaction of electron-rich tetraanisylethylene with iodine monochloride produces the iodine dichloride (ICl_2^-) salt of tetraanisylethylene dication whose structure has been established by X-ray crystallography,[228] i.e.,

$$An_2C{=}CAn_2 + 4\,ICl \longrightarrow An_2\overset{+}{C}\text{-}\overset{+}{C}An_2\ (ICl_2^-)_2 + I_2$$

The ICl_2^- anion arises from the well-established equilibrium between iodine monochloride and chloride.[229]

Time-resolved spectroscopy establishes that the 25-ps laser irradiation of the relatively persistent charge-transfer complex of *p*-bromoanisole with iodine monochloride generates the contact ion pair (see Fig. 15b) in which the metastable $ICl^{-\bullet}$ undergoes mesolytic fragmentation to form the reactive triad, i.e.,

$$[ArH\,,\,ICl]_{EDA} \xrightleftharpoons{h\nu_{CT}} ArH^{+\bullet};\ ICl^{-\bullet} \xrightarrow{\text{fast}} ArH^{+\bullet};\ I^{\bullet};\ Cl^-$$

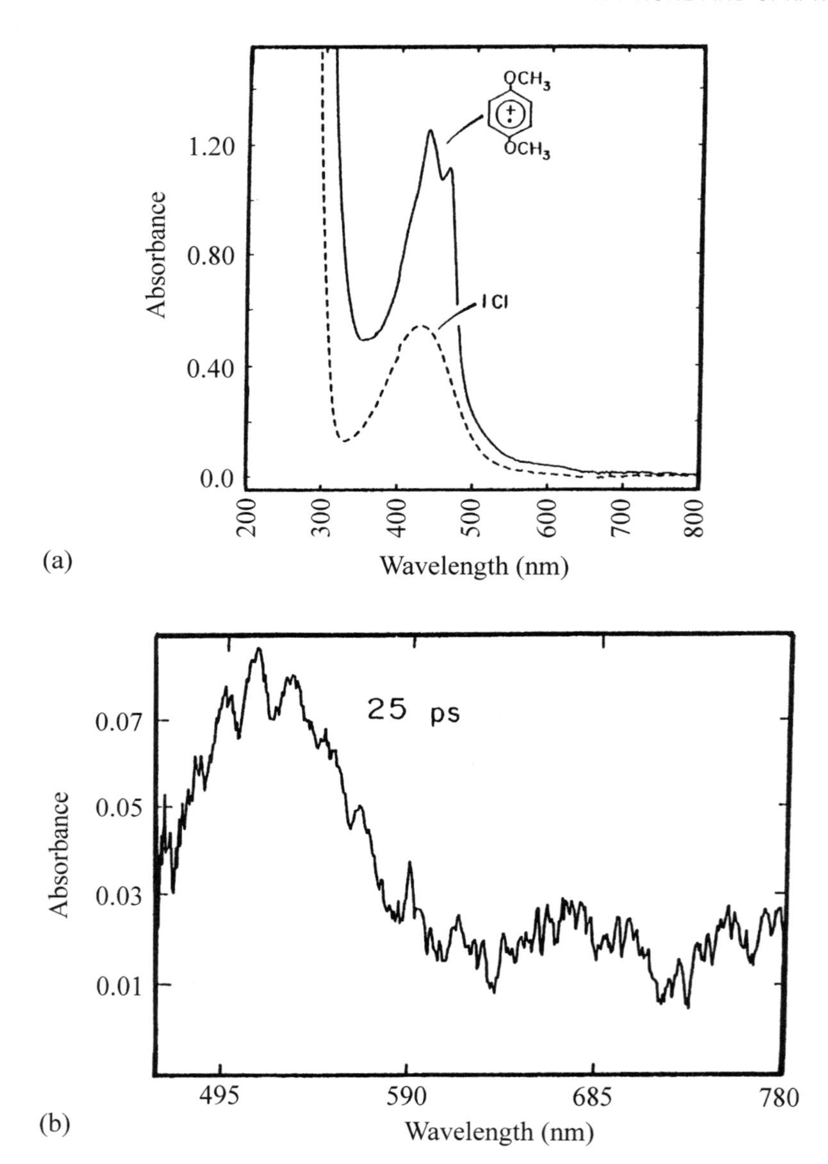

Fig. 15 (a) Absorption spectrum of dimethoxybenzene cation radical (——)obtained upon mixing iodine monochloride (– – –) and dimethoxybenzene at $-78°C$ in dichloromethane. (b) Transient spectrum of the bromoanisole (BA) cation radical generated by photoexcitation of the [BA, ICl] charge-transfer complex. Reproduced with permission from Ref. 225.

It is noteworthy that both thermal and photoinduced electron-transfer activation of the [ArH, ICl] complex leads to the ion-radical triad. Consequently, iodination versus chlorination represents the competition between ion-pair and radical-pair collapse. This is confirmed by reactivity studies of dimethoxybenzene cation radical with chloride and iodine (atom), respectively,[225] i.e.,

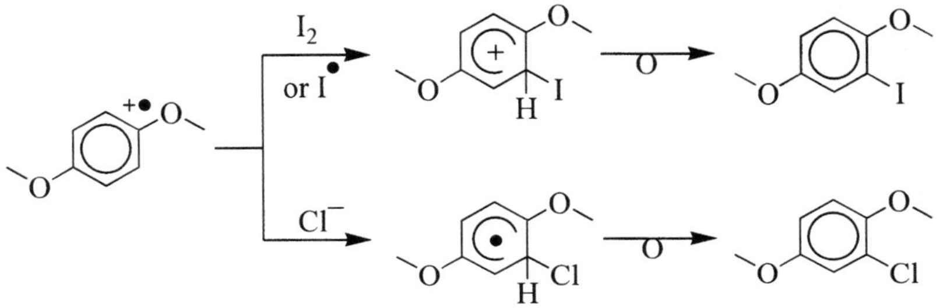

The competition between ion-pair and radical-pair collapse can be predictably modulated by the polarity of the solvent and by the addition of inert salt.

Aromatic nitration

The aromatic nitration with nitric acid represents an electrophilic aromatic substitution in which the nitronium ion (NO_2^+) is the active electrophile.[230] The idea that charge transfer may play a key role in aromatic nitration was first suggested by Kenner in 1945,[231] but this is not well recognized. The independent experimental proof of the reactive intermediates pertaining to the electron-transfer paradigm is not readily forthcoming owing to the rapidity of aromatic nitration with nitronium ion. However, such a kinetic restriction is overcome by use of nitronium ion carriers (NO_2Y) such as nitric acid, acetyl nitrate, dinitrogen pentoxide, *N*-nitropyridinium, tetranitromethane, etc. where Y = OH, OAc, NO_3, Py, and $C(NO_2)_3$ as milder nitrating agents which can also serve as effective electron acceptors.[232] Notably, the most characteristic property of the latter is that they readily form colored electron donor–acceptor complexes with aromatic hydrocarbons,[233,234] and thus allow the charge-transfer activation to effect aromatic nitration as follows.

A. *Nitropyridinium cations.* The spontaneous formation of vividly colored charge-transfer (CT) complexes occurs upon exposure of *N*-nitropyridinium ($PyNO_2^+$) cation to various aromatic donors,[235] i.e.,

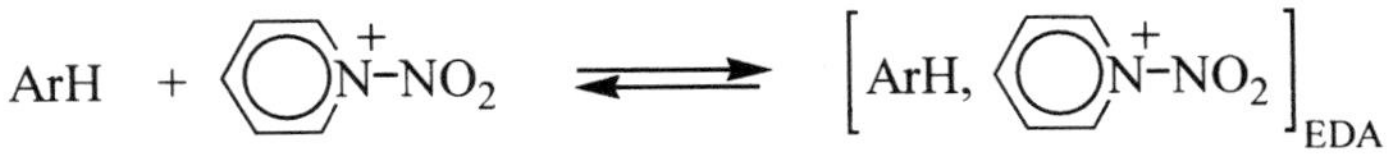

$$ArH + PyNO_2^+ \rightleftharpoons [ArH, PyNO_2^+]_{EDA}$$

Figure 16a shows the progressive bathochromic shift in the CT absorption bands ($h\nu_{CT}$) obtained from $PyNO_2^+$ with aromatic donors with increasing donor strength (or decreasing ionization potential). A similar red shift is observed in the CT absorption bands ($h\nu_{CT}$) of hexamethylbenzene complexes with various *para*-substituted *N*-nitropyridinium cations ($X\text{-}PyNO_2^+$) as shown in Fig. 16b. Such a trend in the $h\nu_{CT}$ is in accord with the increasing acceptor strength of $X\text{-}PyNO_2^+$ in the order X = OMe < Me < H < Cl < CO_2Me < CN.

It is particularly noteworthy that the acceptor strengths of nitropyridinium cations as measured by the oxidation potential (Ep) parallel the acidities of the corresponding hydropyridinium cations (p*K*a) as shown in Fig. 17.[235a] Such a remarkable correlation derives directly from the interconvertibility of oxidants and acids (Chart 6)

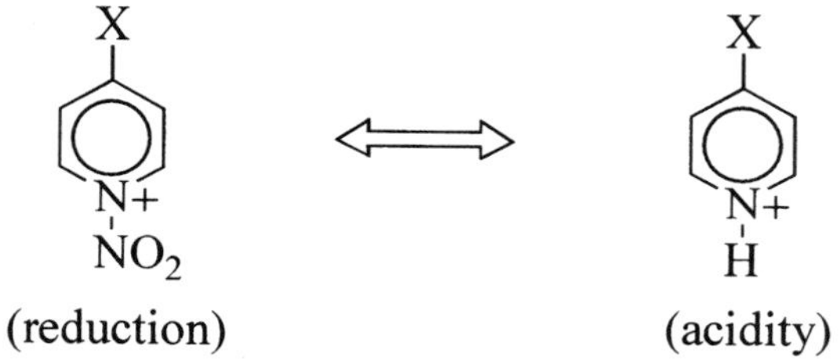

X-substituted N-nitropyridinium (N^+–NO_2) (reduction) ⟺ X-substituted N-hydropyridinium (N^+–H) (acidity)

Chart 6

insofar as the electron-transfer paradigm is the unifying theme (see Chart 1)

Thermal versus photochemical nitration. Activation of the various [$X\text{-}PyNO_2^+$, ArH] complexes can be effected both thermally and photochemically to afford identical isomeric mixtures of aromatic nitration products. For example, the bright lemon color immediately attendant upon the mixing of $PyNO_2^+$ and naphthalene (NAP) in acetonitrile slowly fades within 1 h to afford a mixture of 1- and 2-nitronaphthalene in excellent yield. The same isomeric mixture is obtained when the yellow solution of the [NAP, $PyNO_2^+$] complex (cooled to $-40°C$ to inhibit thermal nitration) is irradiated at the charge-transfer absorption band ($h\nu_{CT}$),[236] i.e.,

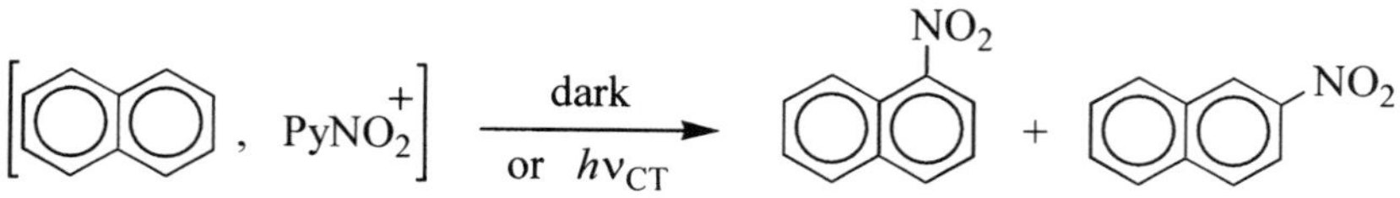

$$[\text{naphthalene}, PyNO_2^+] \xrightarrow[\text{or } h\nu_{CT}]{\text{dark}} \text{1-nitronaphthalene} + \text{2-nitronaphthalene}$$

Electron-transfer activation. Time-resolved spectroscopy shows that the activation of the [ArH, $PyNO_2^+$] complex by the specific irradiation of the CT absorption band results in the formation of transient aromatic cation radical

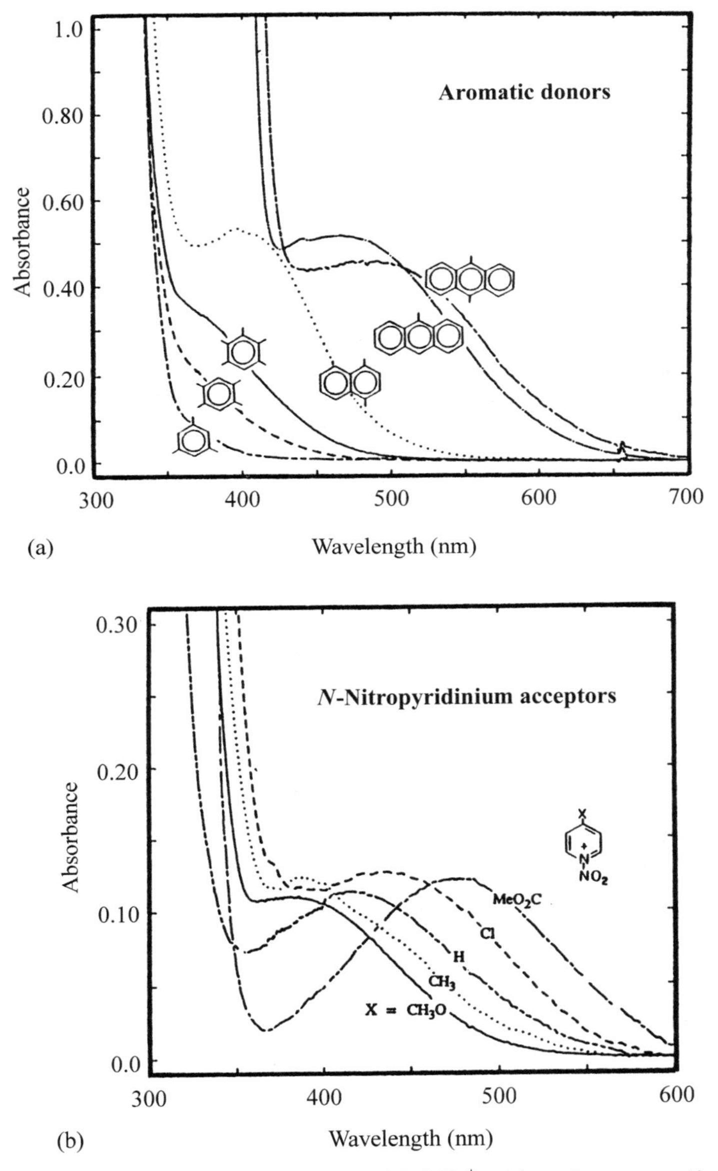

Fig. 16 Charge-transfer spectra of (a) $MeOPyNO_2^+$ with various aromatic donors (as indicated) and (b) hexamethylbenzene with various *N*-nitropyridinium acceptors (as indicated) in acetonitrile. Reproduced with permission from Ref. 235a.

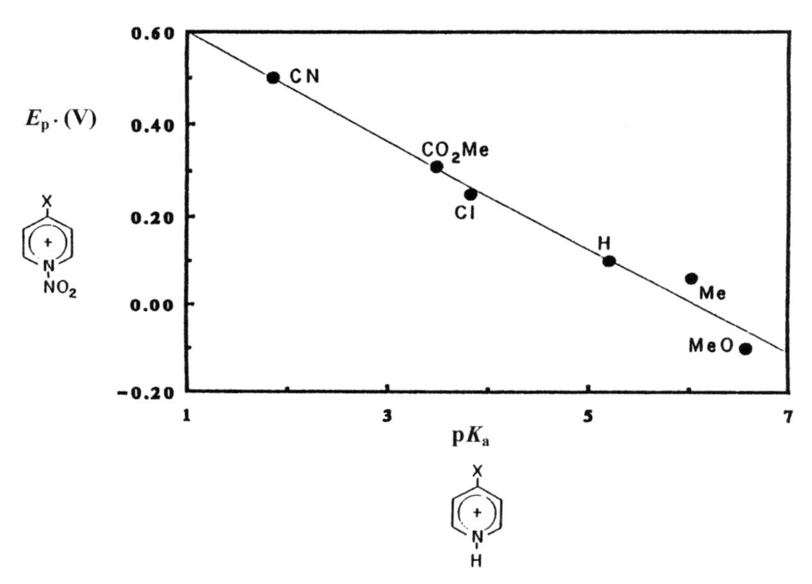

Fig. 17 Comparison of the electron acceptor properties of *N*-nitropyridinium cations ($XPyNO_2^+$) as measured by their E_p^c values in relationship to the thermodynamic acidities of the corresponding hydropyridinium cations ($XPyH^+$) as evaluated by their pK_a values. Reproduced with permission from Ref. 235a.

($ArH^{+\bullet}$) consistent with the photoinduced electron transfer according to equation (4),[235] i.e.,

$$[ArH,\ PyNO_2^+] \xrightleftharpoons{h\nu_{CT}} ArH^{+\bullet}_{,}\ PyNO_2^{\bullet}$$

The subsequent, spontaneous fragmentation of the nitropyridinyl radical and the collapse of the ion-radical pair to the Wheland intermediate (followed by rapid deprotonation) completes the nitration, i.e.,

$$[ArH,\ PyNO_2^+] \underset{k_{BET}}{\overset{h\nu_{CT}}{\rightleftharpoons}} ArH^{+\bullet}_{,}\ PyNO_2^{\bullet} \xrightarrow{k_f} ArH^{+\bullet}_{,}\ NO_2^{\bullet},\ Py$$

Moreover, the thermal nitration of various aromatic substrates with different X-$PyNO_2^+$ cations shows the strong rate dependence on the acceptor strength of X-$PyNO_2^+$ and the aromatic donor strength. This identifies the influence of the HOMO–LUMO gap in the EDA complexes (see Chart 3), and thus provides electron-transfer activation as the viable mechanistic basis for the aromatic nitration. Indeed, the graphic summary in Fig. 18 for toluene nitration depicts the isomeric composition of *o*-, *m*- and *p*-nitrotoluene to be singularly invariant over a wide range of substrate selectivities (k/k_o based on the benzene

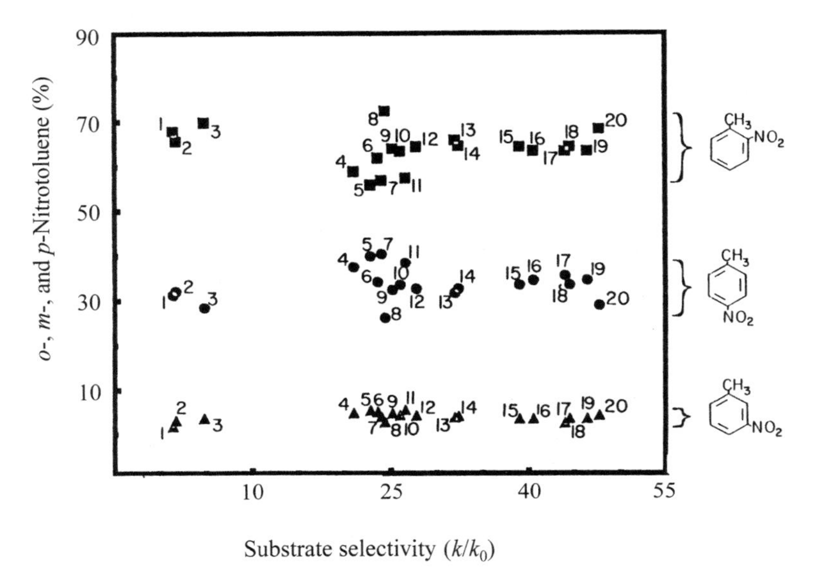

Fig. 18 Invariant composition of nitrotoluene isomers obtained from various nitrating agents, with their (varying) reactivities indicated by the substrate selectivity (k/k_0 for toluene relative to benzene). The nitrating agents are identified by numbers in footnote 49 of ref. 235a. Reproduced with permission from Ref. 235a.

reference[235a]) with different nitrating agents. Thus within minor deviations ($\pm 10\%$) the same distribution of *ortho*, *meta* and *para* isomers is formed irrespective of the nitrating agent (ranging from the very reactive NO_2^+ to the relatively unreactive *p*-MeOPyNO$_2^+$ at the other extreme). In other words, there is a complete decoupling of the product-forming step from the rate-limiting activation of electrophilic aromatic nitration. Such a clear violation of the reactivity/selectivity principle can only arise when there is at least one reactive intermediate such as the ion-radical pair, as predicted by the electron-transfer paradigm (Scheme 1).

B. *Tetranitromethane.* Tetranitromethane forms colored charge-transfer (CT) complexes with a variety of organic donors such as substituted benzenes, naphthalenes, anthracenes, enol silyl ethers, olefins, etc. For example, an orange solution is instantaneously obtained upon exposure of a colorless solution of methoxytoluene (MT) to tetranitromethane (TNM),[237] i.e.,

$$\text{MT} + C(NO_2)_4 \overset{K_{EDA}}{\rightleftharpoons} [\text{MT}, C(NO_2)_4]_{EDA}$$

The [ArH, TNM] complexes persist for prolonged periods at room temperature if the solutions are carefully protected from adventitious light.

Nitration versus alkylation. Upon the CT irradiation of an orange solution of the charge-transfer complex, the color bleaches rapidly, and either an aromatic nitration product (i.e. 3-nitro-4-methoxytoluene) or an aromatic alkylation product (i.e. 3-trinitromethyl-4-methoxytoluene) is obtained in high yield depending on the reaction conditions summarized in Scheme 22.[41c]

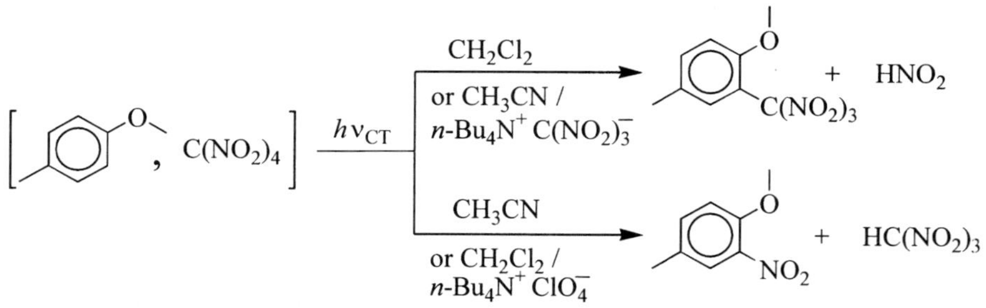

$$[\text{MT}, C(NO_2)_4] \xrightarrow{h\nu_{CT}} \begin{cases} \xrightarrow{CH_2Cl_2 \text{ or } CH_3CN / n\text{-}Bu_4N^+ C(NO_2)_3^-} \text{3-}C(NO_2)_3\text{-MT} + HNO_2 \\ \xrightarrow{CH_3CN \text{ or } CH_2Cl_2 / n\text{-}Bu_4N^+ ClO_4^-} \text{3-}NO_2\text{-MT} + HC(NO_2)_3 \end{cases}$$

Scheme 22

Electron-transfer activation. Time-resolved spectroscopy has established that the irradiation of the [MT, $C(NO_2)_4$] complex generates the contact ion pair,[237] i.e.,

$$[\text{MT}, \ C(NO_2)_4] \underset{}{\overset{h\nu_{CT}}{\rightleftharpoons}} \text{MT}^{+\bullet} C(NO_2)_4^{-\bullet} \tag{80}$$

The rapid (mesolytic) fragmentation of the resulting $C(NO_2)_4^{-\bullet}$ in equation (80) leads to an ion-radical triad[238] (equation 81).

$$\text{MT}^{+\bullet} C(NO_2)_4^{-\bullet} \xrightarrow{\text{fast}} \text{MT}^{+\bullet} NO_2^{\bullet}, C(NO_2)_3^- \tag{81}$$

The explanation for the solvent and salt effect in Scheme 22 lies in the dynamics of the photogenerated ion-radical triad in equation (81). Thus, the ion-pair annihilation is favored in nonpolar solvents such as dichloromethane to afford the alkylation product[237] (equation 82).

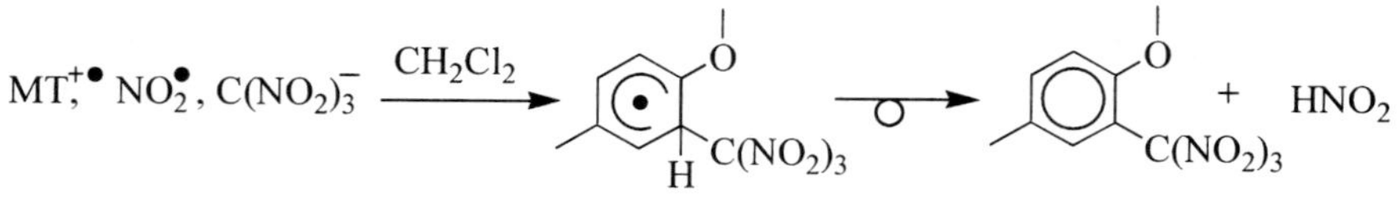

$$\text{MT}^{+\bullet} NO_2^{\bullet}, C(NO_2)_3^- \xrightarrow{CH_2Cl_2} [\text{H, } C(NO_2)_3\text{-adduct radical}] \xrightarrow{-HNO_2} \text{3-}C(NO_2)_3\text{-MT} + HNO_2 \tag{82}$$

On the other hand, in the polar acetonitrile diffusive separation leads to the free cation radical and $NO_2^{\bullet}$ which then undergo (homolytic) collapse. The subsequent deprotonation of the resulting Wheland intermediate leads to the nitration product[237] (equation 83).

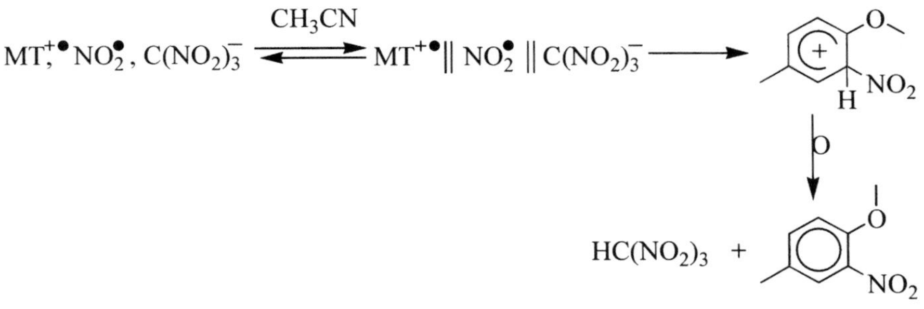

$$MT^{+\bullet}, NO_2^{\bullet}, C(NO_2)_3^- \underset{}{\overset{CH_3CN}{\rightleftharpoons}} MT^{+\bullet} \| NO_2^{\bullet} \| C(NO_2)_3^- \longrightarrow \text{[Wheland intermediate, } NO_2 \text{, H]} \longrightarrow HC(NO_2)_3 + \text{[nitro product]} \quad (83)$$

Such competition between ion-pair collapse of $MT^{+\bullet}$, $C(NO_2)_3^-$ and the radical-pair collapse of $MT^{+\bullet}$, $NO_2^{\bullet}$ is also readily modulated by the addition of inert salt.[14] The description of the solvent and salt effects in equations (82) and (83) is further confirmed by direct kinetics analysis of the decay of the cation radical $MT^{+\bullet}$ on the nanosecond/microsecond timescale.

Nitration versus oxidative dealkylation with nitrogen dioxides. The reaction of various substituted hydroquinone ethers with nitrogen oxides leads to either oxidation (i.e. 1,4-benzoquinones) or nitration (i.e. nitro-*p*-dimethoxybenzenes) depending on the reaction conditions[239] (equation 84).

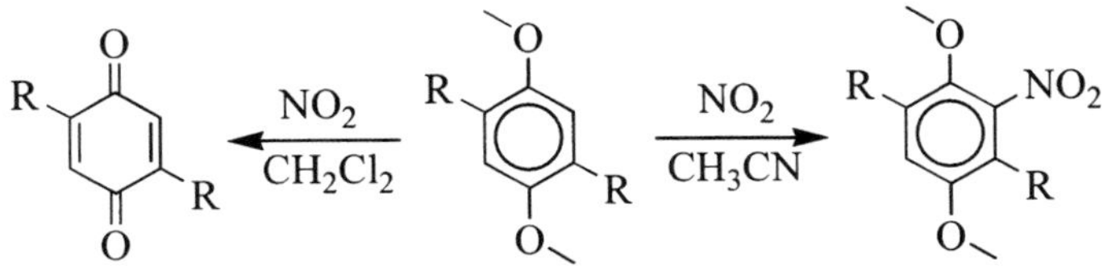

$$\text{[quinone]} \xleftarrow[CH_2Cl_2]{NO_2} \text{[R, R-dimethoxybenzene]} \xrightarrow[CH_3CN]{NO_2} \text{[nitro, R, R-dimethoxybenzene]} \quad (84)$$

The stoichiometries for the nitration and oxidative dealkylation with nitrogen dioxide are established both in dichloromethane and acetonitrile,[239] i.e.,

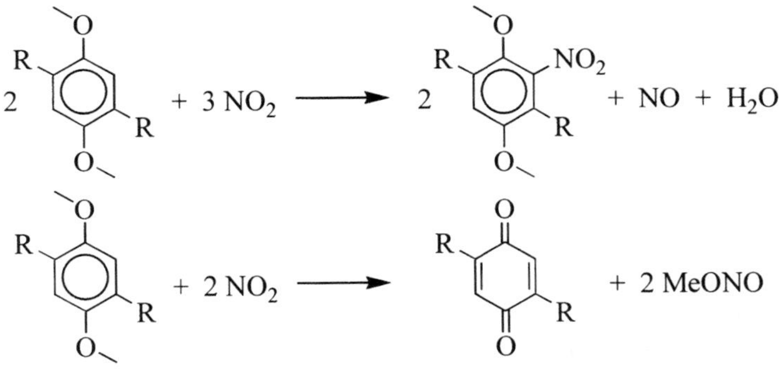

$$2\ \text{[dimethoxybenzene]} + 3\ NO_2 \longrightarrow 2\ \text{[nitro-dimethoxybenzene]} + NO + H_2O$$

$$\text{[dimethoxybenzene]} + 2\ NO_2 \longrightarrow \text{[quinone]} + 2\ MeONO$$

Electron-transfer activation. The observation of intense coloration upon mixing the solutions of hydroquinone ether MA and nitrogen dioxide at low temperature derives from the transient formation of $MA^{+\bullet}$ cation radical, as confirmed by the spectral comparison with the authentic sample. The oxidation of MA to the corresponding cation radical is effected by the nitrosonium oxidant, which is spontaneously generated during the arene-induced disproportionation of nitrogen dioxide,[239] i.e.,

$$\text{(MA)} + 2NO_2 \rightleftharpoons [MA, NO^+]\,NO_3^- \rightleftharpoons MA^{+\bullet}, NO_3^-, NO^\bullet$$

The disproportionation of nitrogen dioxide has been independently confirmed by the direct observation of nitrosonium cation as the EDA complex of hexamethybenzene by spectroscopic means as well as by the spectral comparison with the authentic [HMB, NO^+] complex,[240] i.e.,

$$HMB + 2NO_2 \rightleftharpoons [HMB, NO^+]_{EDA} \rightleftharpoons HMB + NO^+$$

The explanation for the dichotomy between aromatic nitration versus dealkylative oxidation in equation (84) lies in the dynamics of ion-radical triad (which is predictably modulated by solvent polarity and added inert salt). For example, the nonpolar dichloromethane favors aromatic nitration via a radical-pair collapse of $ArH^{+\bullet}$, $NO_2^\bullet$,[239] i.e.,

$$ArH^{+\bullet} + NO_2^\bullet \longrightarrow [Ar(H)NO_2]^+ \xrightarrow{-H^+} ArNO_2$$

whereas the ion-pair annihilation of $ArH^{+\bullet}$, NO_3^- in polar acetonitrile ultimately leads to oxidative dealkylation,[239] i.e.,

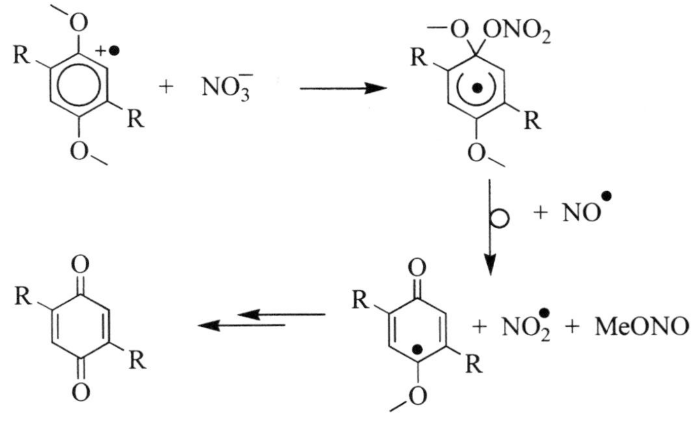

Aromatic nitrosation

Aromatic nitrosation with nitrosonium (NO^+) cation – unlike electrophilic nitration with nitronium (NO_2^+) cation – is restricted to very reactive (electron-rich) substrates such as phenols and anilines.[241] Electrophilic nitrosation with NO^+ is estimated to be about 14 orders of magnitude less effective than nitration with NO_2^+.[242] Such an unusually low reactivity of NO^+ toward aromatic donors (as compared to that of NO_2^+) is not a result of the different electron-acceptor strengths of these cationic acceptors since their (reversible) electrochemical reduction potentials are comparable. In order to pinpoint the origin of such a reactivity difference, let us examine the nitrosation reaction in the light of the donor–acceptor association and the electron-transfer paradigm as follows.

Donor-acceptor association. It is experimentally well established that nitrosonium cation forms vividly colored charge-transfer complexes with a wide variety of aromatic donors[243] (equation 85).

$$\mathrm{ArH} + \mathrm{NO}^+ \underset{}{\overset{K_{\mathrm{EDA}}}{\rightleftharpoons}} [\mathrm{ArH}, \mathrm{NO}^+] \tag{85}$$

For example, benzene, toluene, xylene, mesitylene, durene, and hexamethylbenzene generate yellow to dark purple colors when added to colorless solutions of a NO^+ salt. A careful spectroscopic evaluation establishes that the values of K_{EDA} in equation (85) are highly dependent on the donor strength of the aromatic substrate, increasing dramatically from 0.5 M^{-1} with benzene to 31 000 M^{-1} for hexamethylbenzene.[243] Moreover, exceptionally large formation constants (K_{EDA}) allow ready isolation of crystalline [ArH, NO^+] complexes, and the molecular structures of a number of these highly-colored complexes have been established by X-ray crystallography.[243,244] Indeed, the strong CT interaction is evident in the centrosymmetric structures of [ArH, NO^+] in which the nitrosonium cation lies within the pocket of the van der Waals radius of the arene moiety, as shown by the space-filling representations in Chart 7.

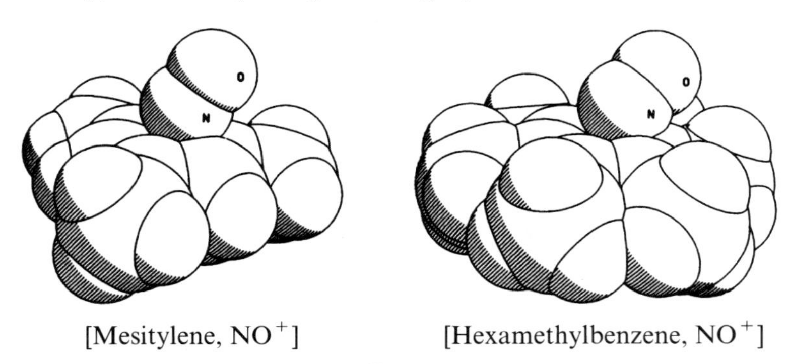

Chart 7
Reproduced with permission from Refs 243 and 244.

A close inspection of the structural data of the [ArH, NO^+] complexes shows considerable lengthening of the aromatic C—C bonds and shortening of the N—O bond compared to those of the single components. This suggests a sizeable degree of charge transfer in the ground-state complexes. Indeed, the comparison of the infrared N—O stretching frequency of the pure [$NO^+PF_6^-$] salt ($\nu_{NO^+} = 2280\ cm^{-1}$), [hexamethylbenzene, NO^+] complex ($\nu_{HMB\text{–}NO^+} = 1880\ cm^{-1}$), and free neutral $NO^{\bullet}$ ($\nu_{NO} = 1876\ cm^{-1}$) confirms a nearly complete charge transfer in the ground state.[243]

Thermal electron transfer. The high degree of charge-transfer character in [ArH, NO^+] complexes is consistent with the fact that a variety of electron-rich aromatic donors undergo reversible electron transfer (in the dark) to form the corresponding cation–radical pair[239] (equation 86).

$$\underset{\text{(CT)}}{[\text{ArH},\ \text{NO}^+]} \ \underset{-78^{\circ}\text{C}}{\overset{25^{\circ}\text{C}}{\rightleftharpoons}} \ \underset{\text{(ET)}}{\text{ArH}^{+\bullet} + \text{NO}^{\bullet}} \tag{86}$$

The thermal electron-transfer (ET) via the charge-transfer (CT) equilibrium depicted in equation (86) is established by temperature dependent (UV-vis) spectral studies. For example, an equimolar mixture of hydroquinone ether MA and NO^+ salt at low temperatures (-78°C) immediately forms the purple [MA, NO^+] charge-transfer complex ($\lambda_{max} = 360$ nm). However, upon warming the solution an orange-red color of the $MA^{+\bullet}$ cation radical ($\lambda_{max} = 518$ nm) develops, and the intensity increases with increasing temperature. Moreover, the identity of liberated $NO^{\bullet}$ is confirmed by the quantitative analysis of the head gas with a diagnostic N—O stretching band at $1876\ cm^{-1}$ in the infrared

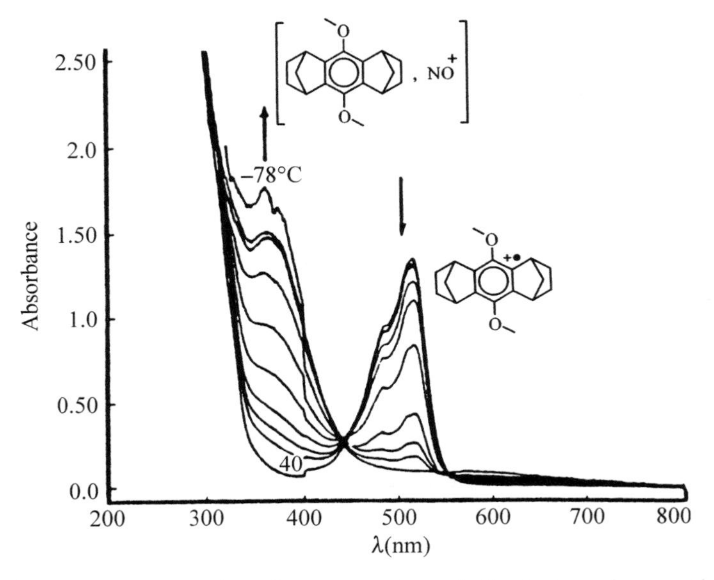

Fig. 19 Temperature-dependent interconversion of the hydroquinone ether (MA) cation radical ($\lambda_{max} = 518$ nm) and its EDA complex with nitrosonium cation ($\lambda_{max} = 360$ nm) according to equation (86) in the temperature range from $+40°C$ to $-78°C$ (incrementally).

spectrum. The quantitative effect of this dramatic color change (which is completely reversible over multiple cooling/warming cycles) is illustrated in Fig. 19. Moreover, the molecular structures of the CT and ET products (i.e., [MA, NO^+] complex and $MA^{+\bullet}$) in equation (86) are determined by X-ray crystallography[245] and clearly demonstrate that thermal activation of the charge-transfer complex yields quantitatively the ion-radical pair (i.e. $MA^{+\bullet}$ and $NO^\bullet$).

Nitrosation. Nitrosonium tetrafluoroborate readily effects the nitrosation of the various polymethylbenzenes and anisoles in a thermal (dark) reaction,[246] i.e.,

$$\text{ArR} + NO^+ \rightleftharpoons [\text{ArR}, NO^+] \longrightarrow \text{R-C}_6\text{H}_4\text{-NO} + H^+$$

Although a detailed kinetic analysis of the nitrosation reaction with $NO^+BF_4^-$ is not available, the time/conversions with various aromatic donors suggest that the reactivity does not follow the variation of the ionization potentials of the aromatic donors as observed in other electrophilic substitutions such as nitration. Nevertheless, the nitrosation of various aromatic donors results from the pre-equilibrium formation of charge-transfer complexes, i.e.,

$$\mathrm{ArH} + \mathrm{NO}^+ \rightleftharpoons [\mathrm{ArH}, \mathrm{NO}^+]_{\mathrm{EDA}}$$

As described in equation (86), the thermal activation of the EDA complex leads to the ion-radical pair, which subsequently collapses to the Wheland intermediate (in a reaction sequence similar to that described for nitration) (equation 87)

$$[\mathrm{ArH}, \mathrm{NO}^+] \rightleftharpoons \mathrm{ArH}^{+\bullet}, {}^{\bullet}\mathrm{NO} \longrightarrow [\mathrm{C_6H_6^+}(\mathrm{H})(\mathrm{NO})] \quad (87)$$

The Wheland intermediate in equation (87) is identified by time-resolved spectroscopy as follows.[247] Laser excitation of the EDA complex of NO^+ with hexamethylbenzene in dichloromethane immediately generates two transient species as shown in the deconvoluted spectrum in Fig. 20. The absorption band at $\lambda_{\max} = 495$ nm is readily assigned to the cation radical of

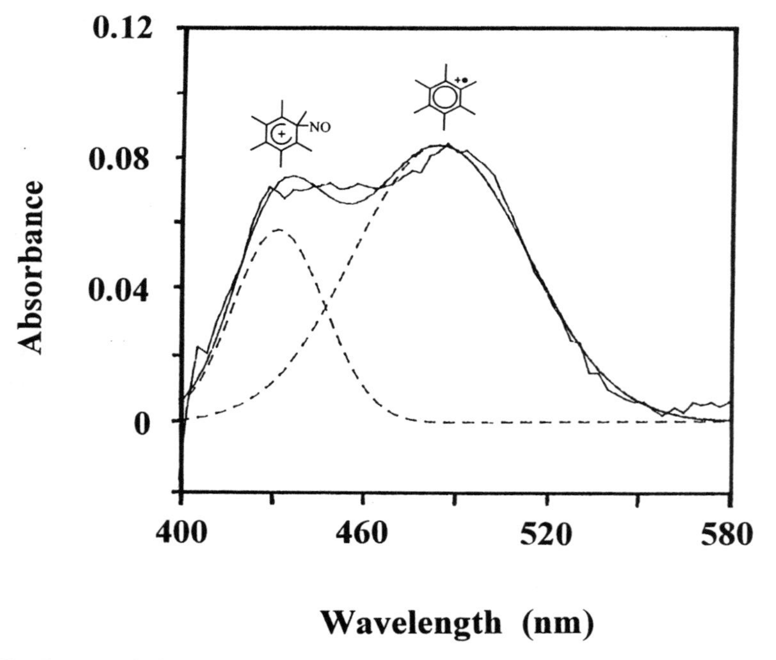

Fig. 20 Deconvolution of the transient spectrum obtained upon the application of a 25-ps laser pulse to a solution of [hexamethylbenzene, NO^+] charge-transfer complex showing the Wheland intermediate (430 nm) and the hexamethylbenzene cation radical (495 nm). Courtesy of S.M. Hubig and J.K. Kochi, unpublished results.

hexamethylbenzene, whereas the 430-nm absorption is ascribed to the Wheland intermediate $ArH(NO)^+$ on the basis of the spectral comparison with other Wheland-type intermediates such as $ArH(H)^+$, $ArH(Cl)^+$, $ArH(Me)^+$, $ArH(NO_2)^+$, etc. Femtosecond time-resolved experiments reveal that the 430-nm component is formed from the cation radical by rapid coupling with $NO^\bullet$,[247] i.e.,

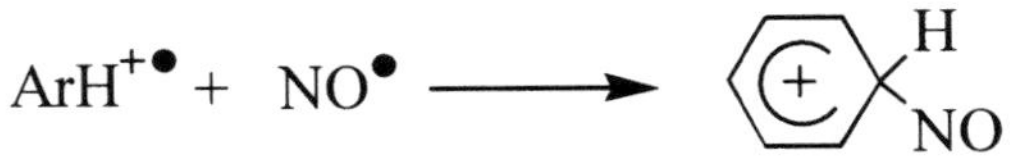

$$ArH^{+\bullet} + NO^\bullet \longrightarrow [\text{Wheland intermediate: } ArH(NO)^+]$$

However, the Wheland intermediate decays on the early nanosecond time scale to restore the original EDA complex. The observation of the EDA complexes, the ion-radical pair, as well as the Wheland intermediate $ArH(NO)^+$ points to the reaction scheme for thermal nitrosations shown in Scheme 23.

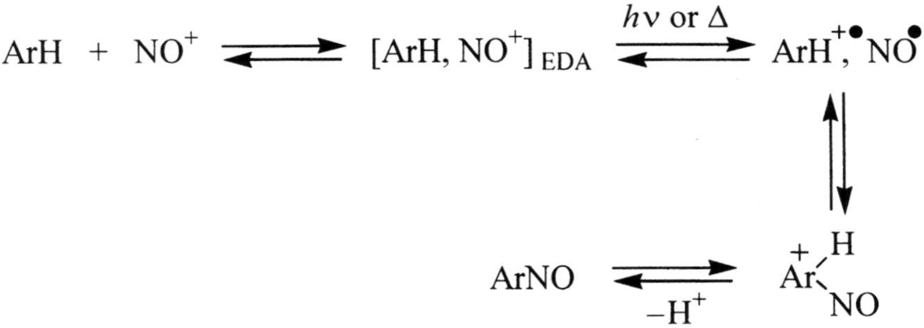

$$ArH + NO^+ \rightleftharpoons [ArH, NO^+]_{EDA} \underset{}{\overset{h\nu \text{ or } \Delta}{\rightleftharpoons}} ArH^{+\bullet}, NO^\bullet$$

$$ArH^{+\bullet}, NO^\bullet \rightleftharpoons {}^{+}Ar(H)(NO)$$

$$ArNO \underset{-H^+}{\rightleftharpoons} {}^{+}Ar(H)(NO)$$

Scheme 23

The electron-transfer formulation in Scheme 23 suggests that the efficiency of nitrosation is the direct result of the competition between deprotonation of the Wheland intermediate versus its breakdown to the original EDA complex via the ion-radical pair, i.e.,

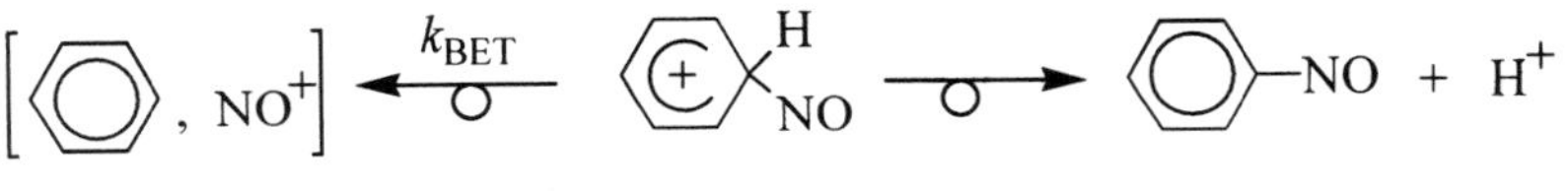

$$[C_6H_6, NO^+] \xleftarrow{k_{BET}} C_6H_6(H)(NO)^+ \longrightarrow C_6H_5\text{–}NO + H^+$$

Note that a fragmentation of Wheland intermediate into the corresponding ion-radical pair has been recently observed with various chloroarenium cations,[248] i.e.,

$$(\text{MA-Cl}^{+}) \rightleftharpoons (\text{MA}^{+\bullet}) + \text{Cl}^{\bullet}$$

The formation of the Wheland intermediate from the ion-radical pair as the critical reactive intermediate is common in both nitration and nitrosation processes. However, the contrasting reactivity trend in various nitrosation reactions with NO^+ (as well as the observation of substantial kinetic deuterium isotope effects) is ascribed to a rate-limiting deprotonation of the reversibly formed Wheland intermediate. In the case of aromatic nitration with NO_2^+, deprotonation is fast and occurs with no kinetic (deuterium) isotope effect. However, the nitrosoarenes (unlike their nitro counterparts) are excellent electron donors as judged by their low oxidation potentials as compared to parent arene.[246] As a result, nitrosoarenes are also much better Brønsted bases[249] than the corresponding nitro derivatives, and this marked distinction readily accounts for the large differentiation in the deprotonation rates of their respective conjugate acids (i.e., Wheland intermediates).

ELECTRON-TRANSFER ACTIVATION IN NITROGEN DIOXIDE REACTIVITY TOWARD ORGANIC DONORS

Nitrogen dioxide ($NO_2^{\bullet}$) reacts with a wide variety of functional groups and it is the reagent of choice for a number of synthetic transformations. For example, the selective oxidation of sulfide to sulfoxide by $NO_2^{\bullet}$ forms the basis for the commercial production of dimethyl sulfoxide (from dimethyl sulfide) via a catalytic procedure (see below).[250] Some representative examples of oxidative transformations carried out with $NO_2^{\bullet}$ are presented in Chart 8.

The organic substrates in Chart 8 can be divided into two main categories in which (i) the oxidation of olefins, sulfides, and selenides involves oxygen atom transfer to yield epoxides, sulfoxides, and selenoxides, respectively, whereas (ii) the oxidation of hydroquinones and quinone dioximes formally involves loss of two electrons and two protons to yield quinones and dinitrosobenzenes, respectively. In order to provide a unifying mechanistic theme for the seemingly disparate transformations in Chart 8, we note that nitrogen dioxide exists in equilibrium with its dimeric forms, namely, the predominant N—N bonded dimer $O_2N—NO_2$ and the minor N—O bonded isomer $ONO—NO_2$[240] (equation 88).

$$O_2N{-}NO_2 \rightleftharpoons 2\,NO_2^{\bullet} \rightleftharpoons ON{-}ONO_2 \tag{88}$$

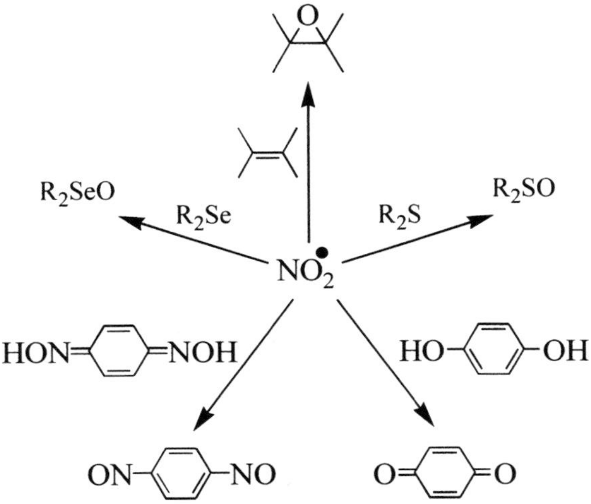

R_2SeO R_2Se R_2S R_2SO

$NO_2^{\bullet}$

HON=⟨⟩=NOH HO–⟨⟩–OH

ON–⟨⟩–NO O=⟨⟩=O

Chart 8

Significantly, the minor isomer in equation (88) undergoes a ready disproportionation to form nitrosonium nitrate, which is stabilized by strong complexation with organic donors[240] (equation 89)

$$\text{ON–ONO}_2 \rightleftarrows \text{NO}^+\,\text{NO}_3^- \overset{\text{D}}{\rightleftarrows} [\text{D, NO}^+]\,\text{NO}_3^- \qquad (89)$$

Indeed, the extent of disproportionation of $NO_2^{\bullet}$ according to equation (89) clearly depends on the donor strength of the aromatic hydrocarbon.[240] For example, hexamethylbenzene which is a strong donor (IP = 7.85 V) promotes the ionization of $NO_2^{\bullet}$ to an extent of 80%; whereas the weaker donor durene (IP = 8.05 V) affords less than 25% ion-pair formation. Furthermore, the resulting NO^+ cation is a powerful electron acceptor (E_{red} = 1.48 V versus SCE) in contrast to $NO_2^{\bullet}$ (E_{red} = 0.25 V versus SCE) and thus readily forms donor/acceptor complexes with a variety of aromatic, olefinic and heteroatom-centered donors. Accordingly, the donor/acceptor complexation and electron-transfer activation are the critical steps in various transformations in Chart 8 as described below.

Oxidation of olefins, sulfides and selenides

The oxidation of olefins,[251] sulfides,[252] and selenides[253] (denoted as D) involves oxygen-atom transfer from nitrogen dioxide to yield epoxides, sulfoxides, and selenoxides, respectively, in excellent yields according to the stoichiometry of equation (90).

$$D + NO_2^{\bullet} \xrightarrow{\text{dark}} DO + NO^{\bullet} \qquad (90)$$

where

$D = R_2S, R_2Se$ or $\rangle\!=\!\langle$ and $DO = R_2SO, R_2SeO$ or (epoxide)

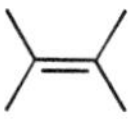

Importantly, the oxidations in equation (90) are considerably accelerated in the presence of added salt such as $n\text{-}Bu_4N^+PF_6^-$ and markedly retarded in nonpolar solvents (such as CCl_4 and hexane) as well as in the presence of added nitrate salts, which suggests the critical role of ionic intermediates (viz. $NO^+NO_3^-$).[251–253] Accordingly, the ionic mechanism for the oxygen-atom transfer from $NO_2^{\bullet}$ to various donors is outlined in Scheme 24.

$$D + 2\,NO_2^{\bullet} \overset{K_{EDA}}{\rightleftharpoons} [D, NO^+]\,NO_3^- \qquad (91)$$

$$[D, NO^+] \overset{k_{ET}}{\rightleftharpoons} D^{+\bullet} + NO^{\bullet} \qquad (92)$$

$$D^{+\bullet} + NO_2^{\bullet} \xrightarrow{\text{fast}} DO + NO^+ \qquad (93)$$

$$NO^+ + NO_3^- \xrightarrow{\text{fast}} N_2O_4\ (2\,NO_2^{\bullet}) \qquad (94)$$

Scheme 24

The donor-induced disproportionation in equation (91) leads to the EDA complex, i.e., $[D, NO^+]NO_3^-$ as the (first) directly observable intermediate. The critical role of the nitrosonium EDA complex in the electron-transfer activation in equation (92) is confirmed by the spectroscopic observation of the cation-radical intermediates (i.e., $D^{+\bullet}$) as well as by an alternative (low-temperature) photochemical activation with deliberate irradiation of the charge-transfer band[252] (equation 95).

$$[R_2S\,,\,NO^+]\,NO_3^- \xrightarrow{h\nu_{CT}} R_2SO\,,\ \text{etc.} \qquad (95)$$

(Note that under these conditions the thermal reaction in equation (90) is too slow to compete.) Finally, the stoichiometry for the oxygen-atom transfer from $NO_2^{\bullet}$ to the donor cation radical in equation (93) is independently established by the reaction of isolated cation radical intermediates with $NO_2^{\bullet}$.[251,252]

Catalytic oxidation with dioxygen. The well-known aerial oxidation of the nitric oxide ($NO^{\bullet}$) produced in the stoichiometric oxidations in equation (90) to nitrogen dioxide ($NO_2^{\bullet}$) forms the basis for the $NO_2^{\bullet}$-catalyzed oxidation of various donors with dioxygen[251–253] (equation 96).

$$D + \tfrac{1}{2}O_2 \xrightarrow{[NO_2^{\bullet}]} DO \tag{96}$$

Moreover, the aerial oxidation in equation (96) can be catalyzed by $NO_2^{\bullet}$, NO^+, or $NO^{\bullet}$ with comparable catalytic efficiency, which confirms the ready interchangeability of the various NO_x species in Scheme 25.

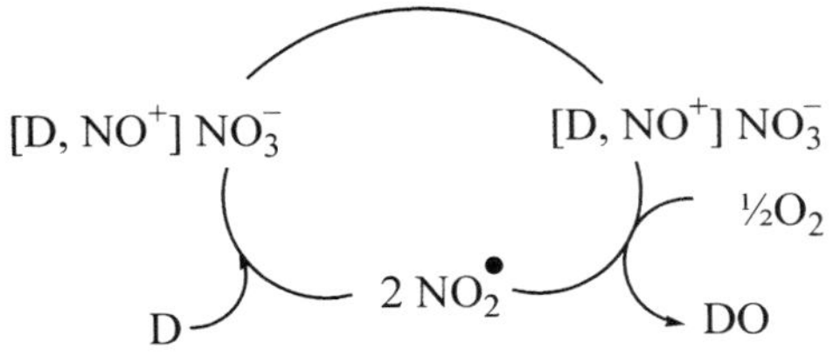

[D, NO$^+$] NO$_3^-$ ⟶ [D, NO$^+$] NO$_3^-$; D, 2 NO$_2^{\bullet}$, ½O$_2$, DO

Scheme 25

Oxidation of hydroquinones and quinone dioximes

The oxidation of hydroquinones[254] and quinone dioximes[255] (denoted as QH_2) involves removal of two electrons and two protons. This redox stoichiometry is experimentally established both in the stoichiometric oxidations with $NO_2^{\bullet}$ and with two equivalents of nitrosonium cation (equations 97a,b).

$$QH_2 + NO_2^{\bullet} \longrightarrow Q + NO^{\bullet} + H_2O \tag{97a}$$

$$QH_2 + 2\,NO^+ \longrightarrow Q + 2\,NO^{\bullet} + 2H^+ \tag{97b}$$

Since the substituted hydroquinones and quinone dioximes are better electron donors than hexamethylbenzene (as established by cyclic voltammetric studies), donor-induced disproportionation (to generate NO^+ NO_3^-) is even more favored. Furthermore, either two successive one-electron oxidations of hydroquinone (or quinone dioxime) by NO^+ followed by the loss of two protons from the dication or two sequential oxidation/deprotonation steps complete the oxidative transformation in equation (97). Importantly, the ready aerial oxidation of $NO^{\bullet}$ to $NO_2^{\bullet}$ provides the basis for the nitrogen oxide catalysis of hydroquinone (or quinone dioxime) autoxidation as summarized in Scheme 26.

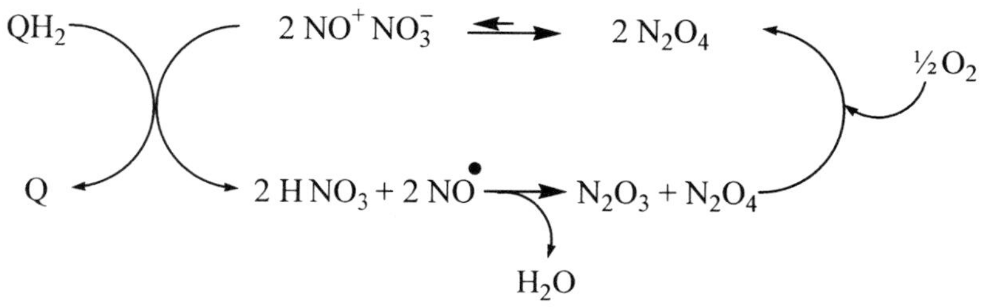

$$QH_2 \rightarrow Q;\quad 2\,NO^+NO_3^- \rightleftharpoons 2\,N_2O_4;\quad 2\,HNO_3 + 2\,NO^{\bullet} \rightarrow N_2O_3 + N_2O_4 + H_2O;\quad N_2O_3 + N_2O_4 + \tfrac{1}{2}O_2 \rightarrow 2\,N_2O_4$$

Scheme 26

Moreover, the efficiency of catalytic autoxidation in Scheme 26 is also attributed to the short lifetime of the hydroquinone cation radical ($t < 10^{-10}$ s^{-1})[254], which renders the sequential electron-transfer/proton-transfer cycles extremely efficient.

10 Quantitative aspects of the electron-transfer paradigm: the FERET

The electron-transfer paradigm in Scheme 1 (equation 8) is subject to direct experimental verification. Thus, the deliberate photoactivation of the pre-equilibrium EDA complex via irradiation of the charge-transfer absorption band ($h\nu_{CT}$) generates the ion-radical pair, in accord with Mulliken theory (equation 98).

$$D + A \underset{}{\overset{K_{EDA}}{\rightleftharpoons}} [D, A]_{EDA} \xrightarrow{h\nu_{CT}} D^{+\bullet}, A^{-\bullet} \qquad (98)$$

The use of short (fs) laser pulses allows even highly transient ion-radical pairs with lifetimes of $\tau \simeq 10^{-12}$ s to be detected, and their subsequent (dark) decay to products is temporally monitored through the sequential spectral changes. As such, *time-resolved (ps) spectroscopy provides the technique of choice for establishing the viability of the electron-transfer paradigm.* This photochemical (ET) mechanism has been demonstrated for a variety of donor–acceptor interactions, as presented in the foregoing section.

The extension of the same mechanistic reasoning to the corresponding thermal process (carried out in the dark) is not generally rigorous. Most commonly, the adiabatic electron-transfer step (k_{ET}) is significantly slower than the fast back electron transfer and follow-up reactions (k_f) described in Section 7, and the pseudo-steady-state concentration is too low for the ion-radical pair to be directly observed (equation 99).

$$D + A \overset{K_{EDA}}{\rightleftharpoons} [D, A]_{EDA} \xrightarrow{k_{ET}} D^{+\bullet} + A^{-\bullet} \xrightarrow{k_f} \qquad (99)$$

As a result, any mechanistic deductions applicable to the electron-transfer paradigm for the thermal reactivities[19a] of the donor/acceptor couple must necessarily be based on indirect methods, as follows.

It is first mandatory that the products (and selectivities) obtained from the donor–acceptor reaction in the dark are the same as obtained by charge-transfer activation at sufficiently low temperatures where the thermal reaction does not compete. Such a parallel between thermal and charge-transfer activation has been established for electrophilic nitration of arenes and enol silyl ethers, aromatic osmylation, etc. presented in the foregoing section. Short of the actual observation of the ion-radical pair, this constitutes the best (indirect) evidence for the electron-transfer pathway in the thermal reaction of a donor/acceptor couple. In the absence of such evidence, it is not possible to present unequivocal evidence for the applicability of the electron-transfer paradigm to thermal reactivities of a given donor/acceptor couple. Accordingly, let us take an alternative approach and consider how the electron-transfer paradigm can be utilized as a working mechanistic tool.

As a 3-step mechanism, the electron-transfer paradigm provides a pair of discrete intermediates [D, A] and $D^{+\bullet}$, $A^{-\bullet}$ for the prior organization and the activation, respectively, of the donor and the acceptor. The quantitative evaluation of these intermediates would allow the overall second-order reaction (k_2) to be determined. Although the presence of [D, A] does not necessarily imply its transformation to $D^{+\bullet}$, $A^{-\bullet}$, a large number and variety of donor/acceptor couples showing transient charge-transfer absorptions associated with [D, A] have now been identified. In each case, the product can be predicted from the expected behavior of the individual ion radicals $D^{+\bullet}$ and $A^{-\bullet}$. Consider for example, the labile 1:1 benzene complex with bromine that has been isolated at low temperatures and characterized crystallographically (Chart 9).[256]

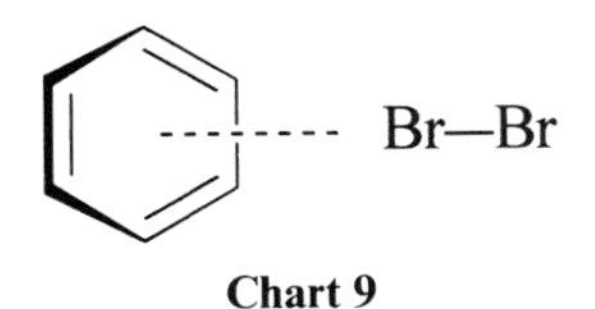

Chart 9

Upon standing at room temperature (in the dark), the charge-transfer absorption (λ_{CT} = 286 nm) bleaches rapidly and bromobenzene (and HBr) is formed in quantitative yields. The electron-transfer paradigm in Scheme 1 is readily invoked since $Br_2^{-\bullet}$ is mesolytically unstable and $C_6H_6^{+\bullet}$ is susceptible to homolytic addition[257] (Scheme 27).

The first question to be raised about the viability of Scheme 27 is that the driving force ($-\Delta G_{ET}$) for the charge-transfer step (equation 100) is insufficient to account for the fast rate of aromatic bromination.

$$[C_6H_6\,,Br_2]_{EDA} \xrightarrow{-\Delta G_{ET}} C_6H_6^{+\bullet} + Br_2^{-\bullet} \quad (100)$$

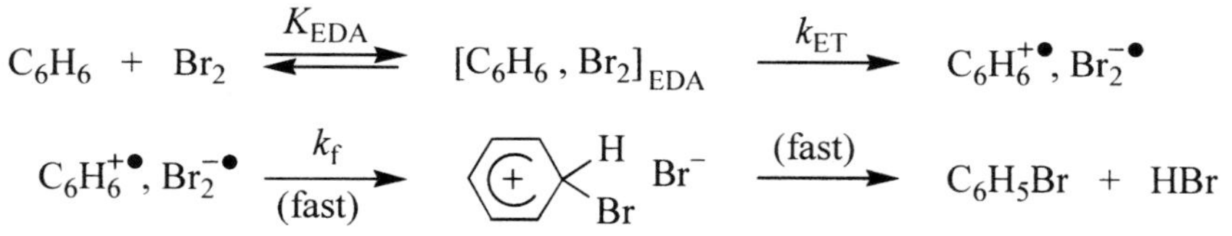

$$C_6H_6 + Br_2 \underset{}{\overset{K_{EDA}}{\rightleftharpoons}} [C_6H_6, Br_2]_{EDA} \xrightarrow{k_{ET}} C_6H_6^{+\bullet}, Br_2^{-\bullet}$$

$$C_6H_6^{+\bullet}, Br_2^{-\bullet} \xrightarrow[\text{(fast)}]{k_f} \text{(Wheland Intermediate: } C_6H_6Br^+\text{)}\ Br^- \xrightarrow{\text{(fast)}} C_6H_5Br + HBr$$

(Wheland Intermediate)

Scheme 27

For such an electron transfer, the free-energy change is related to the oxidation potential of benzene and the reduction potential of bromine that are estimated as $E^0_{ox} = 3.0$ V and $E^0_{red} = 0.4$ V versus NHE, respectively.[8,258] If the free-energy change for equation (100) is taken simply as the difference between these electrode potentials, the magnitude of $-\Delta G_{ET} = -60$ kcal mol^{-1} is clearly too endergonic to be consistent with the fast bromination of benzene. Such an evaluation of the driving force would be (more or less) valid if the electron transfer in equation (100) occurred via an outer-sphere mechanism in which there was minimal (electronic) interaction between C_6H_6 and Br_2 in the transition state.[259] However, the quantitative measurement of the charge-transfer absorption leading to the formation of the 1:1 EDA complex [C_6H_6, Br_2] with the formation constant of $K_{EDA} = 3$ M^{-1} indicates that the donor/acceptor (inner-sphere) pre-organization occurs even in the ground state. As such, the ion radicals $C_6H_6^{+\bullet}$ and $Br_2^{-\bullet}$ are born as a contact (inner-sphere) ion pair with substantial electronic and electrostatic interaction. The free-energy change for the charge-transfer step is accordingly given by[260]

$$-\Delta G_{ET} = -E^0_{ox} + E^0_{red} + V \tag{101}$$

where the ion-pair (orbital) overlap is evaluated by the electronic coupling matrix element V (or H_{DA}).[261] Since charge-transfer forces are a principal contributor, the magnitude of V can be substantial and render the driving force significantly greater than that simply calculated from E^0_{ox} and E^0_{red}. In other words, the electron-transfer paradigm cannot be discounted on the basis of driving forces based on an outer-sphere model for electron transfer. Moreover, in the absence of an (experimental) measurement of V, the inner-sphere electron transfer in equation (99) cannot be quantitatively evaluated by a direct (kinetics) analysis.

A route to the resolution of this conundrum is provided by the photoactivation of the donor–acceptor complex to the ion-radical pair, as described in equation (98). In this case, the close interrelationship between the photochemically-produced or vertical (nonadiabatic) ion pair formed in equation (98) and the thermally accessed or contact (adiabatic) ion pair in equation (99) is illustrated in Scheme 28, where the asterisk identifies the contact ion-pair in the gas phase and s represents the solvation of ion-radical pair. According to

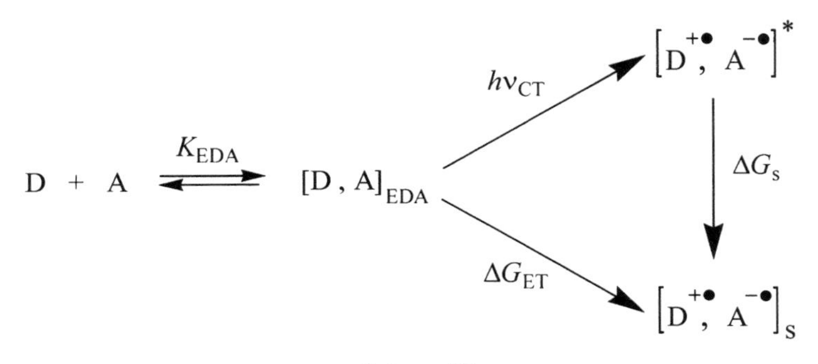

Scheme 28

Mulliken,[6] the energetics of the latter (generated in the course of the photo-irradiation of the charge-transfer absorption band) is given by

$$h\nu_{CT} \cong \mathrm{IP} - \mathrm{EA} + V \tag{102}$$

where the ionization potential (IP) of the donor and electron affinity (EA) of the acceptor have been identified in Section 7.

The energetics in Scheme 28 can be comparatively evaluated in a pair of systems involving either (a) a series of structurally related donors interacting with a single acceptor or (b) a series of structurally related acceptors interacting with a given donor. Under these conditions, the relative free energy change ($\Delta\Delta G_{ET}$) for the adiabatic electron transfer is[262]

$$\Delta\Delta G_{ET} = \Delta h\nu_{CT} + \Delta\Delta G_s \tag{103}$$

and all the changes are taken relative to the reference donor (e.g. C_6H_6 versus a substituted analog ArH). Thus $\Delta h\nu_{CT}$ refers to the spectral shift of the charge-transfer band (e.g., of the [C_6H_6, Br_2] complex relative to [ArH, Br_2] in aromatic bromination), and $\Delta\Delta G_s$ refers to the difference in the solvation energies of $C_6H_6^{+\bullet}$ and $ArH^{+\bullet}$. (Note that the terms due to the electron affinity and anion solvation of Br_2 and $Br_2^{-\bullet}$, respectively, cancel out in this comparison.)

When the energetics of the ion-radical pair are taken as a close approximation to the transition state, the thermodynamic change can be reformulated in terms of relative reactivities, i.e.,

$$\Delta\Delta G_{ET} = -RT \ln k/k_0 = \Delta h\nu_{CT} + \Delta\Delta G_S \tag{104}$$

where $\Delta\Delta G_{ET}$ is the relative activation free energy for the second-order rate constant (k) relative to that (k_0) of the reference donor (or acceptor).

The quantitative formulation of the electron-transfer paradigm in Scheme 1 as the free-energy relationship for electron transfer (FERET)[22] in equation

(104) is illustrated by the linear correlation in Fig. 21a for aromatic bromination according to Scheme 27.[263]

It is also noteworthy that the bromination of various olefins, i.e.,

$$\mathrm{>C{=}C<} \quad + \quad Br_2 \longrightarrow \mathrm{-\overset{Br}{\underset{|}{C}}-\overset{|}{\underset{Br}{C}}-}$$

also adheres to FERET, as shown in Fig. 21a by the separate linear correlation of the rate-limiting charge-transfer activation, i.e.,

$$\mathrm{>C{=}C<} + Br_2 \underset{}{\overset{K_{EDA}}{\rightleftharpoons}} \left[\mathrm{>C{=}C<}, Br_2\right]_{EDA} \xrightarrow{k_{ET}} \mathrm{>C{=}C<}^{+\bullet}, Br_2^{-\bullet}$$

The remarkable generality of FERET to predict the behavior of Br_2 in aromatic bromination as well as in olefin addition is established by the single linear correlation in Fig. 21b when the solvation difference (see equation (103)) between the cation radicals of arene and olefin donors is taken into account.[263]

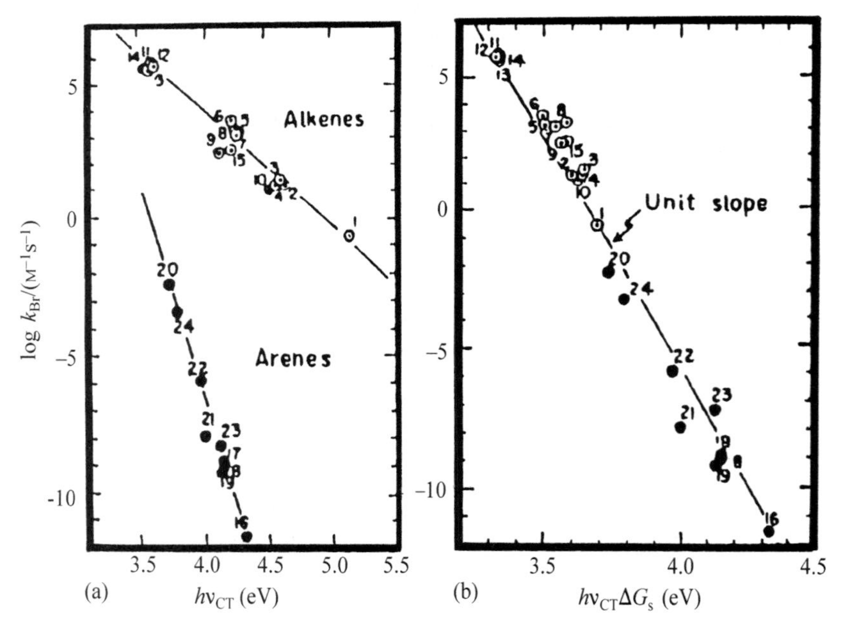

Fig. 21 (a) Correlation of the rates ($\log k_{Br}$) of electrophilic bromination in acetic acid with the CT transition energies ($h\nu_{CT}$) of the bromine complexes of alkenes and arenes identified by the numbers in tables I and II of ref. 263. (b) Unified FERET of electrophilic brominations of arenes and olefins in tables II and III of Ref. 263, with the $h\nu_{CT}$ and cation solvation ΔG_s. Reproduced with permission from Ref. 263.

An even more striking illustration of the electron-transfer paradigm is the comparative behavior of Br_2 and $Hg(OAc)_2$ in olefin addition.[264] Figure 22 shows the divergent and somewhat unsystematic (relative) behavior of these electrophilic reagents. The best way to apply the FERET in this situation is via the direct comparison of the relative reactivity behavior (i.e., $\log k/k_0$) to cancel out the solvation effects. The "buckshot" pattern in Fig. 23a confirms the irregular kinetics in Fig. 22. However, the specific inclusion of the ion-pairing effects of FERET in Fig. 23b shows that the electron-transfer paradigm is the unifying theme in the kinetic behavior of these structurally diverse electrophiles.

The quantitative treatment of the electron-transfer paradigm in Scheme 1 by FERET (equation (104)) is restricted to the comparative study of a series of structurally related donors (or acceptors). Under these conditions, the reactivity differences due to electronic properties inherent to the donor (or acceptor) are the dominant factors in the charge-transfer assessment, and any differences due to steric effects are considered minor. Such a situation is sufficient to demonstrate the viability of the electron-transfer paradigm to a specific type of donor/acceptor behavior (e.g. aromatic substitution, olefin addition, etc.). However, a more general consideration requires that any steric effect be directly addressed,

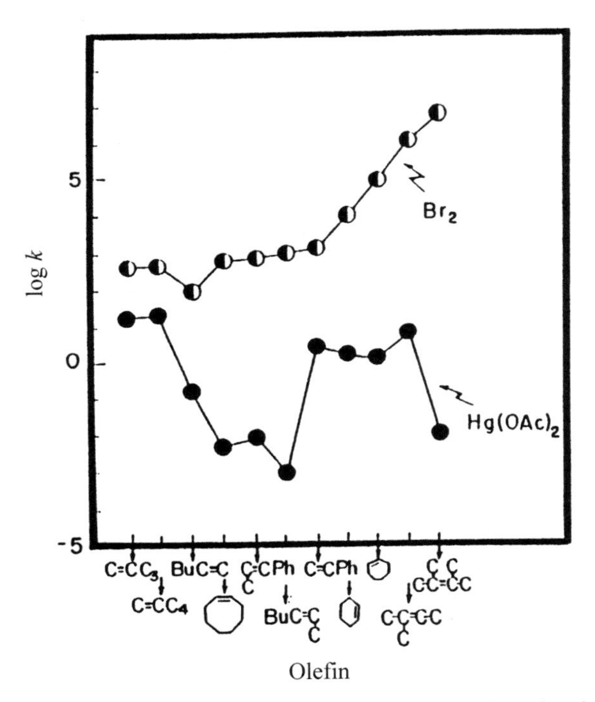

Fig. 22 Divergent trends in the reactivity of various olefins in bromination with Br_2 and in oxymercuration with $Hg(OAc)_2$. Reproduced with permission from Ref. 264.

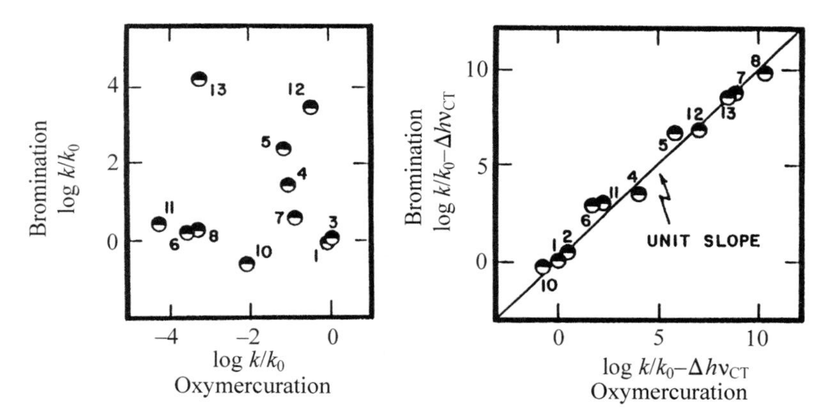

Fig. 23 (a) Comparison of the bromination and oxymercuration data in Fig. 22 plotted as relative reactivities ($\log k/k_0$). (b) Application of FERET according to equation (104) to emphasize that bromination and oxymercuration have common rate-determining activations. Reproduced with permission from Ref. 264.

especially as it affects electron-transfer kinetics. Consider, for example, hexamethylbenzene (HMB) and hexaethylbenzene (HEB) as a pair of aromatic donors with identical E^0_{ox} and IP values.[265] However, HEB is substantially more hindered than HMB owing to an increased van der Waals thickness from the pendant methyl groups that discourage any close cofacial approach to the

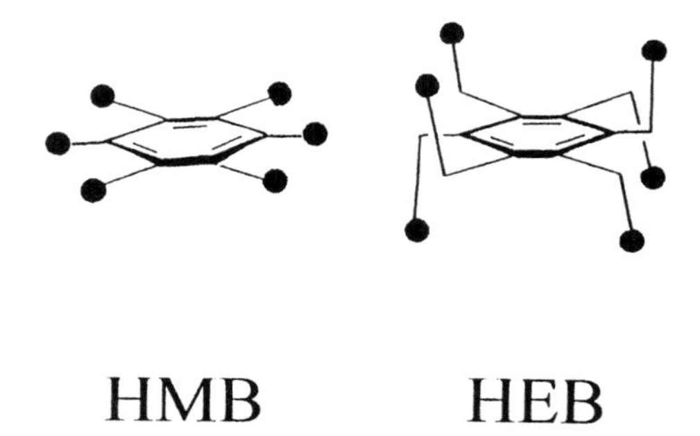

benzenoid (π) chromophore. Direct measurements of the (ET) kinetics via time-resolved spectroscopic techniques show that HEB and HMB behave quite differently toward various (photoactivated) quinone acceptors. Thus the hindered donor HEB is characterized by bimolecular ET rate constants that: (a) are temperature dependent and well-correlated by Marcus theory, (b) show no evidence for the formation of (discrete) encounter complexes, (c) have high dependency of solvent polarity, and (d) have enhanced sensitivity to kinetic salt effects – all diagnostic of outer-sphere electron-transfer mechanisms.[260] In contrast, the unhindered donor HMB is characterized by (a) temperature-independent second-order rate constants that are 10^2 times faster and rather poorly correlated by Marcus theory, (b) weak dependency on solvent polarity, and (c) low sensitivity to kinetic salt effects – all symptomatic of inner-sphere ET mechanisms arising from the pre-equilibrium formation of encounter complexes (K_{EC}) with charge-transfer (inner-sphere) character[260,266] (equation 105)

$$\mathrm{ArH} + \mathrm{Q} \overset{K_{EC}}{\rightleftharpoons} \underset{\text{(Encounter Complex)}}{[\mathrm{ArH}, \mathrm{Q}]_{EC}} \xrightarrow{k_{ET}} \mathrm{ArH}^{+\bullet}; \mathrm{Q}^{-\bullet} \qquad (105)$$

Steric encumbrances that inhibit strong (charge-transfer) coupling between the benzenoid and quinonoid π-systems are critical for the mechanistic change-over. The same applies to the formation of ground-state complexes (i.e., K_{EDA}). For example, HMB and chloranil show an intense CT absorption at $\lambda_{CT} = 532$ nm and readily form a tight donor–acceptor complex with van der Waals separation of 3.5 Å, as revealed by $K_{EDA} = 10\ \mathrm{M}^{-1}$ in solution and the isolation of charge-transfer crystals illustrated in Chart 10.[265] In strong contrast, the closest cofacial approach of chloranil to HEB of more than 4.5 Å precludes complex formation, the characteristic charge-transfer absorption is not observed, and no EDA complex can be isolated. It is especially noteworthy that complex formation with unhindered donors follows the bell-shaped dependence on the ET driving force shown in Fig. 24. The maximum value of $K_{EC} = 67\ \mathrm{M}^{-1}$ is observed in the isoergonic region around $\Delta G_{ET} = 0$ kcal mol^{-1}, and K_{EC} values close to unity are found in the free-energy regions $\Delta G_{ET} > 7$ kcal mol^{-1} and $\Delta G_{ET} < -2$ kcal mol^{-1}.[260]

The quantitative effects of steric encumbrance on the electron-transfer kinetics reinforce the notion that the inner-sphere character of the contact ion pair $D^{+\bullet}, A^{-\bullet}$ is critical to the electron-transfer paradigm in Scheme 1. Charge-transfer bonding as established in the encounter complex (see above) is doubtless an important consideration in the quantitative treatment of the energetics. None the less, the successful application of the electron-transfer paradigm to the

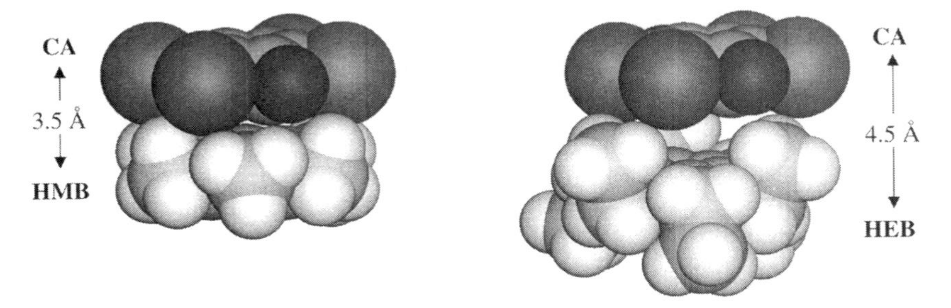

Chart 10
Reproduced with permission from Ref. 260a.

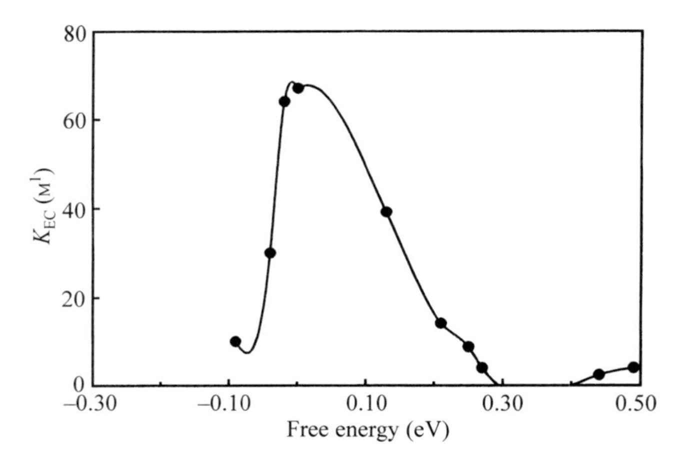

Fig. 24 Bell-shaped free-energy dependence of the formation constants K_{EC} of the encounter complex [ArH, Q*] in dichloromethane. Reproduced with permission from Ref. 260a.

wide spectrum of organic reactions discussed in this article encourages the further development of this mechanistic formulation – particularly by the exploitation of theoretical calculations. By comparison, the conventional electrophilic/nucleophilic formulations involving a direct (one-step) activation process for a second-order (bimolecular) reaction carry with them only limited mechanistic insight of predictive value. Consider, for example, the oxygen-atom transfer from a dioxirane to an olefin,[267]

$$\mathrm{{>}C{=}C{<}} \;+\; \mathrm{{>}C(O_2)} \longrightarrow \mathrm{{>}C(O)C{<}} \;+\; \mathrm{{>}C{=}O}$$

for which various concerted (one-step) mechanisms have been proposed, both experimentally and theoretically.[268] The electron-transfer paradigm for this deceptively simple atom transfer invokes the dioxirane electron acceptor similar to other peroxides with low-lying LUMOs.[269] As such, the formation of the ion-radical pair is followed by a rapid (distonic) ring-opening, and ion-pair collapse[270] (equation 106).

$$\mathrm{C{=}C} + \mathrm{C(O_2)} \rightleftharpoons [\mathrm{C{=}C}, \mathrm{C(O_2)}]_{EDA} \xrightarrow{k_{ET}} \mathrm{C{=}C}^{+\bullet}\ \mathrm{C(O_2)}^{-\bullet} \longrightarrow \mathrm{{}^{+}C{-}C^{\bullet}}\ \ \mathrm{{}^{-}O{-}C{-}O^{\bullet}}, \text{ etc.} \tag{106}$$

The inner-sphere nature of the ion-radical pair may render the biradical (or zwitterionic) intermediate difficult to assess and distinguish from a concerted process. We hope that a systematic study of the olefin-donor properties and dioxirane-acceptor properties, together with an evaluation of the contact ion pair, will provide a more quantitative basis for the applicability of the electron-transfer paradigm to this reaction.

11 Epilogue

The electron-transfer paradigm for chemical reactivity in Scheme 1 (equation 8) provides a unifying mechanistic basis for various bimolecular reactions via the identification of nucleophiles as electron donors and electrophiles as electron acceptors according to Chart 1. Such a reclassification of either a nucleophile/electrophile, an anion/cation, a base/acid, or a reductant/oxidant pair under a single donor/acceptor rubric offers a number of advantages previously unavailable, foremost of which is the quantitative prediction of reaction rates by invoking the FERET in equation (104).

As a mechanistic formulation with a global compass, the virtues of the electron-transfer paradigm can be summarized as follows:

1. Donors (D) and acceptors (A) are primarily considered by a single measure, i.e., IP (or E^0_{ox}) and EA (or E^0_{red}), for the HOMO and LUMO respectively.
2. Donor/acceptor organization as an inherent pre-equilibrium step is rigorously identified by the charge-transfer absorption (λ_{CT}).
3. The energetics of electron transfer is quantitatively evaluated by Mulliken theory ($h\nu_{\mathrm{CT}} = \mathrm{IP} - EA + V$).
4. EDA complex [D, A] and the ion-radical pair $D^{+\bullet}$, $A^{-\bullet}$ are reactive intermediates in a 3-step formulation (Scheme 1).
5. Dynamics of the ion-radical pair $D^{+\bullet}$, $A^{-\bullet}$ are directly observable by time-resolved (fs, ps) spectroscopy during charge-transfer photoactivation.

The presentation of the ion-radical pair $D^{+\bullet}$, $A^{-\bullet}$ as a discrete intermediate is the defining factor that distinguishes the electron-transfer paradigm from a conventional (one-step) mechanism involving a bimolecular (concerted) activation process. There is thus a clear-cut distinction between the electron-transfer paradigm and the concerted mechanism when the ion-radical pair has outer-sphere character that can be spectrally identified via its individual components (i.e., $D^{+\bullet}$ and $A^{-\bullet}$). However when this intermediate has inner-sphere character, the intimate (electronic/electrostatic) interaction in the ion-radical pair can obscure the individual properties of $D^{+\bullet}$ and $A^{-\bullet}$ sufficiently even to make the contact ion pair $D^{+\bullet}$, $A^{-\bullet}$ indistinguishable from a concerted transition-state structure. Even in that case, however, the residual donor/acceptor character should be helpful in qualitatively delineating (or rationalizing) the course of a reaction such as the oxygen-atom transfer in equation (106). Furthermore, the ion-pair dissociation of $D^{+\bullet}$, $A^{-\bullet}$ into kinetically "free" ions provides easy access to free-radical intermediates that often appear as byproducts in an otherwise heterolytic (or concerted) process.[271] In this way, we believe that a more global (and unified) formulation can usefully replace the traditional (and restrictive) heterolytic/homolytic classification of reaction mechanisms.

Acknowledgement

We thank the National Science Foundation and R.A. Welch Foundation for financial support.

References

1. Ingold, C.K. (1969). *Structure and Mechanism in Organic Chemistry*. Cornell, Ithaca, New York
2. See Rathore, R. and Kochi, J.K. (1998). *Acta Chem. Scand.* **52**, 114
3. (a) Edwards, J.O. (1954). *J. Am. Chem. Soc.* **76**, 1540. (b) Edwards, J.O. and Pearson, R.G. (1962). *J. Am. Chem. Soc.* **84**, 16
4. Pearson, R.G. (1963). *J. Am. Chem. Soc.*, **85**, 3533. See also: Ahrland, A., Chatt, J. and Davies, N.R. (1958). *Q. Rev. Chem. Soc.* **12**, 265. Gutmann, V. (1976). *Coord. Chem. Rev.* **18**, 225. For reviews see Pearson, R.G. (1997). *Chemical Hardness*. Wiley-VCH, New York and Ho, T.L. (1975). *Chem. Rev.* **75**, 1
5. See, for example: (a) Klopman, G. (1968). *J. Am. Chem. Soc.* **90**, 223. (b) Klopman, G. (1974). *Chemical Reactivity and Reaction Paths*. Wiley, New York (c) Fleming, I. (1976). *Frontier Orbitals and Organic Chemical Reactions*. Wiley, New York. (d) Jensen, W.B. (1978). *Chem. Rev.* **78**, 1. See also (a) Jensen, W.B. (1980). *The Lewis and Acid–Base Concepts*. Wiley, New York. (f) Jensen, W.B. (1987). In *Nucleophilicity*. (ed. Harris, J.M. and McManus, S.P.), p. 215; *Adv. Chem. Ser.* American Chemical Society, Washington, D.C. (g) Bach, R.D., Winter, J.E. and McDouall, J.J.W. (1995). *J. Am. Chem. Soc.* **117**, 8586
6. (a) Mulliken, R.S. (1952). *J. Am. Chem. Soc.* **74**, 811. (b) Mulliken, R.S. and Person, W.B. (1969). *Molecular Complexes*. Wiley, New York
7. Bockman, T.M. and Kochi, J.K. (1991). *Adv. Organometal. Chem.* **39**, 51
8. Howell, J.O., Goncalves, J.M., Amatore, C., Klasinc, L., Wightman, R.M. and Kochi, J. K. (1984). *J. Am. Chem. Soc.* **106**, 3968
9. Foster, R. (1969). *Organic Charge-Transfer Complexes*. Academic Press, New York
10. Tamres, M. and Strong, R.L. (1979). In *Molecular Association*, Vol. 2. (ed. Foster, R.F.), pp. 332–446. Academic Press, New York
11. Takahashi, Y., Sankararaman, S. and Kochi, J.K. (1989). *J. Am. Chem. Soc.* **111**, 2954
12. (a) Bernasconi, C.F. (1978). *Acc. Chem. Res.* **11**, 147. (b) Bernasconi, C.F., Kliner, D.A.V., Mullin, A.S. and Ni, J.X. (1988). *J. Org. Chem.* **53**, 3342. (c) Bernasconi, C. F. (1992). *Adv. Phys. Org. Chem.* **27**, 119
13. Prout, C.K. and Kamenar, B. (1973). In *Molecular Complexes*, Vol. 1, p. 151ff. (ed. Foster, R.) Russak, Crane, New York
14. (a) Hilinski, E.F., Masnovi, J.M., Amatore C., Kochi, J.K. and Rentzepis, P.M.

(1983). *J. Am. Chem. Soc.* **105**, 6167. (b) Wynne, K., Galli, C. and Hochstrasser, R. (1994). *J. Chem. Phys.* **100**, 4797
15. Bockman, T.M. and Kochi, J.K. (1988). *J. Am. Chem. Soc.* **110**, 1294 and (1989) **111**, 4669. For the behavior of geminate radical pairs, see Kochi, J.K. (ed.) (1973). *Free Radicals*, Vols 1 and 2. Wiley, New York
16. Benesi, H.A. and Hildebrand, J.H. (1949). *J. Am. Chem. Soc.* **71**, 2703
17. Foster, R. (1974). In *Molecular Complexes*, Vol. 2. (ed. Foster, R.), p. 107. Russak, Crane, New York
18. Tamres, M. and Strong, R.L. in ref. 10
19. (a) Colter, A.K. and Dack, M.J.R. (1973 and 1974). In *Molecular Complexes*, Vols 1 and 2 (ed. Foster, R.). Russak, Crane, New York. (b) Kosower, E. (1965). *Prog. Phys. Org. Chem.* **3**, 81. (c) For a recent presentation of the interplay between redox potentials (read: charge transfer) and molecular recognition (read: D/A organization), see Niemz, A. and Rotello, V. M. (1999). *Acc. Chem. Res.* **32**, 44 and Kaifer, A. (1999). *Acc. Chem. Res.* **32**, 62
20. Kochi, J.K. (1990). *Acta Chem. Scand.* **44**, 409
21. For an early treatment of the electron-transfer paradigm, see Kochi, J.K. (1993). *Adv. Phys. Org. Chem.* **29**, 185
22. Kochi, J.K. (1988). *Angew. Chem., Int. Ed. Engl.* **27**, 1227
23. Schmittel, M. (1994). *Top. Curr. Chem.* **169**, 183
24. March, J. (1992). *Advanced Organic Chemistry*, 4th edn. Wiley, New York
25. (a) House, H.O., Czuba, L.J., Gall, M. and Olmstead, H.D. (1969). *J. Org. Chem.* **34**, 2324. (b) Brownbridge, P. (1983). *Synthesis* 1 and 85
26. (a) Stotter, P.L. and Hill, K.A. (1973). *J. Org. Chem.* **38**, 2576. (b) Reuss, R.H. and Hassner, A. (1974). *J. Org. Chem.* **39**, 1785. (c) Hambly, G.F. and Chan, T.H. (1986). *Tetrahedron Lett.* **27**, 2563. (d) Blanco, L., Amice, P. and Conia, J. M. (1976). *Synthesis* 194. (e) Rubottom, G.M., Mott, R.C. and Juve, H.D., Jr. (1981). *J. Org. Chem.* **46**, 2717. (f) Rubottom, G.M. and Mott, R.C. (1979). *J. Org. Chem.* **44**, 1731. (g) Motohashi, S. and Satomi, M. (1982). *Synthesis* 1021. (h) Sha, C.K, Young, J.J. and Jean, T.S. (1987). *J. Org. Chem.* **52**, 3919. (i) Cambie, R.C., Hayward, R.C., Jurlina, J.L., Ruthledge, P.S. and Woodgate, P.D. (1978). *J. Chem. Soc., Perkin Trans. 1*, 126
27. a) Elfehail, F., Dampawan, P. and Zajac, W.W., Jr. (1980). *Synth. Commun.* **10**, 929. (b) Fischer, R.H. and Weitz, H.M. (1980). *Synthesis* 261. (c) Elfehail, F.E. and Zajac, W.W., Jr. (1981). *J. Org. Chem.* **46**, 5151. (d) Evans, P.A. and Longmire, J.M. (1994). *Tetrahedron Lett.* **35**, 8345. (e) Dampawan, P. and Zajac, W.W., Jr. (1983). *Synthesis* 545. (f) Cushman, M. and Mathew, J. (1982). *Synthesis* 597. (h) Feuer, H. and Pivawer, P.M. (1966). *J. Org. Chem.* **31**, 3152. (i) Shavarts, I.S., Yarovenko, V.N., Krayushkin, M.M., Novikov, S.S. and Sevostyanova, V.V. (1976). *Izv. Akad. Nauk. SSSR, Ser. Khim.* 1674. (j) Rathore, R. and Kochi, J.K. in ref. 37. (k) Rathore, R., Lin, Z. and Kochi, J.K. (1993). *Tetrahedron Lett.* **34**, 1859
28. Rasmussen, J.K. and Hassner, A. (1974). *J. Org. Chem.* **39**, 2558
29. (a) Rubottom, G.M., Vazquez, M.A. and Pelegrina, D.R. (1974). *Tetrahedron Lett.* 4319. (b) Brook, A.G. and Macrae, O.M. (1974). *J. Organomet. Chem.* **77**, C19. (c) Rubottom, G.M. and Gruber, J.M. (1978). *J. Org. Chem.* **43**, 1599. (d) Rubottom, G.M., Gruber, J.M., Boeckman, R.K., Jr., Ramaiah, M. and Medwid, J.B. (1978). *Tetrahedron Lett.* 4603. (e) Maume, G.M. and Horning, E.C. (1969). *Tetrahedron Lett.* 343. (f) Hanzlik, R.P. and Hilbert, J.M. (1978). *J. Org. Chem.* **43**, 610. (g) Davis, F.A. and Sheppard, A.C. (1987). *J. Org. Chem.* **52**, 954. (h) Bachman, G.B. and Hokama, T. (1960). *J. Org. Chem.* **25**, 178
30. (a) Bockman, T.M., Perrier, S. and Kochi, J.K. (1993). *J. Chem. Soc., Perkin Trans. 2*, 595. (b) Fukuzumi, S., Fujita, M., Matsubayashi, G. and Otera, J. (1993). *Chem.*

Lett. 1451. (c) Bhattacharya, A., DiMichele, L.M., Dolling, U.-H, Grabowski, E.J.J. and Grenda, V.J. (1989). *J. Org. Chem.* **54**, 6118. (d) Fleming, I. and Paterson, I. (1979). *Synthesis* 736. (e) Tanemura, K., Yamaguchi, K., Arai, H., Suzuki, T. and Horaguchi, T. (1995). *Heterocycles*. **41**, 2165. (f) RajanBabu, T.V., Reddy, G.S. and Fukunaga, T. (1985). *J. Am. Chem. Soc.* **107**, 5473. (g) Kondo, A. and Iwatsuki, S. (1982). *J. Org. Chem.* **47**, 1965. (h) Saraswathy, V.G. and Sankararaman, S. (1995). *J. Org. Chem.* **60**, 5024. (i) Oshima, M., Murakami, M. and Mukaiyama, T. (1985). *Chem Lett.* 1871. (j) Peterson, I. and Price, L.G. (1981). *Tetrahedron Lett.* **22**, 2829. (k) Kopka, I. and Rathke, M.W. (1981). *J. Org. Chem.* **46**, 3771. (l) Magnus, P. and Mugrage, B. (1990). *J. Am. Chem. Soc.* **112**, 462. (m) Sakakura, T., Hara, M. and Tanaka, M. (1994). *J. Chem. Soc., Perkin Trans. 1*, 283 and 289. (n) Sakakura, T. and Tanaka, M. (1985). *J. Chem. Soc., Chem. Commun.* 1309

31. (a) Ito, Y., Konoike, T. and Saegusa, T. (1975). *J. Am. Chem. Soc.* **97**, 649. (b) Moriarty, R.M., Penmasta, R. and Prakash, I. (1987). *Tetrahedron Lett.* **28**, 873. (c) Fujii, T., Hirao, T. and Ohshiro, Y. (1992). *Tetrahedron Lett.* **33**, 5823. (d) Frazier, R.H., Jr. and Harlow, R.L. (1980). *J. Org. Chem.* **45**, 5408. (e) Baciocchi, E., Casa, A. and Ruzziconi, R. (1989). *Tetrahedron Lett.* **30**, 3707. (f) Paolobelli, A.B., Latini, D. and Ruzziconi, R. (1993). *Tetrahedron Lett.* **34**, 721
32. (a) Fukuzumi, S., Fujita, M., Otera, J., Yukihiro, F. and Fujita, Y. (1992). *J. Am. Chem. Soc.* **114**, 10271. (b) Kozikowski, A.P. and Huie, E.M. (1982). *J. Am. Chem. Soc.* **104**, 2923. (c) Paquette, L.A. and Crouse, G.D. (1983). *J. Org. Chem.* **48**, 141. (d) Mukaiyama, T. and Murakami, M. (1987). *Synthesis* 1043. (e) Miura, T. and Masaki, Y. (1994). *J. Chem. Soc., Perkin Trans. 1*, 1659. (f) Heathcock, C.H. (1990). *Aldrichimica Acta* **23**, 99. (g) Sakurai, H., Sasaki, K. and Hosomi, A. (1983). *Bull. Chem. Soc. Jpn.* **56**, 3195. (h) Duhamel, P., Hennequin, L., Poirier, N. and Poirier, J.M. (1988). *Tetrahedron Lett.* **26**, 6201. (i) Kobayashi, S., Murakami, M. and Mukaiyama, T. (1985). *Chem. Lett.* 953. (j) Sainte, F., Serckx-Poncin, B., Hespain-Frisque, A.-M. and Ghosez, L. (1982). *J. Am. Chem. Soc.* **104**, 1428
33. (a) Soga, T., Takenoshita, H., Yamada, M., Han, J.S. and Mukaiyama, T. (1991). *Bull. Chem. Soc. Jpn.* **64**, 1108. (b) Kuroda, Y., Suzuki, Y. and Ogoshi, H. (1994). *Tetrahedron Lett.* **35**, 749
34. Gassman, P.G. and Bottorff, J. (1988). *J. Org. Chem.* **53**, 1097
35. (a) Schmittel, M. in ref. 23. (b) Schmittel, M., Keller, M., Burghart, A., Rappoport, Z. and Langels, A. (1988). *J. Chem. Soc., Perkin Trans. 2*, 869
36. Rathore, R. and Kochi, J.K. (1994). *Tetrahedron Lett.* **35**, 8577
37. Rathore, R. and Kochi, J.K. (1996). *J. Org. Chem.* **61**, 627
38. Bockman, T.M. and Kochi, J.K. (1994). *J. Phys. Org. Chem.* **7**, 325
39. Hilinski, E.F., Masnovi, J.M. and Kochi, J.K. (1984). *J. Am. Chem. Soc.* **106**, 8071
40. Savéant, J.-M. (1987). *J. Am. Chem. Soc.* **109**, 6788
41. (a) Bockman, T.M., Shukla, D. and Kochi, J.K. (1996). *J. Chem. Soc., Perkin Trans. 2*, 1623. (b) Bockman, T.M. and Kochi, J.K. (1996). *J. Chem. Soc., Perkin Trans. 2*, 1633. (c) For picosecond microdynamics (of solvent separation, internal return and special salt effect) in contact ion pairs generated by charge-transfer activation, see Yabe, T. and Kochi, J.K. (1992). *J. Am. Chem. Soc.* **114**, 4491
42. Snider, B.B. and Kwon, T. (1992). *J. Org. Chem.* **57**, 2399
43. (a) Larock, L. (1989). *Comprehensive Organic Transformations*, pp. 93–97. VCH, New York. (b) See March, J. in ref. 24, p. 1225. (c) Guttenplan, J.B. and Cohen, S.G. (1972). *J. Am. Chem. Soc.* **94**, 4040. (d) Fukuzumi, S., Fujita M. and Otera, J. (1993). *J. Org. Chem.* **58**, 5405. (e) Wagner, P.J., Truman, R.J., Puchalski, A.E. and Wake R. (1986). *J. Am. Chem. Soc.* **108**, 7727
44. Schönberg, T. (1968). *Preparative Organic Photochemistry*, pp. 203–217. Springer, New York

45. Baizer, M.M. (1991). *Organic Electrochemistry*, 3rd edn (eds. Lund, H. and Baizer, M.M.), p. 879. Dekker, New York
46. (a) Paterno, E. and Chieffii, G. (1909). *Gazz. Chim. Ital.* **39**, 341. (b) Büchi, G., Inman, C.G. and Lipinsky, E.S. (1954). *J. Am. Chem. Soc.* **76**, 4327
47. (a) Blicke, F.F. and Powers, L.D. (1929). *J. Am. Chem. Soc.* **51**, 3378. (b) (Review) Dagonneau, P. (1982). *Bull. Soc. Chim. Fr. II.* 269. (c) Maruyama, K. and Katagiri, L. (1988). *J. Phys. Org. Chem.* **1**, 21
48. (a) Arbuzov, A.E. and Arbuzova, I.A. (1932). *J. Gen. Chem. USSR.* **2**, 388. (b) Blomberg, C. and Mosher, H.S. (1968). *J. Organomet. Chem.* **13**, 519. (c) Ashby, E.C., Lopp, I. G. and Buhler, J.D. (1975). *J. Am. Chem. Soc.* **97**, 1964
49. (a) Ashby, E.C. (1988). *Acc. Chem. Res.* **21**, 414. (b) Ashby and coworkers have qualitatively shown that a single electron transfer (SET) mechanism is operational in a variety of reactions (such as nucleophilic additions, substitutions, reduction, etc.) based solely on the observation of radical intermediates. However, the quantitative treatment of these results under the unifying mechanistic theme of donor/acceptor association and the electron-transfer paradigm as presented herein provides the definitive mechanistic basis
50. (a) Relles, H.M. (1969). *J. Org. Chem.* **34**, 3687. (b) Becker, H.D. (1967). *J. Org. Chem.* **32,** 4093
51. Holm, T. and Crossland, I. (1971). *Acta Chem. Scand.* **25**, 59
52. Holm, T. (1983). *Acta Chem. Scand.* **B37**, 567
53. Mattay, J., Gersdorf, J. and Buckremer, K. (1987). *Chem. Ber.* **120**, 307
54. Griesbeck, A.G. (1995). *CRC Handbook of Organic Photochemistry and Photobiology*. (eds. Horspool, W.M. and Song, P.S.), p. 522. CRC Press, Boca Raton, FL
55. Sun, D., Hubig, S.M. and Kochi, J.K. (1999). *J. Org. Chem.* **64**, 2250
56. Bosch, E., Hubig, S.M. and Kochi, J.K. (1998). *J. Am. Chem. Soc.* **120**, 386
57. Mulliken, R.S. (1952). *J. Phys. Chem.* **56**, 801
58. Foster, R. in ref. 9
59. Kochi, J.K. (1978). *Organometallic Mechanisms and Catalysis*, p. 445 ff. Academic Press, New York
60. Lee, K.Y. and Kochi, J.K. (1992). *J. Chem. Soc., Perkin Trans. 2*, 1011
61. Bockman, T.M., Lee, K.Y. and Kochi, J.K. (1992). *J. Chem. Soc., Perkin Trans. 2*, 1581
62. Lowry, T.H. and Richardson, K.S. (1981). *Mechanism and Theory in Organic Chemistry*, p. 130 ff. Harper & Row, New York
63. Mann, C. and Barnes, K. (1970). *Electrochemical Reactions in Nonaqueous Systems*. Dekker, New York
64. Yoshida, K. (1984). *Electrooxidation in Organic Chemistry*. Wiley, New York
65. Bard, A.J. and Faulkner, L.R. (1980). *Electrochemical Methods*. Wiley, New York
66. Howell, J.O. *et al.* in ref. 8
67. (a) Howell, J.O. *et al.* in ref. 8. (b) Kiser, R.W. (1965). *Introduction to Mass Spectrometry and Its Applications*, p. 308. Prentice-Hall, Englewood Cliffs, NJ. (c) Zweig, A., Hodgson, W.G. and Jura, W.H. (1964). *J. Am. Chem. Soc.* **86**, 4124. (d) Hammerich. O., Parker, V.D. and Ronlan, A. (1976). *Acta Chem. Scand.* **B30**, 89. (e) Bock, H. and Wagner, G. (1971). *Tetrahedron Lett.* 3713. (f) Rathore, R. and Kochi, J.K. (1995). *J. Org. Chem.* **60**, 4399. (g) Scan rate $v = 1000\ V\ s^{-1}$, unpublished results. (h) Pysh, E.S. and Yang, N.C. (1963). *J. Am. Chem. Soc.* **85**, 2124. (i) Lias, S.G., Bartmess, J.E., Liebman, J.F., Holmes, J.L., Levin, R.D. and Mallard, W.G. (1988). *J. Phys. Chem. Ref. Data* **17**, p. 861. Also see Lias, S.G., Bartmess, J.E., Liebman, J.F., Holmes, J.L., Levin, R.D. and Mallard, W.G. (1990). *NIST Positive Ion Energetic Data Base, Ver 1.1* National Institute of Standards and Technology, Gaithersberg, MD. (j) Siegerman, H. (1975). *Technique of Electroorganic Synthesis*. (Techniques of Chemistry Vol. 5, part II) (ed. Wein-

berg, N.L.), p. 667. Wiley, NY. (k) Zweig, A., Lancaster, J.E., Neglia, M.T. and Jura, W.H. (1964). *J. Am. Chem. Soc.* **86**, 4130. (l) Bosch, E. and Kochi, J.K. (1995). *J. Org. Chem.* **60**, 3172. (m) Miller, L.L., Nordblom, G.D. and Mayeda, E.A. (1972). *J. Org. Chem.* **37**, 916. (n) Kimura, K., Katsumata, S., Achiba, Y., Yamazaki, T. and Iwata, S. (1981). *Handbook of He(I) Photoelectron Spectra of Fundamental Organic Molecules*. Japan Scientific Societies Press, Tokyo. (o) Meites, L. and Zuman, P. (1977). *CRC Handbook Series in Organic Electrochemistry*, Vols I–V. CRC Press, Boca Raton, FL. (p) Gersdorf, J., Mattay, J. and Görner, H. (1987). *J. Am. Chem. Soc.* **109**, 1203. (q) Fukuzumi, S. *et al.* in ref. 43d. (r) Schmittel, M. in ref. 23

68. Holm, T. in ref. 52
69. (a) Fehlner, T.P., Ulman, J., Nugent, W.A. and Kochi, J.K. (1976). *Inorg. Chem.* **15**, 2544. (b) Kochi, J.K. in ref. 59, p. 501 ff
70. Tanner, D.D., Deonarian, N. and Kharrat, A. (1989). *Can. J. Chem.* **69**, 171
71. (a) Maki, A.H. and Geske, D.H. (1961). *J. Am. Chem. Soc.* **83**, 1852. (b) Kebarle, P. and Chowdhury, S. (1987). *Chem. Rev.* **87**, 513. (c) Kortüm, G. and Walz, H. (1955). *Z. Electrochem.* **59**, 184. (d) Mann, C. and Barnes, K. in ref. 63, p. 560. (e) Christodoulides, A.A., McCorkle, D.L. and Christophorou, L.G. (1984). *Electron–Molecule Interactions and Their Applications*, p. 423, Vol. 2. Academic Press, New York. (f) Arnold, D.R. and Maroulis, A.J. (1976). *J. Am. Chem. Soc.* **98**, 5931. (g) Eriksen, J. and Foote, C.S. (1978). *J. Phys. Chem.* **82**, 2659. (h) Mattes, S.L. and Farid, S. (1983). In *Organic Photochemistry*, Vol. 6. (ed. Padwa, A.), p.238. Dekker, New York. (i) Chambers, J.Q. (1988). *The Chemistry of Quinonoid Compound*, Vol. 2. (eds. Patai, S. and Rappoport, Z.), p. 719. Wiley, New York. (j) Rathore, R. *et al.* in ref. 204. (k) Vazquez, C., Calabrese, J.C., Dixon, D.A. and Miller, J.S. (1993). *J. Org. Chem.* **58**, 65. (l) Lee, K.Y., unpublished results. (m) Rehm, D. and Weller, A. (1970). *Isr. J. Chem.* **8**, 259. (n) Birks, J.B. (1970). *Physics of Aromatic Molecules*. Wiley, New York. (o) Meites, L. and Zuman, P. in ref. 67o. (p) Paul, G. and Kebarle, P. (1989). *J. Am. Chem. Soc.* **111**, 464. (q) Barwise, A.J.G., Gorman, A.A., Leyland, R.L., Smith, P.G. and Rodgers, M.A.J. (1978). *J. Am. Chem. Soc.* **100**, 1814. (r) Wagner, P.J., Truman, R.J., Puchalski, A.E. and Wake, R. (1986) *J. Am. Chem. Soc.* **108**, 7727. (s) Grimsrud, E.P., Caldwell, G., Chowdhury, S. and Kebarle, P. (1985). *J. Am. Chem. Soc.* **107**, 4627. (t) Wasielewski, M.R. and Breslow, R. (1976). *J. Am. Chem. Soc.*, **98**, 4222. (u) Saeva, F.D. and Olin, G.R. (1980). *J. Am. Chem. Soc.* **102**, 299. (v) Elofson, R.M. and Gadallah, F.F. (1969). *J. Org. Chem.* **34**, 854. (w) Volz, H. and Lotsch, W. (1969). *Tetrahedron Lett.* 2275. (x) Connelly, N,G. and Geiger, W.E. in ref. 72. (y) Bockman, T.M. and Kochi, J.K. (1989). *J. Am. Chem. Soc.* **111**, 4669. (z) Lee, K.Y., Kuchynka, D.J. and Kochi, J.K. (1990). *Inorg. Chem.* **29**, 4196. (aa) Dinnocenzo, J.P. and Banach, T.E. (1986). *J. Am. Chem. Soc.* **108**, 6063. (bb) Svanholm, U., Hammerich, O. and Parker, V.D. (1975). *J. Am. Chem. Soc.* **97**, 101. (cc) Steckhan, E. (1987). *Top. Curr. Chem.* **142**, 1. (dd) Rathore, R. and Kochi, J.K. in ref. 67f. (ee) Altukhov, K.V. and Perekalin, V.V. (1976). *Russ. Chem. Rev.* **45**, 1052. (ff) Stackelberg, M. and Stracke, W. (1949). *Z. Electrochem.* **53**, 118. (gg) Lopez, L., Calo, V. and Aurora, R. (1986) *J. Photochem.* **32**, 95. (hh) Ross, U., Schulze, T. and Meyer, H.J. (1985). *Chem. Phys. Lett.* **121**, 174. (ii) Song, L. and Trogler, W.C. (1992). *Angew. Chem., Int. Ed. Engl.* **31**, 770. (jj) Kochi, J.K. in ref. 73, p. 854. (kk) Kochi, J.K. in ref. 22
72. Connelly, N.G. and Geiger, W.E. (1996). *Chem. Rev.* **96**, 877
73. Kochi, J.K. (1991). In *Comprehensive Organic Synthesis*, Vol. 7 (eds. Trost, B.M., Fleming, I. and Ley, S.V.), Chapter 7.4, p. 849 ff. New York: Pergamon Press
74. Takahashi, Y. *et al.* in ref. 11
75. Roth, H.D., Schilling, M.L.M. and Ragavachari, K. (1984). *J. Am. Chem. Soc.* **106**, 253

76. (a) Rhodes, C.J. (1988). *J. Am. Chem. Soc.* **110**, 4446. (b) Peacock, N.J. and Schuster, G.B. (1983). *J. Am. Chem. Soc.* **105**, 3632
77. Eaton, D.F. (1981). *J. Am. Chem. Soc.* **103**, 7235
78. Rathore, R. and Kochi, J.K. in ref. 37
79. (a) Chaudhuri, S.A. and Asmus, K.D. (1972). *J. Phys. Chem.* **76**, 26. (b) Masnovi, J.M., Kochi, J.K., Hilinski, E.F. and Rentzepis, P.M. (1986). *J. Am. Chem. Soc.* **108**, 1126
80. (a) Kim, E.K. and Kochi, J.K. (1989). *J. Org. Chem.* **54**, 1692. (b) Brownstein, S., Gabe, E., Lee, F. and Piotrowski, A. (1986). *Can. J. Chem.* **64**, 1616
81. (a) Rappoport, Z. (1966). *J. Chem. Soc.* 4498. (b) Kobashi, H., Funabashi, M.-A., Kondo, T., Morita, T., Okada, T. and Mataga, N. (1984). *Bull. Chem. Soc. Jpn.* **57**, 3557. (c) Hilinski, E.F., Milton, S.V. and Rentzepis, P.M. (1983). *J. Am. Chem. Soc.* **105**, 5193
82. Hammerich, O. and Parker, V.D. (1984). *Adv. Phys. Org. Chem.* **20**, 56
83. Bard, A.J., Ledwith, A. and Shine, H.J. (1976). *Adv. Phys. Org. Chem.* **13**, 155
84. Saeva, F.D. (1990). *Top. Curr. Chem.* **156**, 1990
85. Mizuno, K. and Otsuji, Y. (1994). *Top. Curr. Chem.* **169**, 301
86. Schmittel, M. and Burghart, A. (1997). *Angew. Chem., Int. Ed. Engl.* **36**, 2550
87. Lidman, J. (1975). *J. Am. Chem. Soc.* **97**, 4139
88. (a) Gardner, H.C. and Kochi, J.K. (1976). *J. Am. Chem. Soc.* **98**, 2460. (b) Kochi, J.K. in ref. 22
89. (a) Eaton, D.F. (1984). *Pure Appl. Chem.* **56**, 1191. (b) Eaton, D.F. in ref. 77
90. Akasaka, T., Sato, K., Kako, M. and Anod, W. (1991). *Tetrahedron Lett.* **32**, 6605
91. (a) Engel, P.S., Keys, D.E. and Kitamura, A. (1985). *J. Am. Chem. Soc.* **107**, 4964. (b) Engel, P.S., Robertson, D.M., Scholz, J.N. and Shine, H.J. (1992). *J. Org. Chem.* **57**, 6178. (c) Blackstock, S.C. and Kochi, J.K. (1987). *J. Am. Chem. Soc.* **109**, 2484
92. (a) Bockman, T.M., Hubig, S.M. and Kochi, J.K. (1996). *J. Am. Chem. Soc.* **118**, 4502. (b) Bockman, T.M., Hubig, S.M. and Kochi, J.K. (1997). *J. Org. Chem.* **62**, 2210
93. Bockman, T.M., Hubig, S.M. and Kochi, J.K. (1998). *J. Am. Chem. Soc.* **120**, 6542
94. Arnold, D.R. and Maroulis, A.J. in ref. 71f
95. Adam, W. and Nunez, E.G. (1991). *Tetrahedron.* **47**, 3773
96. (a) Masnovi, J.M. and Kochi, J.K. (1985). *J. Am. Chem. Soc.* **107**, 6781. (b) Wallis, J.M. and Kochi, J.K. (1988). *J. Am. Chem. Soc.* **110**, 8207
97. Albini, A. and Fasani, E. (1988). *J. Am. Chem. Soc.* **110**, 7760
98. (a) Gollas, B. and Speiser, B. (1992). *Angew. Chem., Int. Ed. Engl.* **31**, 332. (b) Rapta, P., Omelka, L., Stasko, A., Dauth, J., Deubzer, B. and Weis, J. (1996). *J. Chem. Soc., Perkin Trans. 2*, 225
99. (a) Toshima, K., Ishizuka, T., Matsuo, G., Nakata, M. and Kinoshita, M. (1993). *J. Chem. Soc., Chem. Commun.* 704
100. Albini, A. and Arnold, D.R. (1978). *Can. J. Chem.* **56**, 2985
101. Schaap, A.P., Siddiqui, S., Gagnon, S.D., Prasad, G., Palmono, E. and Lopez, L. (1984). *J. Photochem.* **25**, 167
102. Kamata, M. and Miyashi, T. (1989). *J. Chem. Soc., Chem. Commun.* 557
103. (a) Abe, M. and Oku, A. (1994). *J. Chem. Soc., Chem. Commun.* 1873. (b) Abe, M. and Oku, A. (1994). *Tetrahedron Lett.* **35**, 3551
104. Miyashi, T., Takahashi, Y., Mukai, T., Roth, H.D. and Schilling, M.L.M. (1985). *J. Am. Chem. Soc.* **107**, 1079
105. (a) Takahashi, Y., Miyashi, T. and Mukai, T. (1983). *J. Am. Chem. Soc.* **105**, 6511. (b) See also, Miyashi, T. *et al.* in ref. 104
106. Takahashi, Y. and Kochi, J.K. (1988). *Chem. Ber.* **121**, 253

107. Majima, T., Pac, C., Nakasone, A. and Sakurai, M. (1981). *J. Am. Chem. Soc.* **103**, 4499
108. (a) Mirafzal, G.A., Kim, T., Liu, J. and Bauld, N.L. (1992). *J. Am. Chem. Soc.* **114**, 10968. (b) Kim, T., Sarker, H. and Bauld, N.L. (1995). *J. Chem. Soc., Perkin Trans. 2*, 577
109. (a) Tanai, T., Mizuno, K., Hashida, I. and Otsuji, Y. (1993). *Tetrahedron Lett.* **34**, 2641
110. Miyashi, T., Konno, A. and Takahashi, Y. (1988). *J. Am. Chem. Soc.* **110**, 3676
111. Pandey, G., Karthikeyan, M. and Murugan, A. (1988). *J. Org. Chem.* **63**, 2867
112. (a) Heidbreder, A. and Mattay, J. (1992). *Tetrahedron Lett.* **33**, 1973. (b) Snider, B.B. and Kwon, T. in ref. 42
113. Schmittel, M., Burghart, A., Malisch, W., Reising, J. and Söllner, R. (1998). *J. Org. Chem.* **63**, 396
114. (a) Baciocchi, E. (1990). *Acta Chem. Scand.* **44**, 645. (b) Amatore, C. and Kochi, J.K. (1991). In *Advances in Electron Transfer Chemistry*. Vol. 1. (ed. Mariano, P.S.), p. 55. JAI Press, London. (c) Baciocchi, E., Del Giacco, T. and Elisei, F. (1993). *J. Am. Chem. Soc.* **115**, 12290
115. Mella, M., Freccero, M. and Albini, A. (1996). *Tetrahedron.* **52**, 5533
116. (a) Zhang, X. and Bordwell, F.G. (1992). *J. Org. Chem.* **57**, 4163. (b) Shaefer, C.G. and Peters, K.S. (1980). *J. Am. Chem. Soc.* **102**, 7566. (c) Dinnocenzo, J.P. and Banach, T.E. (1989). *J. Am. Chem. Soc.* **111**, 8646. (d) Lewis, F.D. (1986). *Acc. Chem. Res.* **19**, 401
117. Parker, V.D. (1973). *Electrochimica Acta* **18**, 519
118. (a) Sankararaman, S., Haney, W.A. and Kochi, J.K. (1987). *J. Am. Chem. Soc.* **109**, 7824. (b) Workentin, M.S., Johnston, L.J., Wayner, D.D.M. and Parker, V.D. (1994). *J. Am. Chem. Soc.* **116**, 8279. (c) Bard, A.J. *et al.* in ref. 83
119. (a) Pac, C., Nakasone, A. and Sakurai, H. (1977). *J. Am. Chem. Soc.* **99**, 5806. (b) Majima, T. *et al.* in ref. 107
120. (a) Maroulis, A.J. and Arnold, D.R. (1979). *Synthesis.* 819. (b) Mizuno, K., Nakanishi, I., Ichinose, N. and Otsuhi, Y. (1989). *Chem. Lett.* 1095
121. Gassman, P.G. and De Silva, S.A. (1991). *J. Am. Chem. Soc.* **113**, 9870
122. (a) Ronlan, A., Hammerich, O. and Parker, V.D. (1973). *J. Am. Chem. Soc.* **95**, 7132. (b) Reynolds, R., Line, L.L. and Nelson, R.F. (1974). *J. Am. Chem. Soc.* **96**, 1087
123. (a) Mizuno, K., Kagano, H., Kasuga, T. and Otsuji, Y. (1983). *Chem. Lett.* 133. (b) Mizuno, K. and Otsuji, Y. (1986). *Chem. Lett.* 683. (c) Carruthers, R.A., Crellin, R.A. and Ledwith, A. (1969). *J. Chem. Soc., Chem. Commun.* 25
124. Bellville, D.J., Wirth, D.D. and Bauld, N.L. (1981). *J. Am. Chem. Soc.* **103**, 718
125. (a) Neunteufel, R.A. and Arnold, D.R. (1973). *J. Am. Chem. Soc.* **95**, 4080. (b) Mattes, S.L. and Farid, S. (1982). *Acc. Chem. Res.* **15**, 80. (c) Caldwell, R.A. and Creed, D. (1980). *Acc. Chem. Res.* **13**, 45
126. (a) Mattes, S.L. and Farid, S. (1986). *J. Am. Chem. Soc.* **108**, 7356. (b) Mizuno, K., Ichinose, N. and Otsuji, Y. (1992). *J. Org. Chem.* **57**, 1855
127. (a) Bauld, N.L. and Pabon, R. (1983). *J. Am. Chem. Soc.* **105**, 633. (b) Bauld, N.L. (1989). *Tetrahedron.* **45**, 5307
128. (a) Sankararaman, S. and Kochi, J.K. (1986). *Recl. Trav. Chim. Pays-Bas.* **105**, 278. (b) Sankararaman, S., Haney, W.A. and Kochi, J.K. (1987). *J. Am. Chem. Soc.* **109**, 5235
129. Masnovi, J.M. and Kochi, J.K. (1986). *Recl. Trav. Chim. Pays-Bas.* **105**, 289
130. Akaba, R., Aihara, S., Sakuragi, H. and Tokumaru, K. (1987). *J. Chem. Soc., Chem. Commun.* 1262
131. (a) Kim, K., Hull, V.J. and Shine, H.J. (1974). *J. Org. Chem.* **39**, 2534. (b) Silber,

J.J. and Shine, H.J. (1971). *J. Org. Chem.* **36**, 2923. (c) Torii, S., Matsuyama, Y., Kawasaki, K. and Uneyama, K. (1973). *Bull. Chem. Soc. Jpn.* **46**, 2912
132. Rathore, R., Lindeman, S.V., Kumar, A.S. and Kochi, J.K. (1998). *J. Am. Chem. Soc.* **120**, 6931
133. Dinnocenzo, J.P. and Banach, T.E. in ref. 71aa
134. Pittman, Jr., C.U., McManus, S.P. and Larsen, J.W. (1972). *Chem. Rev.* **72**, 357
135. Holy, N.L. (1974). *Chem. Rev.* **74**, 243
136. McClelland, B.J. (1964). *Chem. Rev.* **64**, 301
137. Jordan, K.D. and Burrow, P.D. (1987). *Chem. Rev.* **87**, 557
138. (a) Kim, J.K. and Bunnett, J.F. (1970). *J. Am. Chem. Soc.* **92**, 7463. (b) Rossi, R.A. and Bunnett, J.F. (1973). *J. Org. Chem.* **38**, 1407. (c) Rossi, R.A. and Rossi, R.H. de. (1983). *Aromatic Substitution by the $S_{RN}1$ Mechanism.* ACS Monograph 178
139. (a) Kornblum, N., Davies, T.M., Earl, G.W., Holy, N.L., Kerber, R.C., Musser, M.T. and Snow, D.H. (1967). *J. Am. Chem. Soc.* **89**, 725. (b) Fukuzumi, S., Hironaka, K. and Tanaka, T. (1983). *J. Am. Chem. Soc.* **105**, 4722. (c) Fukuzumi, S., Mochizuki, S. and Tanaka, T. (1989). *J. Am. Chem. Soc.* **111**, 1497
140. Holy, N.L. in ref. 135
141. Adam, W. and Schonberger, A. (1992). *Chem. Ber.* **125**, 2149
142. Huyser, E.S., Harmony, J.A.K. and McMillan, F.L. (1972). *J. Am. Chem. Soc.* **94**, 3176
143. Wang, C.-H., Linnell, S.M. and Wang, N. (1971). *J. Org. Chem.* **36**, 525
144. Moëns, L., Baizer, M.M. and Little, R.D. (1986). *J. Org. Chem.* **51**, 4497
145. Mandell, L., Daley, R.F. and Day, R.A., Jr. (1976). *J. Org. Chem.* **41**, 4087
146. Shono, T., Nishiguchi, I., Ohmizu, H. and Mitani, M. (1978). *J. Am. Chem. Soc.* **100**, 545
147. Kise, N., Suzumoto, T. and Shono, T. (1994). *J. Org. Chem.* **59**, 1407
148. (a) Birch, A. J. (1950). *Q. Rev., Chem. Soc.* **4**, 69. (b) Harvey, R.G. (1970). *Synthesis* 161
149. (a) Honzl, J. and Lovy, J. (1984). *Tetrahedron.* **40**, 1885. (b) Also see, Periasamy, M., Reddy, M.R. and Bharathi, P. (1999). *Synth. Commun.* **29**, 677
150. Bock, H., Ruppert, K., Havlas, Z., Bensch, W., Hönle, W. and von Schnering, H.G. (1991). *Angew. Chem., Int. Ed. Engl.* **30**, 1183
151. Hou, Z. and Wakatsuki, Y. (1997). *Chem. Eur. J.* **3**, 1005
152. Ruhlmann, K. (1971). *Synthesis* 236
153. Savéant, J.-M. (1994). *Adv. Phys. Org. Chem.* **4**, 53
154. Angelo, B. (1969). *Bull. Soc. Chim. France* 1710
155. Pandey, G., Krishna, A. and Rao, J.M. (1986). *Tetrahedron Lett.* **27**, 4075
156. (a) Iida, Y. and Akamatu, H. (1979). *Bull. Chem. Soc. Jpn.* **52**, 1875. (b) Iida, Y. (1979). *Bull. Chem. Soc. Jpn.* **53**, 2673
157. Katz, T.J. (1960). *J. Am. Chem. Soc.* **82**, 3785
158. (a) Grzeszczuk, M. and Smith, D. E. (1984). *J. Electroanal. Chem.* **162**, 189. (b) Wolf, M.O., Fox, H.H. and Fox, M.A. (1996). *J. Org. Chem.* **61**, 287
159. Bock, H., Ruppert, K., Näther, C., Havlas, Z., Hermann, H.-F, Arad, C., Göbel, I., John, A., Meuret, J., Nick, S., Rauschenbach, A., Seitz, W., Vaupel, T. and Solukui, B. (1992). *Angew. Chem., Int. Ed. Engl.*, **31**, 550
160. Yoshino, A., Yamasaki, K. and Yonezawa, T. (1975). *J. Chem. Soc., Perkin Trans. 1*, 94
161. (a) Albini, A., Fasani, E. and Sulpizio, A. (1984). *J. Am. Chem. Soc.* **106**, 3562. (b) Albini, A., Fasani, E. and Oberti, R. (1982). *Tetrahedron Lett.* **38**, 1027
162. Borg, R.M., Arnold, D.R. and Cameron, T.S. (1984). *Can. J. Chem.* **62**, 1785
163. Kim, E., Christl, M. and Kochi, J.K. (1990). *Chem. Ber.* **123**, 1209
164. Mattay, J. (1987). *Angew. Chem., Int. Ed. Engl.* **26**, 824
165. Bryce-Smith, D. and Gilbert, A. (1968). *J. Chem. Soc., Chem. Commun.* 1701

166. (a) Farid, S. and Brown, K.A. (1976). *J. Chem. Soc., Chem. Commun.* 564. (b) Kjell, D.P. and Sheridan, R.S. (1985). *J. Photochem.* **28**, 205
167. (a) Kaiser, E.T. and Kevan, L. (1968). *Radical Ions*. Wiley, New York. (b) Davies, A.G. (1966). *Chem. Soc. Rev.* **25**, 299. (c) Courtneidge, J.L. and Davies, A.G. (1987). *Acc. Chem. Res.* **20**, 90. (d) *Electron Spin Resonance, Specialist Periodical Reports*. The Chemical Society, London (1971–1991), Vols. 1–9, 10A–13A. (e) Shiotani, M. (1987). *Magn. Reson. Rev.* **12**, 333. (f) Shida, T., Haselbach, E. and Bally, T. (1984). *Acc. Chem. Res.* **17**, 180. (g) Gerson, F., Lopez, J., Akaba, R. and Nelsen, S.F. (1981). *J. Am. Chem. Soc.* **103**, 6716
168. Shida, T. (1988). *Electronic Absorption Spectra of Radical Ions*. Elsevier, New York
169. (a) DeBoer, J.L. and Vos, A. (1972). *Acta Crystallogr., Section B*. **B28**, 835. (b) Rathore, R. and Kochi, J.K in ref. 67f. (c) Rathore, R. and Kochi, J.K., unpublished results. (d) Rathore, R., Kumar, A.S., Lindeman, S.V. and Kochi, J.K. (1998). *J. Org. Chem.* **63**, 5847. (e) Fritz, H.P., Gebauer, H., Friedrich, P., Ecker, P., Artes, R. and Schubert, U. (1978). *Z. Naturforsch.* **B33**, 498. (f) Sebastiano, R., Korp, J.D. and Kochi, J.K. (1991). *J. Chem. Soc., Chem. Commun.* 1481. (g) Keller, H.J., Nöthe, D., Pritzkow, H., Wehe, D., Werner, M., Koch, P. and Schweitzer, D. (1981). *Mol. Cryst. Liq. Cryst.* **62**, 181. (h) Bockman, T.M. and Kochi, J.K. (1990). *J. Org. Chem.* **55**, 4127. (i) Brown, G.M., Freeman, G.R. and Walter, R.I. (1977). *J. Am. Chem. Soc.* **99**, 6910. (j) Nelsen, S.F., Chen, L.-J., Powell, D.R. and Neugebauer, F.A. (1995). *J. Am. Chem. Soc.* **117**, 11434. (k) Bock, H., Rauschenbach, A., Näther, C., Kleine, M. and Havlas, Z. (1994). *Chem. Ber.* **127**, 2043. (l) Rathore, R., Lindeman, S.V., Kumar, A.S. and Kochi, J.K. in ref. 132. (m) Nishinaga, T., Komatsu, K., Sugita, N., Lindner, H.J. and Richter, J. (1993). *J. Am. Chem. Soc.* **115**, 11642
170. (a) Rathore, R. and Kochi, J.K. in ref. 2. (b) Elson, I.H. and Kochi, J.K. (1973). *J. Am. Chem. Soc.* **95**, 5060. (c) Eberson, L., Hartshorn, M.P. and Persson, O. (1995). *J. Chem. Soc., Chem. Commun.* 1131. (d) Handoo, K.L. and Gadru, K. (1986). *Curr. Sci.* **55**, 920. (e) Sato, Y., Kinoshita, M., Sano, M. and Akamatu, H. (1969). *Bull. Chem. Soc. Jpn.* **42**, 548. (f) Ristagno, C.V. and Shine, H.J. (1971). *J. Am. Chem. Soc.* **93**, 1811. (g) Eberson, L. and Larsson, B. (1987). *Acta Chem. Scand., Ser. B* **41**, 367. (h) Eberson, L., Hartshorn, M.P., Radner, F. and Persson, O. (1996). *J. Chem. Soc., Chem. Commun.* 215. (i) Shine, H.J. and Sullivan, P.D. (1968). *J. Phys. Chem.* **72**, 1390
171. Eberson, L. and Nyberg, K. (1976). *Adv. Phys. Org. Chem.* **12**, 1
172. Baxendale, J.H. and Rodgers, M.A.J. (1978). *Chem. Soc. Rev.* **7**, 235
173. Rathore, R. and Kochi, J.K. in ref. 67f
174. Bandlish, B.K. and Shine, H.J. (1977). *J. Org. Chem.* **42**, 561
175. (a) Rathore, R., Lindeman, S.V. and Kochi, J.K. (1998). *Angew. Chem., Int. Ed. Engl.* **37**, 1585. (b) Rathore, R. and Kochi, J.K. (1998). *J. Org. Chem.* **63**, 8630. (c) Kim, E.K. and Kochi, J.K. (1991). *J. Am. Chem. Soc.* **113**, 4962
176. (a) Bell, F.A., Ledwith, A. and Sherrington, D.C. (1969). *J. Chem. Soc. C.* 2719. (b) Cowell, G.W., Ledwith, A., White, A.C. and Woods, H.J. (1970). *J. Chem. Soc. B.* 227. (c) Luken, E.A.C. (1969). *J. Chem. Soc. C.* 4963
177. Rathore, R. *et al.* in 169d
178. Garst, J. F. (1973). In *Free Radicals*, Vol. 1 (ed. Kochi, J.K.), p. 503. Wiley, New York
179. Angelo, B. (1969). *Bull. Soc. Chim. Fr.* 1710
180. (a) Bock, H., Ruppert, K. and Fenske, D. (1989). *Angew. Chem., Int. Ed. Engl.* **28**, 1685. (b) Bock, H., Näther, C. and Havlas, Z. (1995). *J. Am. Chem. Soc.* **117**, 3869. (c) Bock, H., Claudia, A., Näther, C. and Havlas, Z. (1995). *J. Chem. Soc., Chem. Commun.* 2393. (d) Zanotti, G., Del-Pra, A. and Bozio, R. (1982). *Acta Crystallogr. Sect. B.* **38**, 1225. (e) Konno, M., Kobayashi, H., Marumo, F. and Saito, Y. (1973).

Bull. Chem. Soc. Jpn. **46**, 1987. (f) Bock, M., John, A., Näther, C., Havlas, Z. and Mihokova, E. (1994). *Helv. Chim. Acta* **77**, 41
181. Posner, G.H. (1980). *An Introduction to Synthesis Using Organocopper Reagents.* Wiley, New York
182. (a) House, H.O. and Umen, M.J. (1972). *J. Am. Chem. Soc.* **94**, 5495. (b) House, H.O. and Umen, M.J. (1973). *J. Org. Chem.* **38**, 3893. (c) House, H.O. (1976). *Acc. Chem. Res.* **9**, 59
183. House, H.O., Respess, W.L. and Whitesides, G.M. (1966). *J. Org. Chem.* **31**, 3128
184. Vellekoop, A.S. and Smith, R.A.J. (1994). *J. Am. Chem. Soc.* **116**, 2902
185. Zhu, D. and Kochi, J.K. (1999). *Organometallics.* **18**, 161
186. Lucarini, M., Pedulli, G.F., Alberti, A., Paradisi, C. and Roffia, S. (1993). *J. Chem. Soc., Perkin Trans. 2*, 2083
187. (a) Neumann, W.P. (1987). *Synthesis* 665. (b) Kuivila, H.G. (1970). *Synthesis* 499. (c) Kursanov, D.N., Parnes, Z.N. and Loim, N.M. (1974). *Synthesis.* 633
188. Klingler, R.J., Mochida, K. and Kochi, J.K. (1979). *J. Am. Chem. Soc.* **101**, 6626
189. (a) Peryre, M. and Valade, J. (1965). *Compt. Rend.* **260**, 581. (b) Pommier, J.C. and Valade, J. (1965). *Bull. Soc. Chim. Fr.* 975. (c) Neumann, W.P. and Heyman, E. (1965). *Liebigs Ann. Chem.* **683**, 11
190. Leusink, A.J., Budding, H.A. and Marsman, J.W. (1968). *J. Organomet. Chem.* **13**, 155
191. Perrier, S., Sankararaman, S. and Kochi, J.K. (1993). *J. Chem. Soc., Perkin Trans. 2*, 825
192. Sankararaman, S. and Kochi, J.K. (1989). *J. Chem. Soc., Chem. Commun.* 1800
193. Bockman, T.M. *et al.* in ref. 93
194. Kim, E.K. and Kochi, J.K. (1993). *J. Org. Chem.* **58**, 786
195. (a) Penn, J.H., Deng, D.-L. and Chai, K.J. (1988). *Tetrahedron Lett.* **29**, 3635. (b) Penn, J.H., Lin, Z. and Deng, D.-L. (1991). *J. Am. Chem. Soc.* **113**, 1001
196. Kim, E.K. and Kochi, J.K. in ref. 175c
197. Kolbe, H. (1849). *Ann. Chem. Pharm.* **69**, 257
198. Brown, A.C. and Walker, J. (1891). *Liebigs Ann. Chem.* **261**, 107
199. (a) Serguchev, S. and Beletskaya, D. (1980). *Russ. Chem. Rev.* **49**, 1119. (b) Sheldon, R.A. and Kochi, J.K. (1972). *Org. React.* **19**, 279. (c) Barton, D.H.R., Bridon, D. and Zard, Z. (1989). *Tetrahedron.* **45**, 2615
200. Bockman, T.M. *et al.* in ref. 92
201. (a) Becker, H.D. (1974). In *The Chemistry of Quinonoid Compounds*, Vol. 1 (ed. Patai, S.), p. 335. Wiley, New York. See also Gardner, K.A., Kuehnert, L.L. and Mayer J.M. (1997). *Inorg. Chem.* **36**, 2069. (b) Becker, H.D. and Turner, A.B. (1988). In *The Chemistry of Quinonoid Compounds*, Vol. 2 (eds. Patai, S. and Rappoport, Z.), p. 1351. Wiley, New York
202. (a) Rathore, R., Lindeman, S.V. and Kochi, J.K. (1997). *J. Am. Chem. Soc.* **119**, 9393. (b) Foster, R. in ref. 9
203. (a) Kochi, J.K. in ref. 20. (b) Hubig, S.M., Bockman, T.M. and Kochi, J.K. (1997). *J. Am. Chem. Soc.* **119**, 2926
204. Rathore, R., Hubig, S.M. and Kochi, J.K. (1997). *J. Am. Chem. Soc.* **119**, 11468
205. Bockman, T.M., Hubig, S.M. and Kochi, J.K. (1998). *J. Am. Chem. Soc.* **120**, 2826
206. Woodward, R.B. (1942). *J. Am. Chem. Soc.* **64**, 3058
207. Takahashi, Y. and Kochi, J.K. in ref. 106
208. Kim, E. et al. in ref. 163
209. (a) Scott, L.T. and Brunsvold, W.R. (1978). *J. Am. Chem. Soc.* **100**, 4320. (b) Scott, L.T., Brunsvold, W.R., Kirms, M.A. and Erden, I. (1981). *Angew. Chem., Int. Ed. Engl.* **20**, 274. (c) Scott, L.T., Erden, I., Brunsvold, W.R., Schultz, T.H. Houk, K.N. and Paddon-Row, M.N. (1982). *J. Am. Chem. Soc.* **104**, 3659. (d) Tsuji, T. and Nishida, S. (1984). *Acc. Chem. Res.* **17**, 56

210. Tsuji, T., Shibata, T., Hienuki, Y. and Nishida, S. (1978). *J. Am. Chem. Soc.* **100**, 1806
211. Harada, Y., Ohno, K., Seki, K. and Inokuchi, H. (1974). *Chem. Lett.* 1081
212. Sun, D., Hubig, S.M. and Kochi, J.K. (1999). *J. Photochem. Photobiology.* **122**, 87
213. Schröder, M. (1980). *Chem. Rev.* **80**, 187
214. (a) Collin, R.J., Jones, J. and Griffith, W.P. (1974). *J. Chem. Soc., Dalton Trans.* 1094. (b) Criegee, R. (1936). *Liebigs Ann. Chem.* **522**, 75. (c) Clark, R.L. and Behrman, E.J. (1975). *Inorg. Chem.* **14**, 1425
215. (a) Kolb, H.C., Van Nieuwenzhe, M.S. and Sharpless, K.B. (1994) *Chem. Rev.* **94**, 2483. (b) Corey, E.J., Noe, M.C. and Guzman-Perez, A. (1995) *J. Am. Chem. Soc.* **117**, 10817
216. (a) Corey, E.J. and Noe, M.C. (1996). *J. Am. Chem. Soc.* **118**, 319. (b) Corey, E.J., Sarshar, S., Azimioara, M.D., Newbold, R.C. and Noe, M.C. (1996). *J. Am. Chem. Soc.* **118**, 7851. (c) DelMonte, A.J., Haller, J., Houk, K.N., Sharpless, K.B., Singleton, D.A., Strassner, T. and Thomas, A.A. (1997). *J. Am. Chem. Soc.* **119**, 9907. (d) Nelson, D.W., Gypser, A., Ho, P.T., Kolb, H.C., Kondo, T., Kwong, H.-L., McGrath, D.V., Rubin, A.E., Norrby, P.-O, Gable, K.P. and Sharpless, K.B. (1997). *J. Am. Chem. Soc.* **119**, 1840. (e) The mechanistic distinction between direct [3+2] cycloaddition and [2+2] cycloaddition followed by ligand-assisted ring expansion (to yield the cyclic osmate ester) in Scheme 21 is difficult to establish unambiguously if extensive equilibria amongst the adducts are involved in the osmylation reaction
217. (a) Nugent, W.A. (1980). *J. Org. Chem.* **45**, 4533. (b) Hammond, P.R, Knipe, R.H. and Lake, R.R. (1971). *J. Chem. Soc. A.* 3789
218. (a) Wallis, J.M. and Kochi, J.K. in ref. 96b. (b) Wallis, J.M. and Kochi, J.K. (1988). *J. Org. Chem.* **53**, 1679
219. (a) Motherwell, W.B. and Williams, A.S. (1995). *Angew. Chem., Int. Ed. Engl.* **34**, 2031. (b) (Review) Poli, G. and Scolastico, C. (1996). *CHEMTRACTS. Org. Chem.* **9**, 304
220. Taylor, R. (1990). *Electrophilic Aromatic Substitution.* Wiley, New York
221. (a) Wheland, G.W. (1942). *J. Am. Chem. Soc.* **64**, 900. (b) Schofield, K. (1980). *Aromatic Nitration.* Cambridge University Press, Cambridge. (c) Rathore, R., Hecht, J. and Kochi, J.K. (1998). *J. Am. Chem. Soc.* **120**, 13278. (d) Rathore, R., Loyd, S.H. and Kochi, J.K. (1994). *J. Am. Chem. Soc.* **116**, 8414
222. (a) Kochi, J.K. (1990). *Adv. Free Rad. Chem.* **1**, 53. (b) Strong, R.L., Rand, S.J. and Britt, J.A. (1960). *J. Am. Chem. Soc.* **82**, 5053. (c) Rayner, K.D., Lusztyk, J. and Ingold, K.U. (1989). *J. Phys. Chem.* **93**, 564. (d) Takahashi, Y. *et al.* in ref. 11. (e) Bockman, T.M., Kosynkin, D. and Kochi, J.K. (1997). *J. Org. Chem.* **62**, 5811. (f) Kosynkin, D., Bockman, T.M. and Kochi, J.K. (1997). *J. Chem. Soc., Perkin Trans. 2*, 2003. (g) Kosynkin, D., Bockman, T.M. and Kochi, J.K. (1997). *J. Am. Chem. Soc.* **119**, 4846. (h) Lau, W. and Kochi, J.K. (1986). *J. Am. Chem. Soc.* **108**, 6720
223. Kochi, J.K. (1991). *Pure Appl. Chem.* **63**, 225
224. Fukuzumi, S. and Kochi, J.K. (1981). *J. Am. Chem. Soc.* **103**, 2783
225. Hubig, S. M., Jung, W. and Kochi, J.K. (1994). *J. Org. Chem.* **59**, 6233. Note the competition between ion-pair and radical-pair collapse in halogenation with iodine monochloride on p. 279 is exactly analogous to that in nitration with tetranitromethane in eqns 82/83 (vide infra)
226. Hassel, O. and Stromme, K.O. (1950). *Acta Chem. Scand.* **12**, 1146. (b) Hassel, O. and Stromme, K.O. (1951). *Acta Chem. Scand.* **13**, 1781
227. Mulliken, R.S. in ref. 6a
228. Baenziger, N.C., Buckles, R.E. and Simpson, T.D. (1967). *J. Am. Chem. Soc.* **89**, 3405

229. (a) Andrews, L.J. and Keefer, R.M. (1957). *J. Am. Chem. Soc.* **79**, 1412. (b) Wang, Y.L., Nagy, J.C. and Margerum, D.W. (1989). *J. Am. Chem. Soc.* **111**, 7838
230. Olah, G.A., Malhotra, R. and Narang, S.C. (1989). *Nitration: Methods and Mechanisms*. VCH, New York
231. Kenner, J. (1945). *Nature*. **156**, 369
232. (a) Feuer, H. (Ed.) (1969). *Chemistry of the Nitro and Nitroso Groups*. Wiley, New York. (b) Chowdhury, S., Kishi, H., Dillow, G.W. and Kebarle, P. (1989). *Can. J. Chem.* **67**, 603
233. (a) Andrews, L.J. and Keefer, R.M. (1964). *Molecular Complexes in Organic Chemistry*. Holden-Day, San Francisco
234. Briegleb, G. (1973). *Elektronen-Donator-Acceptor Komplexe*. Springer, Berlin
235. (a) Kim, E.K., Lee, K.Y. and Kochi, J.K. (1992). *J. Am. Chem. Soc.* **114**, 1756. (b) Kim, E.K., Bockman, T.M. and Kochi, J.K. (1992). *J. Chem. Soc., Perkin Trans. 2*, 1879. (c) Kim, E.K., Bockman, T.M. and Kochi, J.K. (1993). *J. Am. Chem. Soc.* **115**, 3091
236. Sankararaman, S. and Kochi, J.K. (1991). *J. Chem. Soc., Perkin Trans. 2*, 1
237. (a) Sankararaman, S., Haney, W.A. and Kochi, J.K. (1987). *J. Am. Chem. Soc.* **109**, 7824. (b) Sankararaman, S. *et al.* in ref. 128b. (c) Masnovi, J.M., Sankararaman, S. and Kochi, J.K. (1989). *J. Am. Chem. Soc.* **111**, 2263
238. See Masnovi, J.M. *et al.* in ref. 79b
239. (a) Rathore, R., Bosch, E. and Kochi, J.K. (1994). *Tetrahedron*. **50**, 6727. (b) Rathore, R., Bosch, E. and Kochi, J.K. (1994). *J. Chem. Soc., Perkin Trans. 2*, 1157
240. (a) Bosch, E. and Kochi, J.K. (1993). *Res. Chem. Intermed.* **19**, 811. (b) Bosch, E. and Kochi, J.K. (1993). *J. Org. Chem.* **59**, 3314
241. Williams, D.L.H. (1988). *Nitrosations*. Cambridge University Press, Cambridge
242. Challis, B.C., Higgins, R.J. and Lawson, A.J. (1972). *J. Chem. Soc., Perkin Trans. 2*, 1831
243. (a) Kim, E.K. and Kochi, J.K. in ref. 175c. (b) Mulliken theory[6] describes the wave function (Ψ_{AD}) of a charge-transfer complex as principally the sum of the dative (bonding) function (Ψ_1) and the "no-bond" function (Ψ_0), i.e. $\Psi_{AD} = a\Psi_0$ (A, D) + $b\Psi_1$ (A^-, D^+) + ... Accordingly, the degree of charge-transfer is defined as the ratio $(b/a)^2$ of the mixing coefficients a and b of the no-bond and the dative wave functions, respectively. For the experimental determination of $(b/a)^2$, see (a) Ketelaar, J.A.A. (1954). *J. Phys. Radium*. **15**, 197. (b) Tamres, M. and Brandon, M. (1960). *J. Am. Chem. Soc.* **82**, 2134. (c) Briegleb, G. in ref. 234
244. Rathore, R. *et al.* in ref. 202a
245. (a) Rathore, R. and Kochi, J.K. in ref. 67f . (b) Rathore, R. and Kochi, J.K. unpublished results
246. Bosch, E. and Kochi, J.K. (1994). *J. Org. Chem.* **59**, 5573
247. Hubig, S.M. and Kochi, J.K. unpublished results
248. Rathore, R. *et al.* in ref. 221b
249. Pearson, R.G. (1987). In *Nucleophilicity*. (eds. Harris, J.M. and McManus, S.P.), *Advances in Chemistry Series* 215. American Chemical Society. Washington, DC
250. Smedslund, T.H. (1949). *Swed. Pat.* **121**, 576 and numerous subsequent patents
251. (a) Bosch, E. and Kochi, J.K. (1996). *J. Am. Chem. Soc.* **118**, 1319. (b) Bosch, E. and Kochi, J.K. (1993). *J. Chem. Soc., Chem. Commun.* 667
252. (a) Bosch, E. and Kochi, J.K. in ref. 67l. (b) Bosch, E. and Kochi, J.K. (1995). *J. Chem. Soc., Perkin Trans 1*, 1057
253. Bosch, E. and Kochi, J.K. (1996). *J. Chem. Soc., Perkin Trans 1*, 2731
254. (a) Bosch, E., Rathore, R. and Kochi, J.K. (1994). *J. Org. Chem.* **59**, 2529. (b) Rathore, R., Bosch, E. and Kochi, J.K. (1994). *Tetrahedron Lett.* **35**, 1335
255. Rathore, R., Kim, J.S. and Kochi, J.K. (1994). *J. Chem. Soc., Perkin Trans. 1*, 2675

256. Hassel, O. and Stromme, O. (1958). *Acta. Chem. Scand.* **12**, 1146
257. Amatore, C. and Kochi, J.K. (1991). *Adv. Electron-Transfer Chem.* **1**, 55
258. Malone, S.D. and Endicott, J.F. (1976). *J. Phys. Chem.* **76**, 2223. See also Henglein, A. (1980). *Radiat. Phys. Chem.* **15**, 151
259. (a) Cannon, R.D. (1980). *Electron-Transfer Reactions*. Butterworth, London. (b) Astruc, D. (1995). *Electron Transfer and Radical Processes in Transition-Metal Chemistry*. VCH, New York
260. (a) Hubig, S.M., Rathore, R. and Kochi, S.K. (1999). *J. Am. Chem. Soc.* **121**, 617. (b) This paper describes the mechanistic distinction between outer-sphere and inner-sphere electron transfer in terms of electron coupling matrix element (V) in the intermolecular process, see Hubig, S.M. *et al.* in ref. 260a
261. Endicott, J.F., Kumar, K., Ramasami, T. and Rotzinger, F.P. (1983). *Prog. Inorg. Chem.* **30**, 141
262. Fukuzumi, S. and Kochi, J.K. (1983). *Bull. Chem. Soc. Jpn.* **56**, 969
263. Fukuzumi, S. and Kochi, J.K. (1982). *J. Am. Chem. Soc.* **104**, 7599
264. Fukuzumi, S. and Kochi, J.K. (1981). *J. Am. Chem. Soc.* **103**, 7240
265. Rathore, R. *et al.* in ref. 202a
266. Rathore, R. *et al.* in ref. 204
267. (a) Murray, R.W. (1989). *Chem. Rev.* **89**, 1187. (b) Compare also oxygen-atom transfer from peracids. See, Koevner, T., Slebocka-Tilt, H. and Brown, R.S. (1999). *J. Org. Chem.* **64**, 196
268. Du, X. and Houk, K.N. (1998). *J. Org. Chem.* **63**, 6480
269. For peroxide anion radicals, see Adam, W. and Schonberger, A. in ref. 141
270. For the lability of alkoxy-type radicals, see Ando, W. (ed.). (1992). *Organic Peroxides*. Wiley, New York. In the same way, olefin epoxidation with peracids can be simply viewed as an electron transfer, followed by mesolytic cleavage of the peracid anion radical to carboxylate and hydroxyl radical, followed by homolytic coupling and proton loss. See also: Nugent, W.A., Bertini, F. and Kochi, J.K. (1974). *J. Am. Chem. Soc.* **96**, 4945
271. For example, see the radical byproducts in the dioxirane reaction in equation (106) as described by Bravo, A. Fontana, F., Fronza, G., Minisci, F. and Zhao, L. (1998). *J. Org. Chem.* **63**, 254

Author Index

Numbers in italic refer to the pages on which references are listed at the end of each chapter

Cumulative Index of Authors

Cumulative Index of Titles

ISBN 0-12-033535-2

9 780120 335350